Simulation Foundations, Methods and Applications

Series Editor

Andreas Tolk, The MITRE Corporation, Charlottesville, VA, USA

Advisory Editors

Rodrigo Castro, Universidad de Buenos Aires, Ciudad Autónoma de Buenos Aires, Argentina

Axel Lehmann, Universität der Bundeswehr München, Neubiberg, Germany

Stewart Robinson, Newcastle University Business School, Newcastle upon Tyne, UK

Claudia Szabo, The University of Adelaide, Adelaide, Australia

Mamadou Kaba Traoré, University of Bordeaux, Talence, France

Bernard P. Zeigler, University of Arizona, Tucson, AZ, USA

Lin Zhang, Beihang University, Beijing, China

Sanja Lazarova-Molnar, Karlsruhe Institute of Technology, Karlsruhe, Germany

The modeling and simulation community extends over a range of diverse disciplines and this landscape continues to expand at an impressive rate. Over recent years, modeling and simulation has matured to become its own discipline, while continuing to provide support to other disciplines. As such, modeling and simulation provides the necessary conceptual insights as well as computational support which has an established record of significantly enhancing the understanding of dynamic system behavior and improving the system design process, as well as providing the foundations for computational sciences and practical applications, from cyber-physical systems to healthcare. Hybrid methods and combinations with artificial intelligence and machine learning open new possibilities as well. The ever-increasing availability of computational power and the availability of quantum computers make applications feasible that were previously beyond consideration. Simulation is pushing back the boundaries of what it can be applied to and what can be solved in practice. Its relevance and applicability are unconstrained by discipline boundaries.

Simulation Foundations, Methods and Applications hosts high-quality contributions that address the various facets of the modeling and simulation enterprise. These range from fundamental concepts that are strengthening the foundation of the discipline to the exploration of advances and emerging developments in the expanding landscape of application areas. The underlying intent is to facilitate and promote the sharing of creative ideas across discipline boundaries.

As every simulation is rooted in a model, which results from simplifying and abstracting the reference of interest to best answer research questions or support the application domain of interest, we understand the model development phase as a prerequisite for any simulation application. There is an expectation that modeling issues will be appropriately addressed in each presentation. Incorporation of case studies and simulation results will be strongly encouraged.

Titles of this series can span a variety of product types, including but not exclusively, textbooks, expository monographs, contributed volumes, research monographs, professional texts, guidebooks, and other references.

These books will appeal to senior undergraduate and graduate students, and researchers in any of a host of disciplines where modeling and simulation has become (or is becoming) an important problem-solving tool. Some titles will also directly appeal to modeling and simulation professionals and practitioners.

Masoud Fakhimi · Navonil Mustafee
Editors

Hybrid Modeling and Simulation

Conceptualizations, Methods and Applications

Editors
Masoud Fakhimi
Surrey Business School
University of Surrey
Guildford, UK

Navonil Mustafee
Centre for Simulation, Analytics
and Modelling
University of Exeter Business School
Exeter, UK

ISSN 2195-2817　　　　　　ISSN 2195-2825　(electronic)
Simulation Foundations, Methods and Applications
ISBN 978-3-031-59998-9　　　ISBN 978-3-031-59999-6　(eBook)
https://doi.org/10.1007/978-3-031-59999-6

© The Editor(s) (if applicable) and The Author(s), under exclusive license to Springer Nature Switzerland AG 2024

This work is subject to copyright. All rights are solely and exclusively licensed by the Publisher, whether the whole or part of the material is concerned, specifically the rights of translation, reprinting, reuse of illustrations, recitation, broadcasting, reproduction on microfilms or in any other physical way, and transmission or information storage and retrieval, electronic adaptation, computer software, or by similar or dissimilar methodology now known or hereafter developed.

The use of general descriptive names, registered names, trademarks, service marks, etc. in this publication does not imply, even in the absence of a specific statement, that such names are exempt from the relevant protective laws and regulations and therefore free for general use.

The publisher, the authors and the editors are safe to assume that the advice and information in this book are believed to be true and accurate at the date of publication. Neither the publisher nor the authors or the editors give a warranty, expressed or implied, with respect to the material contained herein or for any errors or omissions that may have been made. The publisher remains neutral with regard to jurisdictional claims in published maps and institutional affiliations.

This Springer imprint is published by the registered company Springer Nature Switzerland AG
The registered company address is: Gewerbestrasse 11, 6330 Cham, Switzerland

If disposing of this product, please recycle the paper.

"To my parents, Ali and Soheila, whose love and wisdom light my way, and to all my mentors, whose insights into modelling and simulation carved paths in my mind."

Masoud Fakhimi

"To the authors who submitted to the 'Hybrid Simulation' track of the Winter Simulation Conference between 2014 and 2018, to my colleagues and friends who supported the development of the track during its infancy, and above all, to my children, Neelabh and Rudraneel, who make less time seem like more time."

Navonil Mustafee

Foreword

"All Creatures Great and Small" is a BBC television series with iterations since the late 1970s. The series, which takes place in the rolling hills and valleys of the English county of Yorkshire, has one episode that cause me to reflect on this excellent book edited by Masoud Fakhimi and Navonil Mustafee. In this memorable episode, the veterinarian is talking to a boy about his dog. The dog is on the mend, and the veterinarian says that the dog is likely to recover well because it has genetics consistent with "hybrid vigor."

Hybrid vigor is known more scientifically as *heterosis*. Heterosis is defined as improved biological function resulting from hybrid genetics of the parents. Could hybrid modeling and simulation similarly confer a positive outlook? It gets one thinking, but alas, the sorts of models that we create in areas such as operations research are not biological. We may engage in hybrid solutions because it matches our professional interests and inclinations.

What does hybrid modeling and simulation mean? We first acknowledge a difference between modeling and simulation. A model is a language artifact, whose computer-based execution defines a simulation. The model is designed and then the computer executes the model. Because the model is hybrid, the execution can also be hybrid in a computational sense. Continuous models are mathematical and require numerical integration. The models look different when compared with discrete-event models, and the method of solving the models is also different.

Chapter authors take two or more types of approaches. For instance, discrete-event simulation (DES), system dynamics (SD), and agent-based modeling (ABM) are frequently employed. DES, SD, and ABM are different ways of thinking about a problem space. This variety yields a hybrid solution. Decades ago, there were references to Ashby's *Law of Requisite Variety* in cybernetics. This law captures the complexity of state space. This complexity has an analog with information theory: more information in complex systems with many states and events. In the 1980s, simulation researchers constructed the area of "combined modeling." Combined models connect discrete events with continuous time and space modeling, as in fluid dynamics. The collection of scholarly expositions in this new book builds upon early work in cybernetics, systems theory, and combined discrete event/continuous systems.

Even though DES, SD, and ABM reflect our current thinking about hybridity, we need to expand this collection by using other types of models that we routinely use in defining and explaining systems. Information and data modeling are performed. A database and its data are defined with a schema. The knowledge that includes the models extends to semantic web terminology, with the phrase *ontology*. This is also modeling. Packages that we use to aid in our simulation, such as AnyLogic, include 2D and 3D computer-aided design (CAD) models. Even though semantic web models and CAD models do not involve the forward march in time, these models also need to be considered as part of a hybrid modeling system.

To return to our musing on hybrid vigor, our community is left wondering whether there is something akin to biological performance. Can a hybrid model be better or more inclusive than a model that is not hybrid? Is there an argument to be made for hybridity in modeling and simulation? Let's explore two arguments. Both arguments stem from qualities of people.

The first argument is based on education. In education, we observe that every person is different: Different backgrounds and interests. And yet, our school systems are based on mass production through uniform lessons and subject area disciplines. More ideally, we need differentiated learning where there is a tutor who can converse in a wide range of subjects. I was fortunate to explore this idea with my host, Nav, in Exeter during the summer of 2018 with the benefit of a Leverhulme Trust fellowship. We engaged with the Royal Albert Memorial Museum (RAMM) to see how art could be leveraged to understand computer science and scientific modeling. This engagement bridged art, mathematics, and science. The underlying hypothesis is that all objects, including those in museums, have multiple subject area-based interpretations: from art to science. These interpretations have the potential to bring people and subjects together. Curiously, recent research in AI large language models (LLMs) has yielded technologies to assist in this hybrid learning. Khan Academy has been in the forefront with a new type of chatbot tutor.

The second argument is based on management science. This is where I remember parts of my career. When I started my industrial life, I was a small cog in a huge machine that made ships and submarines within the Newport News Shipbuilding company. I then went on to assist in researching CAD models for aircraft at NASA Langley Research Center. Ships and airplanes are very complex, or some would say that they are very complicated. The complications manifest themselves in the assortment of disciplines needed to model and simulate. A variety of people are needed to make the products. The evolving complexity reflects the parts and the people. Large-scale models indicate more disciplines being involved. And this is why this book, and the field of hybrid modeling and simulation, are required. It isn't just that hybridity is "nice to have." It is a "must have" type of research. As we grow our models and simulations, this comes about with increased complexity. This complexity in turn requires more diverse workers, who come with different disciplines—various ways of seeing the world.

I close this short essay with experience I've gained since working for the State of Florida in the modeling and simulation of catastrophic wind and flood risk since the mid-1990s. To understand these risks, one needs the combined, hybrid expertise of different types of science and engineering. The actuarial profession is central to gauging financial risk. This combination includes the ability to deal with people who collaborate across different sectors. Different fields and people mean more complexity leading to a large-scale enterprise. Each player will come knowing different types of models. A hybrid approach is the only possible outcome. This is why this book is a "must read." As we move beyond building small-scale models for ourselves or in our own small labs, we necessarily embrace the hybrid. We are ready to segue to our own version of all models great and small.

Richardson, USA   Paul Fishwick

Preface

Hybrid Modeling and Simulation: Conceptualizations, Methods, and Applications aims to advance our knowledge of mixing methods from the field of modeling and simulation (M&S) and other scientific disciplines. Numerous textbooks and reference bodies of work provide an excellent foundation for discrete and continuous simulation methodologies. However, this is the first book that presents an integrative body of work on hybrid M&S.

Our field has distinct research communities where techniques like agent-based simulation (ABS), system dynamics (SD), and discrete-event simulation (DES) have evolved through decades of research and practice. As these sub-fields have continued to develop, conventional simulation approaches, i.e., using either ABS, DES, or SD, have transitioned to hybrid simulation (e.g., DES+ABS, ABS+SD). The hybrid approach becomes necessary when systems get increasingly complex and individual methods cannot adequately capture their intricacies. Combining methods leverages the strengths of techniques and presents the opportunity to develop a better representation of the system compared to using a single approach.

This book has extended the discussion on model hybridisation by considering methods and approaches developed outside our field. This is referred to as Hybrid Modeling. Thus, the term *Hybrid Modeling and Simulation* in the book's title calls attention to both *Hybrid Simulation* (models predominantly developed within our field) and *Hybrid Modeling* (models that intersect with approaches developed in broader scientific disciplines).

Unlike hybrid simulation, the term hybrid modeling is open to multiple interpretations. As the book is primarily targeted at the M&S community, we define a hybrid model as including at least one simulation technique that is combined with research approaches from a wider array of disciplines.

Further, our definition of hybrid consists of the conjoint application of cross-disciplinary techniques in one or more stages of a simulation study (Chapter 1 provides an integrative taxonomy of hybrid simulation and hybrid modelling). For a model to be classed as a "hybrid model" (as per the definition above), it should include at least one core simulation technique, e.g., DES (it can also be a hybrid simulation), and additionally should have deployed knowledge artefacts from other scientific disciplines in one or more stages of a simulation study (e.g., conceptual modelling stage, validation and verification, experimentation). These artefacts could be disciplinary modelling methods and research paradigms, new ways of framing and answering research questions such as hypothesis testing, use of methodological approaches from hard sciences through formulation of theories and controlled experimentation, new ways of collecting data by adopting ethnographic and other social science methods, novel analysis of primary/secondary data using approaches such as structural equation modelling, and deployment of standards and best practices that have stood the test of time in other disciplines. In the current literature, examples of hybrid models are mostly restricted to those developed in conjunction with Operations Research methods (including various analytics/ML models) and/or applied computing approaches. Examples include hybrid models that combine ABS with machine learning, those that have used DES with distributed computing and studies combining forecasting models with computer simulation.

A Call to the Community!

Our call for book chapters specifically sought contributions to both hybrid simulation and hybrid modeling. Thus, the first objective was to present work that contributed to hybrid simulation conceptualisations, work that applied existing modelling formalisms to hybrid simulation, identified frameworks for mixing methods and developed exemplar hybrid applications. The second objective was to include a collection of chapters that extend this current state-of-the-art in mixing methods and which deployed simulation alongside research approaches from outside M&S; the latter an example of leveraging cross-disciplinary strengths through the use of hybrid models as enablers of multidisciplinary, interdisciplinary, and transdisciplinary research.

Consistent with the intense interest in the M&S community in mixing simulation methods, most of the chapters received were on hybrid simulation. Thus, in relation to objective two, the book includes only a few chapters on hybrid models. Irrespective, we hope our book will broaden the discussion on what hybridity means in M&S and its different facets!

We hope our endeavour will be a call to the community to conduct further research on mixing disciplinary approaches with M&S. This would pave the way for an increasing number of studies on cross-disciplinary hybrid models, which, in the future, (we hope) will follow the same trajectory of growth as we witness for hybrid simulation! Indeed, the challenges that humanity is faced with today, from climate change and the need for climate-resilient regions and infrastructures to the need for sustaining economic growth whilst achieving net zero emission targets, will require the marriage of methods which were theorised, developed, refined and perfected in scientific disciplines, many of which had traditionally existed in isolation. This book is a step towards bridging this divide!

Preface

The book includes a foreword by Paul Fishwick, Chair Emeritus of Arts, Humanities, and Technology at The University of Texas at Dallas, whose work connects Arts and Humanities with Engineering, Mathematics, and Computer Science.

Chapter Summaries

We organised the chapters into three themes: conceptualisations and frameworks; formalisms and methods; applications.

The first four chapters focus on the conceptual aspects of Hybrid M&S, offering readers a foundational understanding of hybrid modeling and hybrid simulation and the use of conceptual modelling approaches. The first chapter is authored by the book's editors. In Chap. 1, Navonil Mustafee and Masoud Fakhimi outline the significance and core concepts of hybrid M&S, constructing a narrative around its research and developing an integrative taxonomy for both hybrid simulation and hybrid modelling. Chapter 2, by William Jones, Kathy Kotiadis, Jesse R. O'Hanley, and Stewart Robinson, introduces a novel method for conceptual modelling in hybrid simulations, emphasising its role in enhancing communication and software development. In Chap. 3, Richard A. Williams presents a semi-systematic literature review on hybrid conceptual modelling within organisations, focusing on social and socio-technical systems. Lastly, Chap. 4 by Andreas Tolk, Jennifer A. Richkus, and Yahya Shaikh underscores the importance of participatory modelling to include diverse community insights, presenting a conceptual framework and open research questions in the field.

Chapters 5 through 10 delve into the methods theme and cover hybrid methodologies and formalisms. Chapter 5 by Fernando J. Barros focuses on model product lines (MPLs) within the HyFlow++ framework, presenting the πHyFlow++ implementation for hybrid simulation. In Chap. 6, Saptaparna Nath and Gabriel A. Wainer use CELL-DEVS modelling to study the influence of social media influencers on follower engagement, employing the Cell-DEVS formalism through the Cell-DEVS Cadmium simulation environment. Chapter 7 by Niclas Feldkamp is on integrating machine learning with simulation. It takes a hybrid modelling perspective, providing guidelines and use case examples for their combined application. Chapter 8, authored by Najiya Fatma, Pranav Shankar Girisha, and Varun Ramamohana, presents a hybrid modelling approach using simulation and machine learning for real-time delay prediction in complex queuing systems; a case study on kidney transplantation waitlists is presented. In Chap. 9, Susan Howick, Itamar Megiddo, Le Khanh Ngan Nguyen, Bernd Wurth and Rossen Kazakov, explore the integration of SD and ABM, offering methodological insights and practical considerations for developing SD-ABM hybrid simulations. Finally, Chap. 10 by Alison Harper, Thomas Monks, and Sean Manzi proposes a hybrid method for enhancing the usability and sharing of simulation models through containerisation with continuous integration, demonstrating this with a Python-based orthopaedic elective recovery planning model.

Chapters 11 through 13 relate to the third theme on applications and case studies in hybrid modelling and simulation, offering in-sights into real-world implementations. Chapter 11, by Anastasia Anagnostou and Simon J. E. Taylor, discusses the application of hybrid simulation in healthcare, particularly in emergency medical services and pandemic crisis management, highlighting its role in holistic analysis and management. In Chap. 12, Vishnunarayan Girishan Prabhu and Kevin M Taaffe present a hybrid modelling approach using machine learning to optimise healthcare operations. They illustrate this through a case study in an emergency department, integrating forecasting, hybrid simulation, and mixed integer linear programming to enhance physician shift scheduling and patient safety. Finally, Chap. 13 by Kavitha Balaiyan, R. K. Amit, Amit Agarwal, and T. V. Krishna Mohan addresses demand forecasting challenges in airline revenue management. The authors propose a sequential two-stage hybrid modeling approach—a simulation-based heuristic algorithm for parameter estimation in joint-forecasting models, using actual airline data in the Airline Planning and Operations Simulator.

As outlined in the chapter summaries, *Hybrid Modeling and Simulation: Conceptualizations, Methods, and Applications* unravels the complexities of hybrid M&S, highlighting its application in multiple areas. The book is designed to foster a deeper comprehension of how various modelling methods can be combined to enhance decision-making and problem-solving in complex environments. The chapter contributions present a call to embrace the complexity of the world around us and to seek out the synergies of hybrid approaches for a deeper understanding of the problem space and engender improved decision-making. For anyone intent on mastering the art of hybrid M&S, this book is an essential companion.

London, UK Masoud Fakhimi
Exeter, UK Navonil Mustafee

Contents

Part I Conceptualisations and Frameworks

1 **Towards an Integrative Taxonomical Framework for Hybrid Simulation and Hybrid Modelling** 3
Navonil Mustafee and Masoud Fakhimi

2 **Using the Modelling Frame in the Conceptual Modelling Activity to Improve the Advantages of Hybridisation** 23
William Jones, Kathy Kotiadis, Jesse R. O'Hanley, and Stewart Robinson

3 **Hybrid Conceptual Modelling of Social and Socio-technical Systems Within Organisations: A Qualitative Semi-systematic Review** 47
Richard A. Williams

4 **Towards Hybrid Modelling and Simulation Concepts for Complex Socio-technical Systems** 73
Andreas Tolk, Jennifer A. Richkus, and Yahya Shaikh

Part II Formalisms and Methods

5 **Defining Families of Hybrid Models with the πHyFlow^{++} Modeling and Simulation Integrative Framework** 103
Fernando J. Barros

6 **CELL-DEVS Modelling of Individual Behaviour Towards Influencers in Social Media** 125
Saptaparna Nath and Gabriel A. Wainer

7 **Application of Machine Learning Within Hybrid Systems Modelling** 159
Niclas Feldkamp

8 Simulation and Machine Learning Based Real-Time Delay
   Prediction for Complex Queuing Systems 185
   Najiya Fatma, Pranav Shankar Girish, and Varun Ramamohan

9 Combining SD and ABM: Frameworks, Benefits, Challenges,
   and Future Research Directions 213
   Susan Howick, Itamar Megiddo, Le Khanh Ngan Nguyen,
   Bernd Wurth, and Rossen Kazakov

10 Deployable Healthcare Simulations: A Hybrid Method
   for Combining Simulation with Containerisation
   and Continuous Integration 245
   Alison Harper, Thomas Monks, and Sean Manzi

Part III Applications

11 Hybrid Simulation in Healthcare Applications 271
   Anastasia Anagnostou and Simon J. E. Taylor

12 Hybrid Modelling Approach Using Reinforcement
   Learning in Conjunction with Simulation: A Case Study
   of an Emergency Department 295
   Vishnunarayan Girishan Prabhu and Kevin M. Taaffe

13 Dependent Demand Forecasting Models in Airline Revenue
   Management: Parametric Estimation Using Simulation 319
   Kavitha Balaiyan, R. K. Amit, Amit Agarwal,
   and T. V. Krishna Mohan

Index 349

Contributors

Amit Agarwal MeraPashu 360, Gurugram, India

R. K. Amit Decision Engineering and Pricing (DEEP) Lab, Department of Management Studies, Indian Institute of Technology Madras, Chennai, India

Anastasia Anagnostou Modelling and Simulation Group, Department of Computer Science, Brunel University London, Uxbridge, Middx, UK

Kavitha Balaiyan Ford Motor Company, Chennai, India

Fernando J. Barros Department of Informatics Engineering, University of Coimbra, Coimbra, Portugal

Masoud Fakhimi Surrey Business School, University of Surrey, Guildford, Surrey, UK

Najiya Fatma Department of Mechanical Engineering, Indian Institute of Technology Delhi, New Delhi, India

Niclas Feldkamp TU Ilmenau, Ilmenau, Germany

Pranav Shankar Girish Department of Mechanical Engineering, Indian Institute of Technology Delhi, New Delhi, India

Vishnunarayan Girishan Prabhu Industrial and Systems Engineering, University of North Carolina at Charlotte, Charlotte, NC, USA

Alison Harper Centre for Simulation, Analytics and Modelling, University of Exeter Business School, Exeter, UK

Susan Howick Strathclyde Business School, Glasgow, UK

William Jones Australian Centre for Field Robotics, University of Sydney, Chippendale, NSW, Australia

Rossen Kazakov Strathclyde Business School, Glasgow, UK

Kathy Kotiadis Kent Business School, University of Kent, Canterbury, UK

T. V. Krishna Mohan Centre for Simulation, Analytics and Modelling, University of Exeter Business School, Exeter, UK

Sean Manzi The National Institute for Health and Care Research Applied Research Collaboration West (NIHR PenARC), Exeter, UK

Itamar Megiddo Strathclyde Business School, Glasgow, UK

Thomas Monks The National Institute for Health and Care Research Applied Research Collaboration West (NIHR PenARC), Exeter, UK

Navonil Mustafee Centre for Simulation, Analytics and Modelling, University of Exeter Business School, Exeter, UK

Saptaparna Nath Carleton University, Ottawa, ON, Canada

Le Khanh Ngan Nguyen Strathclyde Business School, Glasgow, UK

Jesse R. O'Hanley Centre for Logistic and Sustainability Analytics, Kent Business School, University of Kent, Canterbury, UK

Varun Ramamohan Department of Mechanical Engineering, Indian Institute of Technology Delhi, New Delhi, India

Jennifer A. Richkus The MITRE Corporation, McLean, VA, USA

Stewart Robinson Newcastle University Business School, Newcastle Upon Tyne, UK

Yahya Shaikh The MITRE Corporation, Windsor Mill, MD, USA

Kevin M. Taaffe Department of Industrial Engineering, Clemson University, Clemson, USA

Simon J. E. Taylor Modelling and Simulation Group, Department of Computer Science, Brunel University London, Uxbridge, Middx, UK

Andreas Tolk The MITRE Corporation, Charlottesville, VA, USA

Gabriel A. Wainer Carleton University, Ottawa, ON, Canada

Richard A. Williams Department of Management Science, Management School, Lancaster University, Bailrigg, Lancaster, UK

Bernd Wurth Strathclyde Business School, Glasgow, UK

Part I
Conceptualisations and Frameworks

Chapter 1
Towards an Integrative Taxonomical Framework for Hybrid Simulation and Hybrid Modelling

Navonil Mustafee and Masoud Fakhimi

Abstract The modelling and simulation (M&S) literature identifies two forms of hybrid studies—hybrid simulation (HS) and hybrid model (HM). While HS is the combined application of simulation techniques such as discrete-event simulation, agent-based simulation and system dynamics, HM has a cross-disciplinary focus; the objective is to apply research paradigms and approaches, conceptualisations and frameworks, methods, tools and techniques from disciplines such as computer science, engineering, OR and economics in one or more stages of an M&S study. The growing volume of literature on HS evidences the shift from conventional (one-technique) models to HS. It is expected that HM will follow the same trajectory. However, further studies are essential in contextualising hybrid M&S research and identifying opportunities for hybridisation. Towards this, three related themes of research are explored: (a) conceptualisation of HM in the context of the lifecycle of an M&S study, (b) a classification scheme, and (c) mapping existing literature in HS and HM. The themes are based on a series of authors' papers published in the Journal of the Operational Research Society and the Winter Simulation Conference. The chapter concludes with an integrative taxonomical framework that identifies the current advances and opportunities for future research in hybrid M&S.

Keywords Hybrid simulation · Hybrid modelling · Operational research · Taxonomy · Conceptual modelling

---

N. Mustafee (✉)
Centre for Simulation, Analytics and Modelling, University of Exeter Business School, Exeter EX4 4ST, UK
e-mail: n.mustafee@exeter.ac.uk

M. Fakhimi
Surrey Business School, University of Surrey, Guildford, Surrey GU2 7XH, UK
e-mail: Masoud.fakhimi@surrey.ac.uk

1.1 Introduction

M&S techniques such as discrete-event simulation (DES), agent-based simulation (ABS) and system dynamics (SD) enable the development of computational models that aid decision-making. They are among the most frequently used techniques in operations research/management science (OR/MS) [28]. There are numerous examples of their application in manufacturing and business [18], supply chain [37, 50], healthcare [5, 20, 42] and other domains. Many studies now apply multiple M&S techniques to develop hybrid simulations (HS) that combine DES, ABS and SD [6]. HS allows for a better representation of the system being modelled, enables opportunities for analysis at different levels of resolution and provides greater insights into the evolving dynamics of the system.

A parallel but related theme of research is the development of hybrid models (HM) that combine conventional simulations (i.e. single-technique DES, ABS and SD models) or HS, with research approaches, methods and techniques from disciplines such as computer science/applied computing, economics, engineering, information and communications technology, operations research, software engineering and social sciences. Similar to HS, the objective here is to explore the synergies of combining discipline-specific methods in developing M&S studies, studies that go beyond what would otherwise be possible if only approaches developed within our discipline were being used. However, unlike HS, the focus is not limited to combining only simulation techniques, but rather to integrating simulations with theories, frameworks, methods and established research approaches that have been tried and tested and have existed as extant knowledge within distinct academic disciplines [31, 51]. These knowledge artefacts can be seen as the body of knowledge (BOK) that is used by a discipline to guide practice [39], and absorbing them into M&S studies presents the opportunity to complement (rather than supplement) the techniques traditionally used within our field. In social sciences, for example, scientists increasingly use computational experiments through social simulations to explore and test hypotheses concerning aspects of collective action and group dynamics [47].

The book chapter is based on a series of four papers that the authors have published in the *Journal of the Operational Research Society* and the *Winter Simulation Conference* [29–31, 43]. The papers are either conceptual or a mix of conceptual and review papers. The objective of the book chapter is to construct a narrative for HM based on three related themes of hybrid research: (a) conceptualisation of HM in the context of the distinct phases of the lifecycle of an M&S study, which identifies the opportunities for the combined application of cross-disciplinary approaches; (b) a classification scheme for HS and HM; (c) categorisation of the existing work using the classification scheme. These discussions lead to the development of an integrative taxonomical framework for HS and HM. The taxonomy is based on the original HS and HM classification presented by [31] and its extension by [30]. The integrative taxonomical framework is organised in multiple levels of hierarchy.

1.2 Stages of a Simulation Study and Opportunities for Hybrid Modelling

A M&S study begins with the investigation of a real-world problem or a consideration for a future system. A conceptual model is developed and validated, followed by the implementation of a computer model. In the verification stage, the model is checked to ensure that it is a good representation of the conceptual model and is free from errors. Scenarios for experimentation are developed, followed by an analysis of the results of the simulations for possible implementation of the results of the simulation study. These M&S study stages enable us to systematically explore complementary techniques for problem understanding and conceptualisation, model implementation, validation and verification, scenario development and experimentation, results analysis and implementation of the results. Based on the discipline-specific methods and what it has to offer, this added value gained could be mapped to various stages of an M&S study [31]:

- *Problem formulation/Conceptual modelling*: A systems engineering modelling language called systems modelling language (SysML) was proposed by [11] to capture the inception stage of a healthcare service development lifecycle. Soft OR/Problem structuring methods have been widely used to aid the development of conceptual models for DES. Examples include the use of soft systems methodology (SSM) by [21, 23], group model building by Bérard [4] and qualitative systems dynamics (QSD) by Powell and Mustafee [43].
- *Input/Output data analysis*: The use of Hard OR methods with M&S is reported by Mustafee and Bischoff [27], who combined load plan construction heuristics (cutting and packing optimisation) and ABS, with the output of the optimisation algorithm serving as the input for the simulation, and Harper et al. [16] and Harper and Mustafee [15] who used forecasting with DES to model endoscopy services.
- *Model Formalism*: Model-driven engineering and domain-specific modelling languages [58], modelling formalisms such as based on discrete-event system specification (DEVS), e.g. dynamic structure discret-event system specification (DSDEVS) [3] and meta-modelling using UML [52] are examples of deploying complementary methods from fields such as software engineering in the implementation stage of a simulation study.
- *Model Development/Implementation*: Simulation techniques such as ABS, DES and SD can be combined to implement an HS study.
- *Experimentation*: Studies that have used approaches from computer science/ applied computing in the experimentation phase of an M&S study include works by Lendermann et al. [24], who use parallel and distributed simulation for high-fidelity supply chain optimisation, Park and Fishwick [40] who present a graphics processing unit (GPU)-based framework supporting fast DES, and Mustafee and Taylor [33] who developed the "WinGrid" desktop grids to execute simulations over distributed resources.

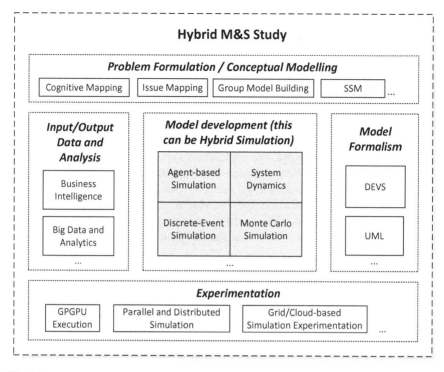

Fig. 1.1 Hybrid modelling extends hybrid simulation by deploying broader disciplinary methods and approaches to different stages of a simulation study; "…" denotes the presence of other methods [43]

Figure 1.1 presents the conceptualisation of an HM study. It presents some examples of disciplinary methods that have been used in various stages of an M&S study (white rectangles) and simulation techniques, including HS, that have been used in the model development/implementation stage (grey boxes). The three dots (…) denote the presence of other methods.

1.3 The Unified Conceptual Representation of Hybrid Simulation and Hybrid Modelling Through a Classification Scheme

The discussions on HS and HM, including Fig. 1.1, were first presented in [43]. It clarified that approaches from the wider OR (including Soft OR) and disciplines such as software engineering and applied computing, when used with M&S techniques, enable opportunities to realise synergies and engender improved insights. Mustafee and Powell [31] developed a unifying conceptual framework and a classification scheme to clarify the hybrid terminologies and to enable the exploration of complementarity between HS and HM. Although the framework furthered the discussion on hybrid from HS to HM, a critique is that the definition of HM mostly considered methods from the broader OR discipline (e.g. analytical modelling, game-theoretic modelling, forecasting). Mustafee et al. [30] added to the original classification by explicitly referring to cross-disciplinary methods and research approaches and the opportunities to combine them with M&S techniques. The extension is consistent with the intention of the authors of the original work who note that "*A classification scheme also has the benefit of being extensible, thus allowing the accommodation of new types of hybrid models that may be realised in future*" [43]. The work by Mustafee et al. [30] thus transitioned the original OR-focused classification of HM in Mustafee and Powell [43] to a classification scheme with disciplinary intersections with M&S, which was how the authors (*ibid.*) originally conceptualised the term HM.

In developing the original classification scheme, Mustafee and Powell [31] used the definitions of paradigm, methodology, technique and tool from Mingers and Brocklesby [25] and adapted them for the hybrid context. They identified five types of hybrid models, namely, Types A, B, C, D and D.1 (Type D.1 is a sub-type of Type D). Mustafee et al. [30] extended this original scheme to Define Type E hybrid models. The authors acknowledged that the definition could be open to multiple interpretations. However, the aim was to provide consistency in using the terms rather than seek a consensus.

The definitional parameters are presented next, followed by a discussion on the various model types defined in the classification scheme.

- **Paradigm:** Paradigms are a "*very general set of philosophical assumptions that define the nature of possible research and interventions*" [25]. Qualitative (interpretive, subjective, soft) and quantitative (positivist, objective, hard) paradigms are well understood. Computer simulations are computational models used for experimentation; they belong to the quantitative paradigm. Models may also be qualitative. As the classification was developed from the standpoint of M&S and models for decision-making, the authors restricted the discussion on paradigms to those relevant to OR.
- **Methodology:** Methodologies develop within a paradigm and usually embody philosophical assumptions [25]. In Mustafee and Powell [31], the authors mainly distinguish between discrete and continuous methodologies in the quantitative paradigm. In the qualitative paradigm, methodologies include soft systems methodology (SSM) and qualitative system dynamics (QSD).
- **Technique:** Techniques exist within the context of methodologies and have well-defined purposes [25]. Mustafee and Powell [31] distinguish between discrete techniques, such as DES and ABS, and continuous techniques, such as stock and flow models developed in SD and numerical computational fluid dynamics (CFD) models.
- **Tool:** Based on Mingers and Brocklesby's [25] definition of tools as an artefact that performs a particular technique, Mustafee and Powell [31] considered

tools as M&S packages and software libraries. Tools are not considered in this classification.

The unified conceptual representation of hybrid models and their classification places the HS and HM literature in context (Fig. 1.2). The classification consists of three forms of HS (Type A, B, C) and three forms of HM (Type D, D.1 and E), respectively; hybrid study serves as an umbrella term for both HS and HM models/studies. The functional definitions of Type A-E are presented next.

- **Type A—Multi-methodology hybrid simulation**: In the quantitative paradigm, Mustafee and Powell [31] distinguish between discrete methodologies, where the system state changes in discrete timesteps, and those which are continuous, where the system state changes continuously based on underlying differential equations. They state that a *multi-methodology HS* comprises simulation techniques with discrete and continuous elements, e.g. SD-DES and SD-ABS.
- **Type B—Multi-technique hybrid simulation**: Mustafee and Powell [31] state that a *multi-technique HS* uses two or more techniques under the same methodology. An example could be a CFD-SD study that uses CFD to model traffic flow and SD to investigate strategic policy related to urban transportation. The combined application of ABS-DES is also Type B HS since the integrated

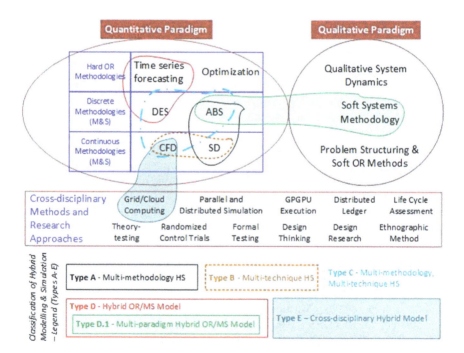

Fig. 1.2 Unified conceptual representation of hybrid M&S with a classification of distinct HS and HM model types [30]. Note that the techniques identified in the figure are not exhaustive; for example, there are numerous OR methods that can be included under Hard OR methodologies

approach combines emergence modelling using ABS with a technique based on queuing theory (DES).

- **Type C—Multi-methodology, multi-technique hybrid simulation**: Multi-methodology, multi-technique HS consist of at least three techniques classified under discrete and continuous methodologies, with the combined mix of techniques spanning both methodologies. Studies that combine the SD technique (continuous) with DES and ABS (both of which are discrete methodologies) fit this definition.
- **Type D—Hybrid operations research/management science (OR/MS) model:** These are HMs that combine M&S with techniques used in the field of OR/MS. An example of this is the combined application of analytical modelling with simulation. The reader is referred to Mustafee and Katsaliaki [28] for OR/MS techniques that are most widely used in the literature.
- **Type D.1—Multi-paradigm hybrid OR/MS model:** Computer simulations are computational models aligned to the quantitative paradigm. However, several studies have identified the use of qualitative/Soft OR approaches in the conceptual modelling phase of an M&S study, e.g. the use of SSM [21, 23]. Type D.1 models are thus a sub-type of Type D and, like the latter, focus on identifying synergies between M&S techniques and methods employed in the wider OR/MS discipline. Mustafee and Powell [31] state that Type D.1 models intersect paradigms! Finally, Type D and D.1 are referred to as HM (rather than HS) since the simulation is only one element of the hybrid model; the other element is from either Soft or Hard OR.
- **Type E—Cross-disciplinary hybrid models:** Mustafee et al. [30] define the Type E *model as one that* combines simulation (SD, DES, ABS, or an HS) with cross-disciplinary techniques from fields such as arts and humanities, economics, computer science/applied computing and systems engineering. *Type E* models go beyond the use of simulation with OR/MS in multiple stages of an M&S study (as is the case with *Model Type D, D.1*). As the realisation of the Type E model generally requires modelling expertise that goes beyond only M&S and OR/MS, it is important that researchers in our field engage with scholars from other scientific disciplines. For example, cloud-based execution of CFD simulations uses theoretical constructs to model agent relationships (e.g. theory of planned behaviour). The cross-disciplinary HMs can thus be considered as enablers of multi-, inter- and transdisciplinary research [51]. In the remainder of the chapter, we use the term HM to refer to both hybrid OR/MS models (Type D, D.1) and cross-disciplinary hybrid models (Type E), unless we need to specifically refer to either type, and in which case we will revert to the original type definitions.

1.4 Mapping Existing Literature to the Classification Scheme for HS and HM

The section builds on the HS-HM classification (Fig. 1.2) by presenting examples of existing hybrid studies and mapping them to distinct model types (i.e. Model Types A, B, C, D, D.1 and E). Mustafee et al. [29] note that although the hybrid classification can help develop a frame for mixing simulation with cross-disciplinary methods in multiple M&S study stages, researchers may continue to experience a gap in translating their conceptual understanding into the development of an empirical HM study. A reflection of the literature on existing hybrid studies (including those that may not have used the term HM but were, in essence, mixing simulation with a wider plethora of non-M&S techniques) and mapping them to the HS-HM classification scheme may help address the gap. Towards this, we use examples presented in Mustafee et al. [29, 30].

1.4.1 Type A Multi-methodology Hybrid Simulation

In OR/MS, Type A HS typically focuses on the combined use of DES or ABS (both discrete methods) with SD as the continuous method. In engineering, computational fluid dynamics (CFD) is widely used to model fluid flow and is a continuous simulation method. Thus, Type A models can be characterised as those that combine techniques from both discrete and continuous methodologies. In the classification scheme, these combinations of techniques are identified as sub-types of the Type A model. Table 1.1 lists examples of studies that could be mapped to specific sub-types. The classification is extensible, and further sub-types may be identified.

1.4.2 Type B Multi-technique Hybrid Simulation

Type B HS employ two or more techniques from either the discrete or the continuous methodology, for example, an HS using DES-ABS. However, there is debate about whether combining two discrete techniques qualifies as a hybrid. In our classification, the combined application of DES-ABM is defined as a sub-type of Type B HS since there are fundamental differences in the execution of the simulation logic, which makes them agreeable to model systems that benefit from adopting both a queuing and an emergence-based approach. Similarly, an HS that combines SD-CFD is identified as a sub-type of Type B HS. Table 1.2 presents some examples. Further sub-types of the Type B model can be identified from the literature.

Table 1.1 Examples of Type A—Multi-methodology HS

Type A model sub-type	Description with emphasis on the use of M&S methods and application area	References
ABS-SD	The authors develop an integrated ABS-SD model to understand behavioural diversities associated with multi-type labourers in multinational projects, revealing the associated impacts and improving project management. ABS was used to model the behaviour of the labourers and estimate their performance, with the SD model using this data to summarise these individual performances and evaluate the deviation in the timelines of the project (**Construction Planning**)	[55]
ABS-SD	An HS was developed to estimate the market share evolution of electric vehicles. Agent-based discrete choice models of consumer choice and awareness were combined with macro-level SD elements that model the interdependencies between consumer choice, technology evolution and available infrastructure for electric vehicles (**Transportation**)	[22]
DES-SD	The authors develop an HS to analyse "schedule risk" in infrastructural projects. DES modelled the construction processes, resource usage and other micro variables, with SD representing the feedback associated with work allocation, rework, etc. and provided a systems perspective (**Construction Planning**)	[56]
DES-SD	The authors investigate total productive maintenance using SD-DES HS. The problem being modelled involved both maintenance scheduling (DES) and considerations for human factors such as attitude (SD) (**Maintenance**)	[36]
ABS-CFD	To demonstrate the feasibility of a hybrid approach for evacuation planning, the authors model the hypothetical case of toxic aerosol release in downtown Los Angeles (using CFD), and simulate the response of a large spatially distributed agent population (ABS) (**Evacuation Planning**)	[12]
DES-CFD	The authors propose an HS consisting of a DES that models the flow of materials through a production line (manufacturing system simulation) with a CFD simulation of a compressed air system. This enables the combined evaluation of the aforementioned systems, with the overall objective of optimising energy consumption per unit of production (**Manufacturing**)	[35]

1.4.3 Type C Multi-methodology, Multi-technique Hybrid Simulation

Type C HS has elements of both Type A (multi-methodology) and Type B (multi-technique). In our classification, Type C multi-methodology, multi-technique HS represent the combined application of three or more simulation techniques, of which at least two techniques employ either continuous or discrete methodologies. Table 1.3 lists two sub-types of Type C, namely DES-ABM-SD and DES-ABM-CFD, however,

Table 1.2 Examples of Type B—Multi-technique HS

Type B model sub-types	Description with emphasis on the use of M&S methods and application area	References
SD-CFD	The authors developed a SD model to simulate vehicle movements with different traffic volumes and a CFD model to simulate the dispersion of pollutants. The objective of the study was to investigate the effects of traffic volume and toll collection methods on the dispersion of pollutants at a toll plaza **(Transportation)**	[17]
DES-ABS	The authors present a case study based on the London Emergency Medical Service where the DES and ABS elements model the hospital processes and first responders/ambulances, respectively **(Healthcare)**	[2]
DES-ABS	The authors implement a Type B hybrid ABS-DES model for the planning of capacity and patient flow in a post-term pregnancy outpatient clinic. The DES models the processes through the clinic, and the ABM models pregnant women as agents **(Healthcare)**	[53]

other sub-types can be identified in the literature. As the classification of hybrid M&S is extensible, it deviates from the definition of HS presented in Brailsford et al. [6] and is restricted to the use of particular combinations of DES, ABM and SD. As the classification presented in Mustafee et al. [30] has cross-disciplinary elements, such extension was necessary in order to incorporate a wider array of simulation techniques, for example, computational fluid dynamics (CFD) is a numerical simulation technique widely used in engineering. Table 1.3 presents examples of DES-ABM-CFD and DES-ABM-SD sub-types. Similar to Type A and Type B, other sub-types of Type C HS may be defined in future.

1.4.4 Type D Hybrid OR/MS Models

Type D HM combines M&S techniques with Hard OR approaches such as forecasting, analytical modelling, mathematical programming and optimisation, metaheuristics, game theory, graph theory, inventory models, multiple-criteria decision-making (MCDM), data envelopment analysis (DEA), process mining and machine learning. The classification can, therefore, include numerous sub-types of Type D models, each identifying a particular combination of M&S and Hard OR methods. Table 1.4 lists examples of a few sub-types of Type D OR/MS HMs.

Table 1.3 Examples of Type C—Multi-methodology, multi-technique HS

Type C model sub-types	Description with emphasis on the use of M&S methods and application area	References
DES-ABS-CFD	The authors developed an HS for evaluating countermeasures for chemical gas emergencies. The gas flow dynamics are modelled in CFD, human movement in ABS and an evacuation model in DEVS (**Evacuation Planning**)	[46]
DES-ABS-SD	The authors combined two discrete methods (DES and ABS) and one continuous method (SD) and applied them to a case study on earthmoving operations. The DES models the process flow of the earthmoving operation; the trucks and drivers are modelled as agents; and SD was used to model agents' physiological processes and decision behaviours (**Construction Planning**)	[13]
DES-ABS-SD	The authors developed an integrated DES-ABS-SD model to complement the standard lifecycle assessment (LCA) methodology. They validated the model using a case study of drink products (e.g. bottled water). SD was used to model the lifecycle of each beverage (e.g. bottled water production and recycled bottles), distribution and energy use; customer behaviour was modelled in ABS. Although the authors claim to have used two discrete methods, the hybrid model has no inherent queuing structures (**Environment**)	[54]
DES-ABS-SD	The authors developed a Type C HS for the assessment of innovative healthcare technologies, namely to evaluate mobile stroke units and prostate cancer screening. DES was used to represent hospital processes, and agents were generated from the SD component of the hybrid model (**Healthcare**)	[7]
DES-ABS-SD	The authors developed an HS for energy efficiency analysis, using SD to model the energy demand of production processes and DES/ABS to map the material flows and logistic processes applied to the mechanical processing of die-cast parts. DES provided meso-level workflow perspective, and ABS modelled micro-level active processes (**Manufacturing**)	[45]

1.4.5 Type D.1 Multi-paradigm Hybrid OR/MS Models

Type D.1 is a multi-paradigm HM (refer to the definition of paradigm in Sect. 1.3) that combines computer simulation with Soft OR techniques such as soft systems methodology (SSM), qualitative system dynamics (QSD) and cognitive mapping. Type D.1 bridges the qualitative and quantitative paradigm and should not be seen merely as a sub-set of the Type D model. Table 1.5 presents examples of Type D.1 models.

Table 1.4 Examples of Type D Hybrid OR/MS Models employing Hard OR methods

Type D model sub-types	Description with emphasis on the use of M&S methods and application area	References
Forecasting with DES	The authors used demographic projections and regression analysis to forecast demand for diagnostic services and used this as inputs into a DES to support long-term capacity planning (**Healthcare**)	[15]
Optimal packing problem with ABS	The authors developed an HM to analyse trade-offs between loading efficiency (using container Loading optimisation algorithms) and various important considerations in relation to the cargo, such as its stability, fragility or possible cross-contamination between different types of items over time (ABS) (**Transportation**)	[27]
Optimal coverage problem with ABS	The authors combine ABS and optimisation model to find the location of wireless sensors that maximises security coverage. The use of ABS is innovative as it allows them to evaluate scenarios in which intruders are intelligent, i.e. they can learn from others (**Security**)	[19]
Process mining with DES	The authors integrated process mining in the conceptual modelling phase of an M&S study to support the development of DES models (**Healthcare**)	[1]
Machine learning with DES	The authors investigated an HM approach that integrates simulation modelling with Machine Learning in an attempt to improve the validity of the simulation model outputs (**Healthcare**)	[10]

1.4.6 *Type E Cross-Disciplinary Hybrid Models*

Distinct from Model Types D and Type D.1, which mainly focus on using simulation with broader OR/MS methods, Type E HM necessitates cross-disciplinary engagement. From the perspective of our research community, exploration of the extant knowledge in disciplines such as engineering and computer science, data science, arts and humanities, medicine and health sciences allow us to identify established research philosophies, methods, techniques and tools, and which could be deployed in conjunction with computer simulation in one or more stages of a M&S study [30]. Table 1.6 presents examples of Type E models.

1 Towards an Integrative Taxonomical Framework for Hybrid Simulation … 15

Table 1.5 Examples of Type D1.1 Hybrid OR/MS Models employing Soft OR methods

Type D.1 model sub-types	Description with emphasis on the use of M&S methods and application area	Reference
Group support with DES	The authors used a collaborative simulation approach combining group support with DES to enhance the convergence of stakeholder viewpoints, acceptance of outcomes and model quality (**Airline industry**)	[8]
Lean with DES	The authors combined lean methodology with DES to improve stakeholder engagement and impact through three use cases (**Healthcare**)	[44]
Cognitive mapping with DES	Cognitive mapping was used to elicit staff perspectives to support DES experiments and guide DES model execution for enhancing surgical capacity (**Healthcare**)	[41]
SSM with DES	The authors use PartiSim, a multi-methodology framework to support participative simulation studies, which combines DES with SSM to engage stakeholders in the M&S study lifecycle through a set of stages and activities (**Healthcare**)	[49]
QSD with DES	The paper discusses the combined application of QSD with DES to aid the understanding of the system in the problem formulation/conceptual modelling stage of a hybrid M&S study (**Healthcare**)	[43]

1.5 Integrative Taxonomical Framework for Hybrid Simulation and Hybrid Modelling

A taxonomy is a scheme of classification, which, in addition to grouping, includes a hierarchy. Similar to a classification scheme, a taxonomy also has the benefit of being extensible. In developing the original unified representation of hybrid M&S and the classification scheme, [30] and [31] used the definitions of paradigm, methodology, technique and tool from Mingers and Brocklesby [25] to identify six distinct types of hybrid model types. The use of these definitions enables us to transition the HS-HM classification scheme (Fig. 1.2) to an integrative taxonomical framework for hybrid studies in M&S (Fig. 1.3). The various forms of hybridity listed in Sect. 1.3 can be considered as having a hierarchy based on the deployment of the body of knowledge (BOK) [39] within distinct fields and sub-fields of study. For example, an HS consisting of a mix of simulation techniques is at a hierarchy distinct from conventional (one-technique) discrete or continuous simulation studies. Similarly, HMs that have combined M&S techniques with the plethora of OR methods and those investigating opportunities for cross-disciplinary HMs have ventured further from their disciplinary confines. Arguably, the HMs also have respective taxonomical hierarchies.

The integrative taxonomical framework for hybrid M&S (Fig. 1.3) has "M&S approaches" as the *root level* and several other sub-levels. The root level defines the scope of the taxonomy and ensures that the hybrid models that refer to it *must include* a simulation component. For example, a hybrid model comprising data envelopment

Table 1.6 Examples of Type E cross-disciplinary Hybrid Models

Type E model sub-types	Description with emphasis on the use of M&S methods and application area	References
DES with desktop grid computing	A desktop-based grid middleware called *WinGrid* was developed and interfaced with a DES and a Monte Carlo simulation package, respectively (two case studies). The objective was to demonstrate faster execution of models using desktop grids (**Automotive; Financial Services**)	[33]
CFD and cloud computing	A cloud-based simulation platform is presented that enables SMEs to use the platform-as-a-service solution to execute CFD simulations (**Engineering**)	[48]
ABS-DES with distributed simulation	The authors present a distributed simulation framework for linking two Type B discrete models developed using ABS and DES. The framework was deployed in a case study related to the London emergency medical service. In their hybrid model, DES simulated the hospital processes, whereas ABS modelled the ambulance service. (**Healthcare**)	[2]
DES with distributed simulation	The IEEE 1516 HLA standard and DMSO RTI1.3-NG were used to investigate the speed-up of DES models using *Time Advance Request* and *Next Event Request*. The DES model was on the supply chain of blood. The objective of this study was to enable faster execution of DES models using distributed simulation (**Healthcare**)	[34]
ABS with parallel computing	The authors developed ABS of population dynamics in a certain demography. To make the simulation faster, they used a parallel computing technique. The experiment was done on a supercomputer (**Demographic study**)	[26]
DES with real-time data and machine learning	Symbiotic simulation is a technology that enables interaction between physical systems and their digital twins. The authors propose an HM architecture for symbiotic simulation, which includes data acquisition to receive data from the physical system, simulation/optimisation/ML models, scenario manager and decision maker/ actuator that relays the results to the physical system (**Methodology**)	[38]
DES with real-time data and forecasting model	The authors investigate proactive service recovery in ED by interfacing their model with real-time data steam from NHSquicker [32] and identifying approaches to trigger real-time DES experiments. It is an example of both the Type D Hybrid OR/MS Model (as it uses forecasting with DES) and Type E (as it employs real-time technologies that have been developed in applied computing) (**Healthcare**)	[14]

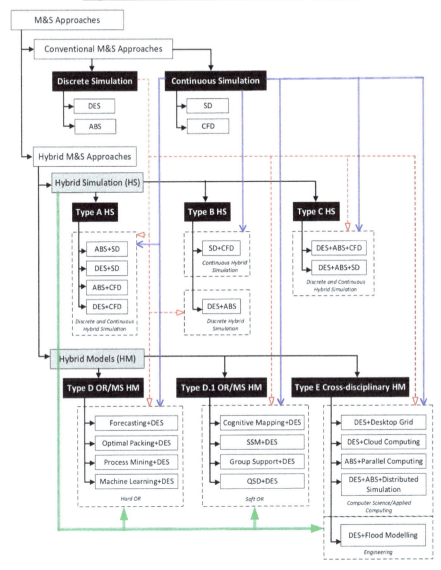

Fig. 1.3 Integrative taxonomical framework for hybrid M&S. The framework is based on the unified conceptual representation of hybrid simulation (HS) and hybrid modelling (HM) and the classification scheme presented in [31] and its extension in [30]. Note that the techniques identified in the figure are not exhaustive; for example, based on the requirement of a study, there are numerous OR/MS and cross-disciplinary methods that can be potentially used with discrete and continuous techniques and hybrid simulation of Model Types A, B and C, to realise Model Types D, D.1 and E. The taxonomy is thus extensible!

analysis and game theory [57] is outside the scope of the taxonomy. However, a study that uses game theory with system dynamic simulation (e.g. [9]) is within scope. "Conventional M&S" and "hybrid M&S" are the two elements in *Level-1* of the taxonomy. These are discussed next.

Level-1 Conventional M&S approach has "discrete" and "continuous" as the two *Level-2* elements, with "DES" and "ABS" as *Level-3* for discrete and "SD" and "CFD" as *Level-3* for continuous simulation, respectively. The taxonomy is extensible, and Fig. 1.3 only includes techniques discussed earlier in the chapter. For example, a discrete-time Markov chain simulation can be shown in *Level-3* under "Discrete Simulation" (along with DES and ABS). In relation to *Level-1* Hybrid M&S approaches, as per the contribution of our existing work to the literature on hybrid M&S, we differentiate between "hybrid simulation (HS)" and "hybrid models (HM)", both of which are *Level-2* elements in our taxonomy.

As per the original classification of HS-HM [31] and its extension [30], *Level-3* of "hybrid simulation (HS)" identifies the mix of simulation techniques—from both discrete (ABS, DES) and continuous methodologies (SD, CFD)—and introduces the following three elements—**Type A HS, Type B HS** and **Type C HS**. Type A is multi-methodology HS and the two arrows from the *Level-2* elements under *Root > Conventional M&S Approaches*, namely a red dotted arrow from discrete simulation and a double blue arrow from continuous simulation to the *Level-3* elements (e.g. ABS + SD, DES + SD) signify that these models have both discrete and continuous elements. Similarly, the arrows from *Root > Conventional M&S Approaches > continuous/discrete simulation* to the two Level-3 elements of Type B Multi-Technique HS, namely SD + CFD and DES + ABS, signify that these models may take the form of either continuous or discrete HS. Finally, *Level-2* Type C HS is multi-methodology, multi-technique HS and includes a minimum of three techniques, two of which will belong to either discrete or continuous simulation methodologies (this is represented through the dotted lines from *Root > Conventional M&S Approaches*). Two examples of Type 3 illustrated in Fig. 1.3 are DES + ABS + CFD and DES + ABS + SD.

Next, we focus on *Level-2* "Hybrid Models (HM)" (*Root > Hybrid M&S Approaches > Hybrid Models*). Based on the classification by Mustafee et al. [30], we define three *Level-3* elements: **Type D** Hybrid Operations Research/Management Science (OR/MS) Model, **Type D.1** Multi-paradigm Hybrid OR/MS Model and **Type E** Cross-disciplinary HM. Type D and Type D.1 models use simulation techniques with Hard OR and Soft OR methods, respectively. Thus, the examples of Type D *Level-3* elements include "forecasting + DES" and "Machine Learning + DES", whereas "Cognitive Mapping + DES" and "SSM + DES" are examples of Type D.1. Similarly, examples of *Level-3* Type E cross-disciplinary HM include uses of DES with desktop grid computing ("DES + Desktop Grid"), "ABS + Parallel Computing" and "DES + Flood Modelling". The taxonomy establishes that the simulation element of all the three *Level-3* elements under *Level-2* HM (*Root > Hybrid M&S Approaches > Hybrid Models*) must comprise techniques from either the discrete or continuous methodologies. This is shown using the arrows from the two *Level-2* elements *Root > Conventional M&S Approaches > discrete /continuous simulation*. Finally, the green double arrow from *Level-2* HS (*Root > Hybrid M&S Approaches*

> *Hybrid Simulation*) to *Level-3* elements under Types D, D.1 and E signify that the simulation models can also be hybrid simulations.

1.6 Conclusion

The book chapter introduced an integrative taxonomical framework for hybrid M&S. It built on the authors' previous research on hybrid simulation (HS) and hybrid modelling (HM). There is a growing volume of literature on HS, and numerous examples of HS Model Types A, B, and C exist. Similarly, as identified in Sect. 1.4, several studies have implemented HM Model Types D and D.1 by combining various simulation techniques with Hard and Soft OR methods and approaches. These models, which in our taxonomy align with HM Type D and D.1, have been referred to as multi-methodology models, analytical-simulation models, etc. However, the literature lacked a structured assessment that considered how different authors used the underpinning methodologies with simulation and its possible extensions in terms of the deployment of combined methods. The taxonomy provides precision in terms of definitions and can be used to identify hybrid approaches using a common reference frame. The taxonomy clarifies the distinction of HS from an HM but recognises that HS models can be a part of HM.

In reference to studies that have used simulation with cross-disciplinary methods (HM Model Type E), there are numerous studies published in journals and conferences that focus on simulation from a computer science perspective, for example, *ACM Transactions on Computer Modelling and Simulation* (ACM TOMACS), *Simulation Modelling Practice and Theory* (SMPT), *IEEE/ACM Symposium on Distributed Simulation and Real-Time Applications (DS-RT)* and *ACM SIGSIM Conference on Principles of Advanced Discrete Simulation (PADS)*. Thus, the taxonomy prominently identifies the parallel and distributed simulation (PADS) community's work (Fig. 1.3). However, in a similar vein, it also identifies opportunities for combining techniques with engineering, economics and other disciplines. The structure of the taxonomy also makes us wonder what other distinct areas of cross-disciplinary M&S research exist and whether there are sub-communities already undertaking work we classify as Type E HM? Indeed, such studies may be published in journals and conferences which are not necessarily on M&S. Our integrative and extensible taxonomy for hybrid M&S may thus evolve to identify new forms of HMs in future years!

References

1. Abohamad W, Ramy A, Arisha A (2017) A hybrid process-mining approach for simulation modelling. In: Proceedings of the 2017 winter simulation conference. IEEE, pp 1527–1538

2. Anagnostou A, Taylor SJE (2017) A distributed simulation methodological framework for OR/MS applications. Simul Model Pract Theory 70:101–119
3. Barros FJ (1995) Dynamic structure discrete event system specification: a new formalism for dynamic structure modeling and simulation. In: Proceedings of the 27th winter simulation conference. IEEE, pp 781–785
4. Bérard C (2010) Group model building using system dynamics: an analysis of methodological frameworks. Electron J Bus Res Methods 8(1):35–46
5. Brailsford SC, Harper PR, Patel B, Pitt M (2009) An analysis of the academic literature on simulation and modelling in healthcare. J Simul 3(3):130–140
6. Brailsford SC, Eldabi T, Kunc M, Mustafee N, Osorio AF (2019) Hybrid simulation modelling in operational research: a state-of-the-art review. Eur J Oper Res 278(3):721–737
7. Djanatliev A, German R (2013) Prospective healthcare decision-making by combined system dynamics, discrete-event and agent-based simulation. In: Proceedings of the 2013 winter simulation conference. IEEE, pp 270–281
8. den Hengst M, de Vreede GJ, Maghnouji R (2007) Using soft OR principles for collaborative simulation: a case study in the Dutch Airline Industry. J Oper Res Soc 58(5):669–682
9. Dehghan H, Nahavandi N, Chaharsooghi SK, Zarei J, Amin-Naseri MR (2022) A hybrid game theory and system dynamics model to determine optimal electricity generation mix. Comput Chem Eng 166:107990
10. Elbattah M, Molloy O (2016) Coupling simulation with machine learning: a hybrid approach for elderly discharge planning. In: Proceedings of the 2016 ACM SIGSIM conference on principles of advanced discrete simulation, pp 47–56
11. Eldabi T, Jun T, Clarkson J, Connell C, Klein J (2010) Model-driven health care: disconnected practices. In: Proceedings 2010 winter simulation conference, Baltimore, pp 2271–2282
12. Epstein JM, Pankajakshan R, Hammond RA (2011) Combining computational fluid dynamics and agent-based modeling: a new approach to evacuation planning. PLoS ONE 6(5):e20139
13. Goh YM, Ali MJA (2016) A hybrid simulation approach for integrating safety behavior into construction planning: an earthmoving case study. Accid Anal Prev 93:310–318
14. Harper A, Mustafee N (2019) A hybrid modelling approach using forecasting and real-time simulation to prevent emergency department overcrowding. In: Proceedings of the 2019 winter simulation conference. IEEE, pp 1208–1219
15. Harper A, Mustafee N (2023) Strategic resource planning of endoscopy services using hybrid modelling for future demographic and policy change. J Oper Res Soc 74(5):1286–1299
16. Harper A, Mustafee N, Feeney M (2017) A hybrid approach using forecasting and discrete-event simulation for endoscopy services. In: Proceedings of the 2017 winter simulation conference. IEEE, pp 1583–1594
17. He J, Qi Z, Hang W, King M, Zhao C (2011) Numerical evaluation of pollutant dispersion at a toll plaza based on system dynamics and CFD. Transp Res Part C Emerg Technol 19(3):510–520
18. Jahangirian M, Eldabi T, Naseer A, Stergioulas LK, Young T (2010) Simulation in manufacturing and business: a review. Eur J Oper Res 203(1):1–13
19. Karatas M, Onggo BSS (2019) Optimising the barrier coverage of a wireless sensor network with hub-and-spoke topology using mathematical and simulation models. Comput Oper Res 106:36–48
20. Katsaliaki K, Mustafee N (2011) Applications of simulation within the healthcare context. J Oper Res Soc 62(8):1431–1451
21. Kotiadis K, Tako A, Vasilakis C (2014) A participative and facilitative conceptual modelling framework for discrete event simulation studies in healthcare. J Oper Res Soc 65(2):197–213
22. Kieckhäfer K, Volling T, Spengler TS (2014) A hybrid simulation approach for estimating the market share evolution of electric vehicles. Transp Sci 48(4):651–670
23. Lehaney B, Paul RJ (1996) The use of soft systems methodology in the development of a simulation of out-patient services at Watford General Hospital. J Oper Res Soc 47(4):864–870
24. Lendermann P, Gan BP, McGinnis LF (2001) Distributed simulation with incorporated aps procedures for high-fidelity supply chain optimisation. In: Proceedings of the 33rd winter simulation conference. IEEE, pp 1138–1145

25. Mingers J, Brocklesby J (1997) Multi-methodology: towards a framework for mixing methodologies. Omega 25(5):489–509
26. Montañola-Sales C, Onggo BSS, Casanovas-Garcia J, Cela-Espín JM, Kaplan-Marcusán A (2016) Approaching parallel computing to simulating population dynamics in demography. Parallel Comput 59:151–170
27. Mustafee N, Bischoff EE (2013) Analysing trade-offs in container loading: combining load plan construction heuristics with agent-based simulation. Int Trans Oper Res 20(4):471–491
28. Mustafee N, Katsaliaki K (2020) Classification of the existing knowledge base of OR/MS research and practice (1990–2019) using a proposed classification scheme. Comput Oper Res 118:104920
29. Mustafee N, Harper A, Fakhimi M (2022) From conceptualization of hybrid modelling & simulation to empirical studies in hybrid modelling. In: Proceedings of the 2022 winter simulation conference (WSC), 11th–14th Dec 2022, Singapore. IEEE, pp 1199–1210
30. Mustafee N, Harper A, Onggo BS (2020) "Hybrid modelling and simulation (M&S): driving innovation in the theory and practice of M&S. In: Proceedings of the 2020 winter simulation conference, December 14–18, virtual conference. IEEE, pp 3140–3151
31. Mustafee N, Powell JH (2018) From hybrid simulation to hybrid systems modelling. In: Proceedings of the 2018 winter simulation conference (WSC), December 9–12, Gothenburg, Sweden. IEEE, pp 1430–1439
32. Mustafee N, Powell J (2020) Providing real-time information for urgent care. Impact 2021(1):25–29
33. Mustafee N, Taylor SJE (2009) Speeding up simulation applications using WinGrid. Concurr Comput Pract Exp 21(11):1504–1523
34. Mustafee N, Taylor SJE, Katsaliaki K, Brailsford SC (2009) Facilitating the analysis of a UK national blood service supply chain using distributed simulation. SIMULATION 85(2):113–128
35. Nagasawa N, Hibino H, Hashimoto M, Kase N (2017) Hybrid simulation method by cooperating between manufacturing system simulation and computational fluid dynamics simulation. In: Proceedings of the 2017 IEEE International conference on industrial engineering and engineering management. IEEE, pp 1616–1620
36. Oleghe O, Salonitis K (2019) The application of a hybrid simulation modelling framework as a decision-making tool for TPM improvement. J Qual Maint Eng 25(3):476–497
37. Oliveira JB, Lima RS, Montevechi JAB (2016) Perspectives and relationships in supply chain simulation: a systematic literature review. Simul Model Pract Theory 62:166–191
38. Onggo BS, Mustafee N, Smart A, Juan AA, Molloy O (2018) Symbiotic simulation system: hybrid systems model meets big data analytics. In: Proceedings of the 2018 winter simulation conference. IEEE, pp 1358–1369
39. Ören TI (2005) Toward the body of knowledge of modeling and simulation. In: Interservice/industry training, simulation, and education conference (I/ITSEC), pp 1–19
40. Park H, Fishwick PA (2010) A GPU-based application framework supporting fast discrete-event simulation. SIMULATION 86(10):613–628
41. Pessôa LAM, Lins MPE, da Silva ACM, Fiszman R (2015) Integrating soft and hard operational research to improve surgical centre management at a University Hospital. Eur J Oper Res 245(3):851–861
42. Philip AM, Prasannavenkatesan S, Mustafee N (2023) Simulation modelling of hospital outpatient department: a review of the literature and bibliometric analysis. SIMULATION 99(6):573–597
43. Powell JH, Mustafee N (2017) Widening requirements capture with soft methods: an investigation of hybrid M&S studies in healthcare. J Oper Res Soc 68(10):1211–1222
44. Robinson S, Radnor ZJ, Burgess N, Worthington C (2012) SimLean: utilising simulation in the implementation of lean in healthcare. Eur J Oper Res 219(1):188–197
45. Roemer AC, Strassburger S (2019) Hybrid system modeling approach for the depiction of the energy consumption in production simulations. In: Mustafee N, Bae K-HG, Lazarova-Molnar S, Rabe M, Szabo C, Haas P, Son Y-J (eds) Proceedings of the 2019 winter simulation conference. IEEE, Piscataway, NJ, pp 1366–1377

46. Seok MG, Kim TG, Choi C, Park D (2016) A scalable modeling and simulation environment for chemical gas emergencies. Comput Sci Eng 18(4):25–33
47. Suleiman R, Troitzsch KG, Gilbert N (eds) (2012) Tools and techniques for social science simulation. Springer
48. Taylor SJ, Anagnostou A, Kiss T, Terstyanszky G, Kacsuk P, Fantini N et al (2018) Enabling cloud-based computational fluid dynamics with a platform-as-a-service solution. IEEE Trans Indus Inf 15(1):85–94
49. Tako AA, Kotiadis K (2015) PartiSim: a multi-methodology framework to support facilitated simulation modelling in healthcare. Eur J Oper Res 244(2):555–564
50. Terzi S, Cavalieri S (2004) Simulation in the supply chain context: a survey. Comput Ind 53(1):3–16
51. Tolk A, Harper A, Mustafee N (2021) Hybrid models as transdisciplinary research enablers. Eur J Oper Res 291(3):1075–1090
52. Traoré MK (2003) Foundations of multi-paradigm modeling and simulation: a meta-theoretic approach to modeling and simulation. In: Proceedings of the 2003 winter simulation conference. IEEE
53. Viana J, Simonsen TB, Faraas HE, Schmidt N, Dahl FA, Flo K (2020) Capacity and patient flow planning in post-term pregnancy outpatient clinics: a computer simulation modelling study. BMC Health Serv Res 20(1):1–15
54. Wang B, Brême S, Moon YB (2014) Hybrid modeling and simulation for complementing lifecycle assessment. Comput Ind Eng 69:77–88
55. Wu C, Chen C, Jiang R, Wu P, Xu B, Wang J (2019) Understanding laborers' behavioral diversities in multinational construction projects using integrated simulation approach. Eng Constr Arch Manag 26(9):2120–2146
56. Xu X, Wang J, Li CZ, Huang W, Xia N (2018) Schedule risk analysis of infrastructure projects: a hybrid dynamic approach. Autom Constr 95:20–34
57. Zare H, Tavana M, Mardani A, Masoudian S, Kamali Saraji M (2019) A hybrid data envelopment analysis and game theory model for performance measurement in healthcare. Health Care Manag Sci 22:475–488
58. Zschaler S, Polack FA (2023) Trustworthy agent-based simulation: the case for domain-specific modelling languages. Softw Syst Model 22(2):455–470

Chapter 2
Using the Modelling Frame in the Conceptual Modelling Activity to Improve the Advantages of Hybridisation

William Jones, Kathy Kotiadis, Jesse R. O'Hanley, and Stewart Robinson

Abstract Conceptual modelling (CM) is fundamental to the complex activity of simulation modelling, yet it remains widely misunderstood. The emergence and growing adoption of hybrid simulation (HS) have added layers of complexity due to the multitude of methods and the diverse expertise of modellers. This chapter introduces a novel representation method to elucidate the modelling frame for HS studies. This method, designed to complement existing CM techniques, guides modellers in pinpointing and communicating the best possible amalgamation of modelling methods tailored for specific projects. The introduced approach pivots on five core components: the combination of simulation techniques, the combination of simulation with analytic techniques, the modelling environment, the experimentation approach, and study outputs. By enhancing the conceptual clarity of HS models, our representation method paves the way for improved model quality, more effective stakeholder engagement, and expedited project timelines. Furthermore, it facilitates a smoother translation from system descriptions to operational computer models.

W. Jones (✉)
Australian Centre for Field Robotics, University of Sydney, 8 Little Queen Street, Chippendale, NSW 2008, Australia
e-mail: william.jones@sydney.edu.au

K. Kotiadis
Kent Business School, University of Kent, Canterbury CT2 7FS, UK
e-mail: K.Kotiadis@kent.ac.uk

J. R. O'Hanley
Centre for Logistic and Sustainability Analytics, Kent Business School, University of Kent, Canterbury CT2 7NB, UK
e-mail: J.Ohanley@kent.ac.uk

S. Robinson
Newcastle University Business School, Newcastle Upon Tyne NE1 4SE, UK
e-mail: stewart.robinson@newcastle.ac.uk

The chapter underscores the pivotal role of our method in bridging communication divides, fostering understanding, and enabling agile software development, ultimately aiding in streamlining the creation of simulation models.

Keywords Conceptual modelling · Modelling frame · Stakeholder engagement · Representation method

2.1 Introduction: The Conceptual Model Activity in Hybrid Simulation

Simulation modelling is a complex process and conceptual modelling (CM) is a critical activity of it. Despite its importance, CM is often poorly understood [1]. While there is no universally accepted definition for a conceptual model [2], the term broadly encompasses "a non-software specific description of the computer simulation model (that will be, is or has been developed), describing the objectives, inputs, outputs, content, assumptions and simplifications of the model" [3] [p. 380].

CM is considered one of the most challenging aspects in simulation model development and usage [4, 5]. While coding and testing can also be difficult, there are established methods to follow. CM, on the other hand, is often referred to as an "art" [6, 7]. This challenge is exacerbated in the case of hybrid simulation (HS), where the expertise and experience of modellers employing a wide range of possible methods will vary. Their perspective on the purpose of the model and, therefore, their assessment of the suitability and benefits of incorporating different methods will be greatly influenced by their prior experience [3]. Models are both "frame and picture" [8] [p. 51], capturing not only what the modellers and stakeholders can see, but also their way of seeing it.

To create a hybrid model, a modeller must not only understand *how* to hybridise the model but also *why* hybridisation is necessary and the advantages it can provide. The *how* presents a technical challenge of implementation. But at the CM activity, defining *why* (i.e., the advantages or insights hybridisation may bring to the problem at hand) is key. *How* must be considered, since implementation must after all be feasible. Likely adding hybridisation adds complexity to the model, so *how* it will be coded must be defined. However, an assessment of whether the benefits justify the additional effort can only be properly determined once the *why* is determined.

A model is typically created in response to a specific need and it evolves as the understanding of the problem improves. By applying a methodology at the CM stage, modellers can determine the most suitable model for the study, which aids in identifying a good model [7]. However, a review of HS research [9] reveals that the CM activity is often overlooked. A HS CM methodology could assist other modellers and stakeholders in comprehending the modeller's view of the system, as well as understanding both *how* the model can address the system problems and *why* hybridisation was incorporated and the advantages it brings. This would enable other modellers and stakeholders to evaluate whether the model or methodology is appropriate for the study at hand.

In this chapter, we present a straightforward method for representing the *modelling frame*. The term *modelling frame* is defined as 'the choice and combination of modelling approaches' for addressing the problem of interest in HS studies. It can be thought of as the lens through which a modeller sees a problem and the set of tools the modeller has identified for addressing it. The representation method, which was first introduced by Jones et al. [10], takes into account the various definitions for the term *hybrid* used in the literature from different fields. These definitions broadly consider HS "as a single simulation method combined with any or multiple of the following; another simulation method, an analytic technique or another method that enhances the simulation study" [10]. See the literature review of [10] for a detailed review of the variety of existing definitions of HS. In recent literature, a distinction is increasingly made between HS, simulation combined with simulation, and hybrid modelling, simulation combined with another technique [11, 12]. To our knowledge, our proposed modelling frame is the only example of a technique for concisely conveying HS approaches in the existing literature.

The key benefit of our method is that it provides modellers with a tool to clearly convey *how* and, more importantly, *why* hybridisation should be introduced in a particular study. Our method can be utilised at any stage of the modelling process, such as pre, during, or post-model development. Modellers can also use our method as a conceptual tool to aid their thinking about the structure of their study, the opportunities to hybridise, and the benefits of hybridising, in addition to explaining, communicating, and documenting proposed or existing models.

The representation method is not meant to replace existing CM approaches that are based on a single simulation paradigm [3, 7, 13–16]. Rather, it is designed to complement those methods and assist modellers who wish to go beyond a single paradigm when developing a HS. By using our method, modellers can determine the nature of the HS (i.e., the combination of methods) and communicate it effectively. The representation method enables modellers to move from a system description of what is known about the world, through a defined modelling frame that outlines the choice and combination of modelling approaches, all the way to a complete conceptual model that describes the objectives, inputs, outputs, content, assumptions, and simplifications of the model [3]. The individual conceptual models can then be expressed using appropriate representations for each simulation paradigm adopted.

The use of hybrid simulation (HS) has increased over the past two decades [9]. However, more research is needed to be able to measure the value added by hybridisation. Robinson [17] identifies model representation as a significant challenge in CM. There is a need for further research to explore methods that can facilitate the incorporation of hybridisation at the earliest stages of CM. Adopting our representation method during the CM activity can realise significant benefits that will be discussed in this chapter such as improved model quality, stakeholder engagement, and project timeframes.

The remainder of this chapter proceeds as follows. Section 2.2 presents our method for representing the modelling frame, providing a detailed description of each component and the structure. Section 2.3 uses the representation method to illustrate an example hybrid study. Section 2.4 explains *how* and *why* the modelling frame representation method should be used, including a discussion of the benefits it provides in determining the nature of a HS and its communication, planning and documentation. Section 2.5 summarises the key take-home messages of the chapter.

2.2 Representation of the Modelling Frame

In Jones et al. [10], we introduced a technique for illustrating the modelling frame (i.e., the choice and combination of modelling approaches). This representation method can provide a conceptual-level description of a modelling study [18] and demonstrate how the proposed model addresses the problem at hand. It reveals the combination of simulation paradigms/methods and other elements that come together to form a HS. The representation method serves as a link between the system description, which encompasses all available information about the real system and the conceptual model (i.e., a description of the computer simulation model [3, 7]). It can assist modellers in defining the blend of methods forming the hybrid model. Thus, it helps modellers move from the system description to the modelling frame and eventually to the full conceptual models of each modelling paradigm for abstractly representing the real system. Additionally, the method can be utilised to depict the modeller's vision of the modelling frame at any point during, or after, the iterative model development process.

The representation method can be used in conjunction with any existing HS frameworks. HS often brings together techniques developed by groups with diverse expertise. Tolk et al., framework [11], for example, is useful in identifying the level of abstraction and for demonstrating compatibility. This representation method can help modellers communicate the reasons for selecting specific components of the hybrid modelling study and how they are suited to solve the problem at hand. Unlike other methods that provide a set of steps to develop a HS [19–22], our representation technique acknowledges the highly iterative nature of creating a conceptual model and the simulation lifecycle [3, 23, 24]. Therefore, it accommodates and encourages the ad hoc integration of hybridisation, which is commonly seen in HS when the limitations of single-method approaches are encountered [9, 25]. It is a loose representation that the user is free to modify to meet the specific needs of their project.

As Fig. 2.1 shows, our representation method can be progressively assembled from its five components: (1) HS representation, (2) HS and analytic model representation, (3) hybrid modelling environment representation, (4) hybrid experimental approach representation, and (5) hybrid outputs representation, each of which is described in detail below. This dissection of a hybrid study is unique to this representation and the name of each component aims to describe what it may capture. As noted above, components 2–4 may all be categorised as hybrid modelling, our representation divides that category into a more granular view. In addition to the core HS (component 1) and HS analytic model (component 2), both hard and soft methods can be incorporated alongside one another within components 3–5 and are considered equally significant. It is important to note that our representation method is software independent. The components of the representation can be approached in any order and can evolve in any way that is deemed appropriate. Further, it is important to note that the specific techniques mentioned in the following description are merely examples and the representation method is no way limited to those.

Figure 2.2 illustrates how the framework can be populated with some example hybridised methods and outcomes. More details on each section of the framework and how it may be populated are given in the following subsections, one for each component of the framework.

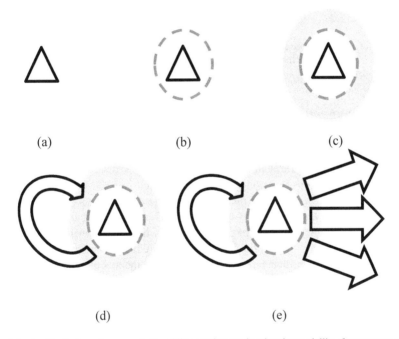

Fig. 2.1 An illustration of progressively adding sections to develop the modelling frame representation, **a** HS representation, **b** HS and analytic model representation, **c** hybrid modelling environment representation, **d** hybrid experimental method representation, **e** hybrid outputs representation. See how sections can be populated to illustrate a model in Fig. 2.2. The authors sometimes refer to this simple representation of the modelling frame as the 'teapot', with the specific benefits important to capture and motivate the hybridisation being the 'tea' that comes out of the pot

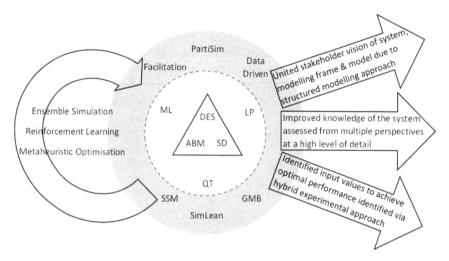

Fig. 2.2 Full example hybrid modelling frame (teapot) representation reproduced from Jones et al. [10]. The specific techniques/benefits stated here are only used as examples to illustrate the representation

2.2.1 Component 1: Hybrid Simulation Representation

A HS model in the context of operational research (OR) consists of a combination of at least two out of the three main simulation paradigms: discrete-event simulation (DES), system dynamics (SD), and agent-based modelling (ABM). The HS triangle, shown in Fig. 2.1a is a subjective measure based on the modeller's perception of the relative importance and complexity of the DES, SD, and ABM elements in a model. The user of the framework can mark with an 'x' where they consider their model sits within the triangle. This definition of a HS model assumes genuine integration of the paradigms, where their interaction is integral to the model's functioning. This is distinct from a scenario where two separate models simply exchange data in a sequential manner [26]. While this simpler form of HS can still be represented using our representation method, it would be clearer to show the sequence (e.g., DES → SD) or the use of a single paradigm instead of the hybrid triangle at the centre of the representation. Alternatively, other simulation methods such as Monte Carlo, could replace the suggested depictions in the hybrid triangle.

2.2.2 Component 2: Hybrid Simulation and Analytic Model Representation

The simulation triangle can be expanded (Fig. 2.1b) and a model can be considered a hybrid simulation and analytic (HSA) model when it combines one or more simulation paradigms or methods with one or more analytic techniques, such as queuing theory (QT), linear programming (LP), or machine learning (ML) [27, 28]. To be considered a HSA model in this representation, the model must use both simulation and analytic techniques in combination when the model is executed, differentiating it from a model that uses hybrid experimentation (as described in Sect. 2.2.4d). Clearly defining the introduction of analytic techniques (i.e., as a HSA model or through hybrid experimentation) improves the ability of our representation method to communicate the purpose of the hybridisation.

In Fig. 2.1b, the position of the analytic techniques relative to the simulation triangle is not meant to imply any particular relationship, but users of the representation method may choose to place the analytic techniques near the simulation method they are linked to in the design of their conceptual model (e.g., DES and LP), or use arrows to connect the two. The specific placement of the example techniques in the representation method (Fig. 2.2) should not be viewed as meaningful.

2.2.3 Component 3: Hybrid Modelling Environment Representation

A modelling approach can be considered hybrid if it incorporates a mixture of different techniques during its development. Specific techniques for data-driven analysis of trends, common in the development of ML models [29], may be used to define the model parameters from existing data. Alternatively, soft methods, such as group model building (GMB) [30], soft systems methodology (SSM) [31], stakeholder facilitated modelling [32], PartiSim [33], and SimLean [34], which aim to engage stakeholders or provide a structured approach to building a model may be used, thus also hybridising model development. The model can be thought of as being contained within a hard and or soft hybrid modelling environment (see Fig. 2.1c).

2.2.4 Component 4: Hybrid Experimentation Approach Representation

Once a simulation model has been created, the modeller will then run experiments with it. Simply running the model with certain inputs to produce outputs is not considered a hybrid experimentation method. However, the experimentation can be considered hybrid if it incorporates additional techniques such as ensemble simulations to explore a range of parameters [35, 36], metaheuristics such as genetic algorithms [37] or tabu search [38] to find an optimal set of input parameters, or methods like reinforcement learning [39], where the model is repeatedly run to determine the best combination of settings to achieve a desired outcome. These examples and other experimental methods often require many iterations of the model, sometimes needing high-performance computing resources or the ability to run the simulations in parallel. Our representation method can be extended (as seen in Fig. 2.1d) to accommodate instances where a model is hybridised through a hybrid experimentation approach. It is worth noting that some of the techniques mentioned here may also be found in the HSA model component of the representation method. However, to be considered a HSA model with a feature to be captured in component 2 of the framework, the simulation and analytic techniques must be integrated and used within a single run of the model. Here, in the hybrid experimentation component of the representation method, introduced techniques are used outside the core HS to run the complete simulation model multiple times. Clearly differentiating between these two methods of hybridising simulation with analytic techniques enhances the ability of our representation method to communicate the design of a model.

2.2.5 *Component 5: Hybrid Outputs Representation*

The final component of our representation method (Fig. 2.1e) highlights the advantages of combining different techniques from the viewpoint of the individuals or stakeholders involved in developing the model. The results showcase the reasons behind incorporating different approaches and the benefits that these combinations bring to the simulation study, as deemed important by the modelling team. Introducing hybridisation and incorporating additional methods may offer a broader perspective on the situation of interest and capture previously overlooked features. Integrating analytic techniques into the model can also provide similar benefits [27, 28]. Developing the model in collaboration with stakeholders using techniques such as GMB and then experimenting with large ensemble simulations and metaheuristic optimisation techniques can offer numerous advantages some of which may even be qualitative in nature. Engaging stakeholders, for instance, can lead to learning throughout the model development process and bring the team together around a shared vision of the system [40, 41]. Additionally, combining the model with heuristic optimisation techniques can reveal relationships between outputs and inputs [35, 36] or determine the best inputs for the model [37, 38]. Hybridising the model may become necessary when the limitations of a single simulation method are reached due to the complexity of the problem [9].

2.3 Example Hybrid Study Using the Modelling Frame

Figure demonstrates how the representation may be used to illustrate the modelling frame for a specific study. In this example, all components of the frame are populated with at least some text, indicating that the study uses each type of hybridisation to some extent, note however, populating all components is not a requirement of the modelling frame. The figure captures a HS methodology developed to support freight rail operations in the mining industry. Details of the study are described in Jones and Gun [42]. The work automated operations planning by providing an on-demand tool for determining train destinations and generating a feasible timetable capable of meeting operational needs (component 5—hybrid outputs).

Work by Jones and Gun [42] addresses a common challenge face by mining organisations. When extracting ore from a mine site, the quality of the material produced fluctuates over time. To maintain a consistent ore percentage, mine planners (staff managing onsite operations) mix higher and lower quality material from different areas of the site. With a network of mine sites, materials from multiple sites are combined so as to ensure that the blend of material received at the ports still meets requirements. In the case, Jones and Gun [42] describe, which is typical for large mining organisations, a rail network connects individual mine sites to the ports, with mined material transported by large freight trains. The rail operation is in effect used to balance the individual mine site fluctuations in ore quality.

Jones and Gun [42] present a HS method for generating schedules for a mining freight rail fleet that meets port requirements for material quality and move trains in an efficient way within the constraints of the network. At any given time, each mine can load a number of trains with a quality of material at a known rate. Operational requirements dictate the quantity and quality of material needed to reach each ship in the port. Hence, demand (i.e., the number of trains required to visit each mine site in any period in order to achieve the required mix of material at the ports) is known. The challenge is to determine which trains should be directed to which mine sites and in what order to best satisfy demand and to generate efficient timetables that can achieve this in a limited amount of time. Due to the operational complexity of mining operations and the uncertain nature of the problem, which includes changing external conditions, requirements may be revised at any moment, thereby necessitating updates to the demand at each site and train destinations. Some using the frame may wish to note the objective on the representation, e.g. as a teapot lid.

The problem Jones and Gun [42] seek to address of choosing optimal destinations for a fleet of trains in a large network is a significant computational challenge (NP-hard in the general case). To manage the computational challenge, they devised a method that combines DES and ABM (Component 1–HS) with heuristics (Component 2–HSA model) and an ensemble of simulation runs (Component 4–HS experimental method) to significantly reduce the parameter space. The model was developed by collaborating closely with a stakeholder with relevant domain expertise (Component 3–hybrid modelling environment). Regular collaboration was required over the long period of iterative development and data sharing among the parties to define parameters and validate the methodology.

In line with the categorisation presented in Mustafee et al. [43], a purely ABM and DES model would be categorised as a multi-technique hybrid model (Type-B). When a simulation model is combined with other OR techniques, such as a heuristic, as in Jones and Gun [42], it becomes a hybrid systems model (Type-D). However, this categorisation gives no indication of how the methods were combined or why the combination was suitable for addressing the problem at hand. In evaluating a simulation methodology's suitability, importance should be given to both the selected modelling frame and the design of the conceptual model. Depicting the modelling frame helps explain how and why the combination of modelling approaches is suitable for a given problem. The suitability of the conceptual model and the overall study can be evaluated only if the modelling frame's suitability is justified.

The heuristics incorporated within the modelling frame are essential for addressing the problem of Jones and Gun [42]. The heuristic algorithms developed govern the individual movements of train agents between intermediate nodes within the network and their destinations. Due to the large number of agents and their interactions, their behaviours are difficult to predict such that their movements can only be observed by running the simulation. The adopted modelling frame includes an ensemble experimental method for generating a feasible timetable, which is essential for addressing the task at hand (Fig. 2.3). Since the diagrammatic representation may quickly become overcrowded with text, more detail can be captured in Table 2.1. In Jones et al. [10], two further case studies illustrate the flexibility of the representation method in conveying different types of hybrid models.

2.4 How and Why to Use the Modelling Frame

We have outlined a new method for representing the modelling frame in hybrid simulation studies that can assist modellers in conveying the relationship between the study design and the research questions of interest. Any stage of model development (e.g. pre, during, or post) can benefit from the representation method. At the beginning of a study, as a conceptual model is emerging, the representation provides a guide to aid modellers' thinking. As a conceptual model is being further formed, it provides a critical lens with which to view the model and question if hybridisation is being used to best effect. Subsequently, as the conceptual model is being expanded, it provides a tool to communicate the model to clients, stakeholders, or other modellers. Once a complete conceptual model has been developed, the representation provides a guide to begin thinking about how the conceptual model can be coded into a computer model. After a computer model is developed, it can provide a useful contribution to model documentation by illustrating a model's structure.

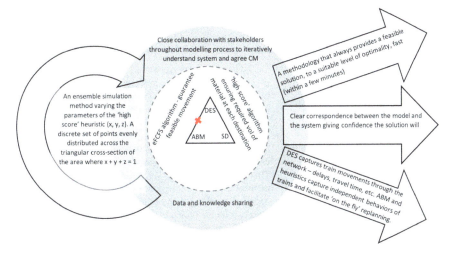

Fig. 2.3 Example modelling frame (teapot) representation illustrating Jones and Gun's [42] HS methodology combining ABM, DES, heuristics and an ensemble simulation methodology to aid planning of a mining freight rail operation

2.4.1 To Aid in Determining the Nature of the HS

CM requires a deep understanding of the system or process being modelled, including its structure, behaviour, and underlying principles. This can be a challenge because real-world systems are often complex and dynamic, with numerous interdependent variables and feedback loops that can be difficult to model accurately. It is a particularly relevant challenge for HS, which is often applied to highly complex systems where a single paradigm approach is not able to capture the system to the level of fidelity the modeller requires [9]. One of the main challenges of CM for HS is selecting the appropriate simulation techniques and integrating them effectively into a cohesive and viable model. This requires a sound grasp of each technique and how they can be combined to accurately represent the system being modelled, while simultaneously balancing model accuracy and complexity.

HS offers the advantage of allowing modellers to examine a situation of interest from different philosophical perspectives [18, 46]. When developing a model, modellers must clearly articulate the benefits that the model can bring to the problem at hand. The proposed conceptual model should be valid, credible, feasible, and useful [7]. Although models aim to reflect the real world, simplifications are always necessary. Many of these simplifications are hidden in model design assumptions and software implementation [47]. Each form of hybridisation has its own assumptions and simplifications.

The representation method we propose outlines the role that each form of hybridisation plays in addressing the problem. In this way, it encourages a process of identifying the components of the model and outlining their relationships and impacts.

Table 2.1 Example modelling frame giving additional details of the high-level description captured in Fig. 2.3

Component	Details
(1) HS	• DES–Given the core rail network has well-defined parameters (travel times, loading/unloading times, average train speed, etc.), it is appropriate to utilise a DES approach to model the system in a comprehensive manner • ABM–The adopted modelling approach considers each terminal and train agent as independent and autonomous entities, which closely aligns with an ABM perspective. This perspective focuses on modelling the entities themselves and the interactions among them. The approach used to generate timetables in the model aligns closely with the philosophy of ABM; relying on the micro-decisions of individual agents. The authors recognise that a central controller capable of optimising the movement of all trains would be the preferred approach, however, due to the scale of the system they are modelling applying mathematical optimisation would not be computationally tractable and so use an agent-based approach representing the network infrastructure as the environment (constraints) in which the trains operate and prescribing rules governing movements of individual trains and their interactions with each other and the network infrastructure [44] • Note the 'x' in Fig. 2.3 falls on the ABM/DES edge of the triangle, indicating no SD component in this model
(2) HSA model	• An extended first come first served (eFCFS) policy governs train movements to try to avoid grid-lock situations where trains need to reverse. It works by requiring trains to reserve sections of track up to the next available passing point before proceeding. The deterministic logic of eFCFS is repeatedly applied over time as trains progress along their journey, generating a timetable as the stochastic simulation runs • The method for selecting an optimal destination for a train involves the use of a 'high-score' methodology, which calculates three metrics for each possible destination option. The three metrics include destination priority, traffic density on the route, and estimated waiting time. An aggregate score can be calculated based on these metrics, indicating the best (highest scoring) destination option at the time of calculation. The three metrics reflect the network state and are regularly recalculated at major junction nodes to ensure the destination selected remains the best choice for each train (adjusted in response to stochastic events)

(continued)

Table 2.1 (continued)

Component	Details
(3) Hybrid modelling environment	• Due to the complexity of the system being modelled, the high-fidelity model was developed parsimoniously [45] over an extended period while working closely with stakeholders who have expert knowledge of the system. A process of regular collaboration between domain and modelling experts using an agile approach that adapts and responds to feedback and new information as they arrive was adopted. This process was crucial in appropriately constraining the train agents' behaviours within the model and ensuring that emergent timetables were suitably accurate and usable in real-world scenarios • Significant data sharing was required
(4) Hybrid experimental approach	• The ensemble method identifies the best train timetable by finding the input values to the high-score method (x, y, and z) that ensure demand at each mine site is met. Varying the relative input values adjusts the outcome of high-score method. Simulating different combinations of the triangular cross section $x + y + z = 1$ of the three-dimensional parameter space produces different timetables that prioritise each parameter in the high-score method at different levels. The best combination of inputs can be identified by considering both demands satisfied and fairness across destinations. The method does not require multiple repetitions of simulations. The identified high-score method input values produce the best timetable (i.e., the one that satisfies operational requirements in the shortest amount of time)
(5) Hybrid outputs (numbers shown relate the output back to the hybridisation method 1–4)	(1) The hybrid approach captures the system in a way that reduces the communication gap between the modeller and stakeholder and provides clear face validation (2) The incorporation of heuristics within the modelling frame is essential for addressing the problem at hand. Predicting the behaviours of a large number of agents is very challenging due to their numerous interactions and the stochastic nature of some elements of the simulation. The eFCFS and destination selection heuristics guarantee a feasible timetable is produced for every simulation run (3) Regular stakeholder engagement promoted buy-in and the model that was developed captured modellers' and stakeholders' shared vision of the system. The high-fidelity model incorporated the domain expertise of key stakeholders (4) The ensemble approach enables a good-quality timetable to be produced in a computationally efficient manner. Where standard mathematical optimisation would not be computationally tractable, this method provides a sub-optimal answer, but, one that is good enough for meeting operational requirements and likely better than a solution developed manually

Additionally, it can serve as a reference for modellers during the CM activity, helping them to evaluate if they are making the most of hybridisation and whether further hybridisation would add value to the model and the problem being addressed. Accordingly, it can assist modellers and stakeholders in evaluating the suitability of the model's design which is a crucial step in the validation process [40].

2.4.2 To Communicate Hybrid Simulation Studies

Multiple examples of HS exist within the current literature. However, the reasoning behind hybridisation and the advantages of the selected modelling frame are often unclear. This may be due to the difficulty in communicating the modelling frame and its rationale. Defining the modelling frame before developing a model can determine when HS would be advantageous and communicate the benefits to stakeholders for HS studies. Our representation method allows modellers to convey how their chosen modelling frame is suitable for addressing the problem in question by identifying the hybrid outputs the modelling frame will produce. Modellers can then evaluate if these outputs are suitable for achieving the modelling objectives of the problem. The representation method also highlights the limitations of the selected modelling frame and helps with identifying issues not directly addressed by the chosen modelling approaches. Aligning the model to the representation method can enable modellers to convey their understanding of the system and how their proposed model can address the problem, explaining why hybridisation is necessary and its benefits. This, in turn, enables other modellers and stakeholders to determine if the model and methodology are appropriate for the study.

Key to developing a suitable conceptual model is engaging collaboratively with stakeholders to adequately understand the system in question. The benefits of a collaborative approach are clear [33, 48–54] and the risks of failing to engage (i.e., results not being accepted or acted upon [49, 51, 55]) are well known. Stakeholder-facilitated modelling [32] is common practice in certain fields with a variety of existing tools/frameworks for designing and documenting the system to be modelled that directly lend themselves to conceptual modelling. Combining DES with soft systems methodology (SSM) [31] is well explored. Kotiadis et al. [56] present a framework modellers can follow for collaborating with stakeholders to develop a conceptual model and throughout the SM development lifecycle. Their approach advocates engaging stakeholders through structured workshops, often using SSM, to inform the conceptual model design. For SD, group model building (GMB) is a common tool for engaging diverse stakeholders in a process of jointly understanding and addressing complex issues, while increasing engagement in systems practices and confidence in the use of systems ideas [57]. GMB has a long association with SD model development [58] to capture required knowledge of the mental models of the stakeholder group and increase the chances model results are implemented [49, 51, 55]. Surprisingly, there is little, if any, literature formally exploring the benefits of involving stakeholders in ABM studies. However, ABM, by its nature, provides a highly descriptive representation of a system that enables clear correspondence between model and reality, thereby providing a form of face validation [59] and facilitating stakeholder participation in model development and validation [60]. A caveat to note is that some authors [61] argue that while face validation may be slightly easier to obtain for ABMs, quantitative external validation is typically more of a challenge than for DES or SD, especially due to higher levels of output uncertainty due to uncertainty operating at the individual agent level. An inability to quantitatively validate a model post-development will likely lose stakeholder engagement gained early in the CM process due to the design's correspondence with reality. Modellers should consider this risk when designing their model.

These collective CM procedures involve recognising and comprehending the issues, setting objectives, identifying outputs and inputs, and determining the model's content [7]. The outcome is often documented through a paradigm-specific representation method, such as stock and flow, causal loop, process flow, or state diagrams [13, 15, 16]. Additionally, there may be supplementary documentation that details the objectives, inputs, outputs, content, assumptions, and simplifications of the model. These methods capture and visualise the conceptual model developed such that it can be communicated and provide a record of the outcome of the CM activity. In the context of a HS, these methods may still be utilised to visualise the SD, DES, and ABM components of the model. However, there is no standard method for demonstrating how these components come together to form a HS. The modelling frame representation we present helps fill this gap. Used together with single paradigm representation methods, it can explain how they combine to address the question of interest.

A hybrid modelling approach has been found to potentially reduce the communication gap [62, 63] between the modeller and stakeholders by providing a clearer link between the observable system and the model [59]. Our modelling frame representation can help bridge the gap between stakeholder questions and their understanding of how the model can address those questions. Considering the example shown in Fig. 2.3, the model alone may not fully answer stakeholders' questions, though the model combined with a hybrid experimental approach may, as the example shows. Alternatively, it may be only when the model is combined with a collaborative hybrid modelling environment that the study is able to address the key questions of interest. The representation conveys how the proposed hybrid study can address the stakeholders' questions of interest and the impact of the model's various components, making it easier to understand why a hybrid approach should be introduced.

The representation method allows modellers to assess the suitability of the proposed modelling frame for addressing the problem and achieving the modelling objectives. It also highlights the limitations of the selected modelling frame. It is important to note that the representation method does not delve into issues that are not directly addressed by the combination of modelling approaches. Nonetheless, using this method can help modellers effectively communicate their perception of the system, their proposed solution, and why a hybrid approach is necessary. As such, this can provide other modellers and stakeholders with a clear understanding of the model and its methodology.

2.4.3 To Plan the Development of the HS

Introducing hybridisation into a model, almost always, adds complexity to the model code. Many software packages facilitate simulation development via 'drag and drop' through a graphical user interface. However, even for those packages that support HS, typically when hybridisation is introduced, at least some additional coding is required. Complex model development should, therefore, be preceded by planning the model structure and how it will be built.

The representation can help modellers convey how the model will be built. The hybrid frame representation aids clear thinking about a model's structure and its development [64] regardless of the software package or programming language chosen by the modelling team. An HS is often modular in nature [64]. The boundaries of hybridisation are divided within the full-system conceptual model. For many HS models, the modular structure will naturally generate a set of distinct model components to be coded, some of which can be decomposed further. Often, divisions can be made in line with the components of the modelling frame.

Defining the hybrid frame provides a structure that guides model development. It helps modellers conceptualise a full-system conceptual model and consider how hybridisation can be used to maximum effect by encouraging modellers to clearly define the purpose of introducing hybridisation. Once clarified, it can help define the minimum requirements of the model code to satisfy the modelling frame and objectives of the study (i.e., the minimum viable product (MVP) to be developed). An MVP approach ensures an efficient model development process with minimal wasted time/code. Defining the modelling frame can aid with the different MVP stages of development (i.e., components to develop, tasks to do, and the order in which to do them [65]). Defining MVP stages can allow for agile software development [66, 67] and enable the model coding to be divided across a team working in parallel, a benefit noted in Jones and Gun [42]. Developing models collaboratively within a team has clear advantages, particularly in the case of complex examples requiring significant labour-hours to develop. Improved planning of how the model will be developed using the modelling frame representation can potentially reduce the overall time of a study by minimising wasted effort.

Figure 2.4 illustrates how the processes of defining the modelling frame via our representation method fit into the cycle where artefacts emerge during the model development process. The modelling frame aids in abstracting the real world and planning the model development process, bridging the gap between the problem domain and the model domain [3].

Modellers can utilise the representation method to transition from a system description, which covers their understanding of the world, to a well-defined modelling frame that specifies the selection and combination of modelling approaches, and eventually to a comprehensive conceptual model and software implementation design. The modelling frame representation facilitates abstraction of the real world to a conceptual model that can be coded.

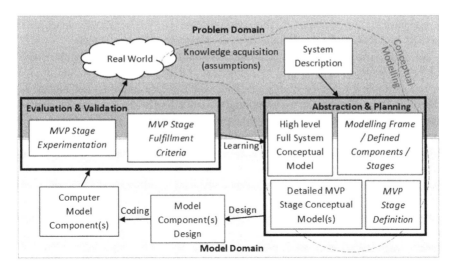

Fig. 2.4 The Artefacts of simulation modelling. Adapted from [65], originally adapted from [3]

2.4.4 To Document Hybrid Simulation Studies

Proper documentation ensures that the model can be replicated, modified, and validated by others, which is critical for scientific rigour and reproducibility. The lack of clarity in reporting simulation studies can make it difficult to reproduce or reuse findings [68]. To address this, Monks et al. [68] propose the Strengthening the Reporting of Empirical Simulation Studies (STRESS) guidelines as a set of good practices for documenting simulation models. For HS, they suggest using specific guidelines for each paradigm.

There are standard ways to illustrate models using a single simulation paradigm [13, 15, 16]. Our representation method alone is not suitable for fully documenting a model for reproduction. Instead, it primarily conveys the suitability of the proposed modelling frame for addressing the problem at hand. By combining with single simulation paradigm representation methods and incorporating the representation method into STRESS guidelines, one can ensure that the hybridisation is reported in a clear and concise manner, providing the necessary information to understand its added value. Our representation method can be combined with existing methods of documentation to provide a clear description of a HS and ensure the reproducibility of simulation studies.

A detailed description of a model from which the model can be reproduced necessitates a highly complex document be produced. Likely, it will be overly technical or difficult to understand for non-experts, which can limit its usefulness to a wider audience. This is particularly important to consider when simulation models are used in decision-making processes that involve stakeholders with varying levels of expertise, not just simulation experts. Our representation method can enhance documentation by summarising complex models, their chosen combination of modelling approaches adopted, and why each adds value into a concise communicable format.

An often-overlooked benefit of simulation studies, to both modellers and stakeholders, is the learning generated along the way [40, 41]. Simulation modelling is a process that is known to be highly iterative, with the conceptual model and model code often undergoing revisions throughout the modelling lifecycle [3, 23, 24], particularly as understanding of the system and problem in question evolves. At any point during or after the iterative development process, the modelling frame representation method can be employed to capture the modeller's perspective of the modelling frame. As the study progresses, the modelling frame and conceptual model may change [23, 24]. Our representation method provides a format for documenting and communicating these changes and the learning generated throughout the modelling lifecycle. If the choice of hybridisation changes, the representation method will also change to reflect this. Documenting each evolution of the representation method helps to track the learning generated by the study.

2.5 Summary

CM is a crucial but often poorly understood activity in the complex process of simulation modelling. HS is increasingly popular for modelling complex systems that present further challenges for CM due to the range of methods available and the varying expertise and experience of modellers. This chapter details a straightforward method for representing the modelling frame in HS studies that can aid modellers in determining the most suitable combination of modelling approaches for a particular project and communicating effectively their approach to stakeholders. This method is not meant to replace existing CM approaches, but rather to complement them and assist modellers in developing a HS. The use of this method can improve model quality, stakeholder engagement, and project timeframes.

The chapter describes a method for representing the modelling frame and understanding HS models. It consists of five components:

- HS representation: which combination of simulation paradigms is used?
- Hybrid simulation and analytic model representation: which analytic techniques are combined with the simulation?
- Hybrid modelling environment representation: what techniques are used during model development?
- Hybrid experimentation approach representation: which additional techniques are used when experimenting with the model?
- Hybrid outputs representation: what advantages does each type of hybridisation bring?

The method aims to improve communication and understanding of hybrid simulation models, which can offer broader perspectives and capture previously overlooked features in complex systems. Defining and illustrating the modelling frame for a HS via the representation method can help modellers in the conceptual modelling activity move from a system description to full computer model. The representation method can be used in conjunction with any existing HS frameworks and accommodates and encourages the ad hoc integration of hybridisation.

Selecting the appropriate techniques and integrating them effectively is essential for achieving HS model accuracy and complexity. The proposed representation method provides a conceptual-level description of the modelling frame, including the choice and combination of modelling approaches, and assists modellers in evaluating the suitability of the model's design. The method can aid modellers with diverse expertise in communicating the hybridised methods adopted and why they are suitable for addressing the problem of interest. Further, it can bridge the communication gap between modellers and stakeholders, making it easier for non-experts to understand why a hybrid approach should be introduced.

The hybrid frame representation aids initially in conceptualising the full-system conceptual model and further in planning the structure of the model to be coded. It facilitates the transition from a system description to a well-defined modelling frame that can be coded. The nature of hybridisation often generates distinct model components that can be coded separately. Developing a clear modelling frame can help define these components and, in turn, enable agile software development and collaborative work. This may reduce the overall time and effort required to develop complex simulation models. The representation method presented can summarise complex models in a concise format. The method provides a clear and concise way to document the choice and combination of modelling approaches adopted and why each adds value to the study.

Acknowledgements This is chapter derived in part from an article published in the Journal of the Operational Research Society 2021 © Operational Research Society, available online: https://doi.org/10.1080/01605682.2021.2018368.

References

1. Tako AA, Eldabi T, Fishwick P, Krejci CC, Kunc M (2019) Panel—towards conceptual modeling for hybrid simulation: setting the scene. In: 2019 Winter simulation conference (WSC). IEEE, pp 1267–79 [Internet]. Available from: https://ieeexplore.ieee.org/document/9004838/
2. Robinson S (2015) A tutorial on conceptual modeling for simulation. In: 2015 Winter simulation conference (WSC). IEEE, pp 1820–1834 [Internet]. Available from: http://ieeexplore.ieee.org/document/7408298/
3. Robinson S (2013) Conceptual modeling for simulation. In: 2013 Winter simulations conference (WSC). IEEE, pp 377–88 [Internet]. Available from: http://ieeexplore.ieee.org/document/6721435/
4. Law A (2007) Simulation modeling and analysis. McGraw-Hill, Boston
5. Brooks RJ, Tobias AM (1996) Choosing the best model: level of detail, complexity, and model performance. Mathl Comput Modell 24
6. Robinson S (2007) The future's bright the future's…Conceptual modelling for simulation! J Simul 1(3):149–52 [Internet]. Available from: https://doi.org/10.1057/palgrave.jos.4250026

7. Robinson S (2008) Conceptual modelling for simulation part II: a framework for conceptual modelling. J Oper Res Soc 59(3):291–304
8. Matos JF, Houston S, Blum W, Carreira SP (2001) Modelling and mathematics education: ICTMA 9-Applications in science and technology, 1st edn. Elsevier
9. Brailsford SC, Eldabi T, Kunc M, Mustafee N, Osorio AF (2018) Hybrid simulation modelling in operational research: a state-of-the-art review. Eur J Oper Res 278(3):721–37 [Internet]. Available from: https://doi.org/10.1016/j.ejor.2018.10.025
10. Jones W, Kotiadis K, O'Hanley JR, Robinson S (2021) Aiding the development of the conceptual model for hybrid simulation: representing the modelling frame. J Oper Res Soc 1–19 [Internet]. Available from: https://doi.org/10.1080/01605682.2021.2018368
11. Tolk A, Harper A, Mustafee N (2020) Hybrid models as transdisciplinary research enablers. Eur J Oper Res [Internet]. Available from: https://linkinghub.elsevier.com/retrieve/pii/S0377221720308869
12. Mustafee N, Powell JH (2018) From hybrid simulation to hybrid systems modelling. In: 2018 Winter simulation conference (WSC). IEEE, pp 1430–9 [Internet]. Available from: https://ieeexplore.ieee.org/document/8632528/
13. Triebig C, Klügl F (2009) Elements of a documentation framework for agent-based simulation models. Cybern Syst 40(5):441–474
14. Robinson S (2008) Conceptual modelling for simulation part I: definition and requirements. J Oper Res Soc 59(3):278–290
15. Coyle RG (1997) System dynamics modelling: a practical approach. J Oper Res Soc 48(5):544–544
16. Oscarsson J, Moris MU (2002) Documentation of discrete event simulation models for manufacturing system life cycle simulation. In: Proceedings of the winter simulation conference. IEEE, pp 1073–1078
17. Robinson S (2020) Conceptual modelling for simulation: progress and grand challenges. J Simul 14(1):1–20
18. Zulkepli J, Eldabi T (2015) Towards a framework for conceptual model hybridization in healthcare. In: 2015 Winter simulation conference (WSC). IEEE, pp 1597–608 [Internet]. Available from: http://ieeexplore.ieee.org/document/7408280/
19. Hwang S, Park M, Lee HS, Lee S (2016) Hybrid simulation framework for immediate facility restoration planning after a catastrophic disaster. J Constr Eng Manag 142(8):04016026
20. Venkateswaran J, Son YJ (2005) Hybrid system dynamic—discrete event simulation-based architecture for hierarchical production planning. Int J Prod Res 43(20):4397–4429
21. Linnéusson G, Ng AHC, Aslam T (2020) A hybrid simulation-based optimization framework supporting strategic maintenance development to improve production performance. Eur J Oper Res 281(2):402–414
22. Nasirzadeh F, Khanzadi M, Mir M (2018) A hybrid simulation framework for modelling construction projects using agent-based modelling and system dynamics: an application to model construction workers' safety behavior. Int J Constr Manag 18(2):132–143
23. Balci O (1994) Validation, verification, and testing techniques throughout the life cycle of a simulation study. Ann Oper Res 53(1):121–173 [Internet]. Available from: https://doi.org/10.1007/BF02136828
24. Willemain TR (1995) Model formulation: what experts think about and when. Oper Res 43(6):916–932 [Internet]. Available from: https://doi.org/10.1287/opre.43.6.916
25. Eldabi T, Balaban M, Brailsford S, Mustafee N, Nance RE, Onggo BS, et al. Hybrid Simulation: Historical lessons, present challenges and futures. In: 2016 Winter Simulation Conference (WSC) [Internet]. IEEE, pp 1388–1403. Available from: http://ieeexplore.ieee.org/document/7822192/
26. Chahal K, Eldabi T (2008) Applicability of hybrid simulation to different modes of governance in UK healthcare. In: 2008 Winter simulation conference. IEEE, pp 1469–1477 [Internet]. Available from: https://ieeexplore.ieee.org/document/4736226/
27. Powell J, Mustafee N (2014) Soft OR approaches in problem formulation stage of a hybrid M&S study. In: Proceedings of the winter simulation conference 2014. IEEE, pp 1664–1675 [Internet]. Available from: http://ieeexplore.ieee.org/document/7020017/

28. Eldabi T, Tako AA, Bell D, Tolk A (2019) Tutorial on means of hybrid simulation. In: 2019 Winter simulation conference (WSC). IEEE, pp 33–44 [Internet]. Available from: https://ieeexplore.ieee.org/document/9004712/
29. Rausch T, Hummer W, Muthusamy V (2020) PipeSim: trace-driven simulation of large-scale AI operations platforms. Available from: http://arxiv.org/abs/2006.12587
30. Peck S (1998) Group model building: facilitating team learning using system dynamics. J Oper Res Soc 49(7):766–7 [Internet]. Available from: https://doi.org/10.1057/palgrave.jors.2600567
31. Checkland P (2000) Soft systems methodology: a thirty year retrospective. Syst Res Behav Sci 17(S1):S11-58
32. Franco LA, Montibeller G. Facilitated modelling in operational research. Eur J Oper Res [Internet]. 2010 Sep;205(3):489–500. Available from: https://doi.org/10.1016/j.ejor.2009.09.030
33. Tako AA, Kotiadis K (2015) PartiSim: a multi-methodology framework to support facilitated simulation modelling in healthcare. Eur J Oper Res 244(2):555–64 [Internet]. Available from: https://doi.org/10.1016/j.ejor.2015.01.046
34. Robinson S, Radnor ZJ, Burgess N, Worthington C (2012) SimLean: utilising simulation in the implementation of lean in healthcare. Eur J Oper Res 219(1):188–197
35. Jones W, Willson RE, Sooriyabandara M, Doufexi A (2016) Wireless network MAC layer performance evaluation with full-duplex capable nodes. In: Proceedings of the 12th ACM symposium on QoS and security for wireless and mobile networks—Q2SWinet '16 [Internet]. New York, New York, USA: ACM Press, pp 111–8. Available from: http://dl.acm.org/citation.cfm?d=2988272.2990294
36. Jones W, Wilson RE, Doufexi A, Sooriyabandara M (2019) A pragmatic approach to clear channel assessment threshold adaptation and transmission power control for performance gain in CSMA/CA WLANs. IEEE Trans Mob Comput (9):1
37. Davis L (1991) Handbook of genetic algorithms. Van Nostrand Reinhold Company, New York
38. Glover F (1990) Tabu search: a tutorial. Interfaces (Providence) 20(4):74–94 [Internet]. Available from: https://doi.org/10.1287/inte.20.4.74
39. Gosavi A (2009) Reinforcement learning: a tutorial survey and recent advances. INFORMS J Comput 21(2):178–92 [Internet]. Available from: https://doi.org/10.1287/ijoc.1080.0305
40. Robinson S (2004) Simulation: the practice of model development and use. Wiley, Chichester
41. Monks T, Robinson S, Kotiadis K (2009) Model reuse versus model development: effects on credibility and learning. In: Proceedings of the 2009 winter simulation conference (WSC) [Internet]. IEEE, pp 767–78. Available from: http://ieeexplore.ieee.org/document/5429691/
42. Jones W, Gun P (2022) Train timetabling and destination selection in mining freight rail networks: a hybrid simulation methodology incorporating heuristics. J Simul 1–14 [Internet]. Available from: https://doi.org/10.1080/17477778.2022.2056536
43. Mustafee N, Harper A, Onggo BS (2020) Hybrid modelling and simulation (M&S): driving innovation in the theory and practice of M&S. In: 2020 Winter simulation conference (WSC). IEEE, pp 3140–51 [Internet]. Available from: https://ieeexplore.ieee.org/document/9383892/
44. Siebers PO, Macal CM, Garnett J, Buxton D, Pidd M (2010) Discrete-event simulation is dead, long live agent-based simulation! J Simul 4(3):204–10 [Internet]. Available from: https://doi.org/10.1057/jos.2010.14
45. Vandekerckhove J, Matzke D, Wagenmakers EJ (2015) Model comparison and the principle of parsimony. In: Busemeyer JR, Wang Z, Townsend JT, Eidels A (eds) Oxford library of psychology, vol 1. Oxford University Press [Internet]. Available from: https://doi.org/10.1093/oxfordhb/9780199957996.001.0001/oxfordhb-9780199957996-e-14
46. Shanthikumar JG, Sargent RG (1983) A unifying view of hybrid simulation/analytic models and modeling. Oper Res 31(6):1030–52 [Internet]. Available from: http://www.jstor.org/stable/170837
47. Crooks A, Castle C, Batty M (2008) Key challenges in agent-based modelling for geo-spatial simulation. Comput Environ Urban Syst 32(6):417–30 [Internet]. Available from: https://linkinghub.elsevier.com/retrieve/pii/S0198971508000628

48. Wilson JCT (1981) Implementation of computer simulation projects in health care. J Oper Res Soc 32(9):825–32 [Internet]. Available from: https://doi.org/10.1057/jors.1981.161
49. Young T, Eatock J, Jahangirian M, Naseer A, Lilford R (2009) Three critical challenges for modeling and simulation in healthcare. In: Winter simulation conference, pp 1823–1830 [Internet]. (WSC '09). Available from: http://dl.acm.org/citation.cfm?id=1995456.1995709
50. Lowery JC, Hakes B, Lilegdon WR, Keller L, Mabrouk K, McGuire F (2005) Barriers to implementing simulation in health care. In: Proceedings of winter simulation conference IEEE, pp 868–875 [Internet]. Available from: http://ieeexplore.ieee.org/document/717447/
51. Fone D, Hollinghurst S, Temple M, Round A, Lester N, Weightman A et al (2003) Systematic review of the use and value of computer simulation modelling in population health and health care delivery. J Public Health (Bangkok) 25(4):325–35 [Internet]. Available from: https://doi.org/10.1093/pubmed/fdg075
52. Gunal MM, Pidd M (2005) Simulation modelling for performance measurement in healthcare. In: Proceedings of the winter simulation conference. IEEE, pp 2663–2668 [Internet]. Available from: http://ieeexplore.ieee.org/document/1574567/
53. Eldabi T, Paul RJ, Young T (2007) Simulation modelling in healthcare: reviewing legacies and investigating futures. J Oper Res Soc 58(2):262–70 [Internet]. Available from: https://doi.org/10.1057/palgrave.jors.2602222
54. Robinson S, Worthington C, Burgess N, Radnor ZJ (2014) Facilitated modelling with discrete-event simulation: reality or myth? Eur J Oper Res 234(1):231–40 [Internet]. Available from: https://doi.org/10.1016/j.ejor.2012.12.024
55. Brailsford S, Vissers J (2011) OR in healthcare: a European perspective. Eur J Oper Res 212(2):223–34 [Internet]. Available from: https://doi.org/10.1016/j.ejor.2010.10.026
56. Kotiadis K, Tako AA, Vasilakis C (2014) A participative and facilitative conceptual modelling framework for discrete event simulation studies in healthcare. J Oper Res Soc 65(2):197–213
57. Lane DC, Rouwette EAJA (2022) Towards a behavioural system dynamics: exploring its scope and delineating its promise. Eur J Oper Res
58. Vennix JAM, Forrester JW. Group model-building: tackling messy problems The evolution of group model building. Vol. 15, Dyn. Rev. 1999.
59. Edmonds B, Moss S (2005) From KISS to KIDS—an 'anti-simplistic' modelling approach. In: Davidsson P, Logan B, Takadama K (eds) Lecture notes in computer science, vol 3415. Springer, Heidelberg, pp 130–144 [Internet]. Available from: https://doi.org/10.1007/b106991
60. Barreteau O, Bousquet F, Attonaty JM (2011) Role-playing games for opening the black box of multi-agent systems : method and lessons of its application to Senegal River Valley irrigated systems. J Artif Soc Soc Simul 4(2) [Internet]. Available from: http://jasss.soc.surrey.ac.uk/4/2/5.html
61. Manzo G (2014) Potentialities and limitations of agent-based simulations. An introduction. Rev Fr Sociol 55(4):653–88 [Internet]. Available from: https://www.cairn.info/revue-francaise-de-sociologie-2014-4-page-653.htm?ref=doi
62. Brailsford S. Modeling Human Behaviour - An (ID)entity Crisis? Proceedings of the 2014 Winter Simulation Conference [Internet]. 2014;1539–48. Available from: http://dl.acm.org/citation.cfm?id=2694013.2694045
63. Jahangirian M, Borsci S, Shah SGS, Taylor SJE. Causal factors of low stakeholder engagement: a survey of expert opinions in the context of healthcare simulation projects. Simulation [Internet]. 2015 Jun 23;91(6):511–26. Available from: http://journals.sagepub.com/doi/https://doi.org/10.1177/0037549715583150
64. Kreuger LK, Weichen Qian, Osgood N, Choi K (2016) Agile design meets hybrid models: using modularity to enhance hybrid model design and use. In: 2016 Winter simulation conference (WSC) [Internet]. IEEE, pp 1428–38. Available from: http://ieeexplore.ieee.org/document/7822195/
65. Jones W, Gun P, Foumani M (2022) An MVP approach to developing complex hybrid simulation models. In: 2022 Winter simulation conference (WSC) [Internet]. IEEE. pp 1176–87. Available from: https://ieeexplore.ieee.org/document/10015460/

66. Poppendieck M, Poppendieck T (2003) Lean software development: an agile toolkit. Addison-Wesley
67. Highsmith JA (2002) Agile software development ecosystems. Addison-Wesley Professional
68. Monks T, Currie CSM, Onggo BS, Robinson S, Kunc M, Taylor SJE (2019) Strengthening the reporting of empirical simulation studies: introducing the STRESS guidelines. J Simul 13(1):55–67 [Internet]. Available from: https://doi.org/10.1080/17477778.2018.1442155

Chapter 3
Hybrid Conceptual Modelling of Social and Socio-technical Systems Within Organisations: A Qualitative Semi-systematic Review

Richard A. Williams

Abstract An often-overlooked activity within the software engineering of computational models for social and socio-technical systems is the development of a comprehensive conceptual model (Williams in J Simul, 2022 [1]). With specific reference to modelling these systems within their organisational environment, we not only have to define the scope of the resultant simulation with respect to actors/technical resources, individual and system-level behaviours/dynamics, environment and abstraction level but also have to cater for the added challenge of dealing with a variety of qualitative and quantitative data. This introduces the need to utilise a number of analytical techniques that are specific to the different types of data and provide complementary views of the complex system, resulting in a hybrid conceptual model. This chapter presents a qualitative semi-systematic literature review of the analytical techniques that have been used in the extant literature for hybrid conceptual modelling of social and socio-technical systems within organisations.

Keywords Hybrid conceptual modelling · Semi-systematic review · Social systems · Socio-technical systems

3.1 Introduction

The modelling and simulation techniques used to investigate social and socio-technical systems have significantly advanced over the last 20–30 years, moving on from mere quantitative data analysis to computational techniques that allow us to model and simulate at the level of individual social actors [1]. Within the operational research (OR) community, this has meant a move away from purely mathematical approaches where the equations are solved, to computational ones, where the

R. A. Williams (✉)
Department of Management Science, Management School, Lancaster University, Bailrigg, Lancaster LA1 4YX, UK
e-mail: r.williams4@lancaster.ac.uk

system is simulated. Although this move has been facilitated by the advances in computing power and computational processing, it is also due to the fact that unlike engineered/built physical systems that can be definitively measured and quantified, social and socio-technical systems are modelled descriptively and validated qualitatively [2], meaning empirical data and evidence collected through fieldwork is often a mix of qualitative, quantitative, and network data. This mix of empirical data is a direct consequence of the fact that social systems within an organisational context (e.g. working group, resources within a department, or a project team) are complex and comprise a range of characteristics at the individual-level (e.g. role type, personal demographics, education, seniority, etc.), the system-level (e.g. interactions between resources, be they people or with technologies being used), and with the wider organisational environment in which the system of interest is situated.

We have recently argued that an often-overlooked activity within the software engineering of computational models for social and socio-technical systems is the development of a comprehensive conceptual model [1]. With specific reference to modelling these systems within their organisational environment, we not only have to define the scope of the resultant simulation with respect to actors/technical resources, individual and system-level behaviours/dynamics, environment and abstraction level but also have to cater for the added challenge of dealing with a variety of qualitative and quantitative data. This introduces the need to utilise a number of analytical techniques that are specific to the different types of data and provide complementary views of the complex system, resulting in a *hybrid conceptual model*.

With the above in mind, it is pleasing that conceptual modelling has once again become a focus for the OR community in recent years, with special consideration being applied to the software engineering principles, processes, activities, and tools associated with developing conceptual models that will be used as specification for subsequent computational models. Whereas the use of multiple techniques for development of computational models is termed *hybrid simulation* [3], the use of multiple tools and approaches for other parts of the simulation lifecycle (e.g. requirements analysis, model design, validation, etc.) is termed hybrid modelling [4, 5], meaning that the use of multiple tools and approaches for conceptual modelling results in a *hybrid conceptual model*. As such, a hybrid conceptual model is one that combines different types of models or approaches to create a more comprehensive representation of a complex system or phenomenon. This allows modellers to take advantage of the strengths of multiple tools/techniques, whilst also compensating for any weaknesses that they might have.

Although modellers have often used a variety of approaches to develop conceptual models, it has only recently been acknowledged that these were hybrid conceptual models following its relatively recent conception. As such, extant literature on conceptual modelling, for all but the most recent, does not clearly identify whether a hybrid approach has been taken. Our aim in this chapter is therefore to review the literature in order to understand the various analytical approaches that have been used for developing hybrid conceptual models of social systems within organisations. Due to the nature of the extant literature on conceptual modelling, we have performed a qualitative, standalone, descriptive review [6], which has also been termed a semi-systematic literature review [7]. As per Kraus et al. [8], we consider *domain, theory,* and *method* as three substantive focuses that can be applied to standalone literature reviews. Specifically, this review takes a method-focused hybrid approach, where the primary focus is on the *method* of hybrid conceptual modelling, which is applied to the *concept* of social and socio-technical systems and situated within the *context* of work-based organisations.

This semi-systematic review provides a comprehensive picture of the various approaches used in conceptual modelling of social and socio-technical systems within organisations. In doing so, this study identifies the complementary approaches that have been used by modellers when developing hybrid conceptual models.

3.2 Method

We conducted a qualitative, semi-systematic literature review [7], an approach intended to gather and analyse a heterogeneous literature with the aim of identifying the analytical tools and techniques used for hybrid conceptual modelling of social and socio-technical systems in organisations. This approach has been used within the management/business discipline due to its focus on providing an overview of the topic using a broad range of literature, identifying research themes and implications for the topic, and to synthesise these to develop a meta-narrative instead of using quantitative analysis to measure effect-size differences between literature findings [7]. We followed the literature review protocol of Xiao and Watson [6]. First, we developed our search strategy and collected the literature. Second, we coded the literature for domain, research findings, conceptual modelling tools/techniques. Finally, we interpreted the coded content of the literature in the context of hybrid conceptual modelling.

We followed the updated Preferred Reporting Items for Systematic Review and Meta-Analysis (PRISMA) guidelines for our semi-structured review of the literature [9]. The IEEE Xplore database, along with the Scopus and Web of Science search platforms, was used, to ensure relevant conference papers along with journal articles and book chapters were returned. We began by searching for titles, keywords and abstracts of English-language literature published up to April 2023. We used the search terms "conceptual model" and "organisation", along with associated synonyms (Table 3.1), using Boolean Operators as per Rowley and Slack [10]. Due to the focus of our study, we constrained the disciplines to computer science, engineering, information systems, and management. The citation data was exported from the three databases and imported into the online Rayyan.ai literature review and analysis tool [11]. The inclusion and exclusion selection criteria for population, interventions, setting, and study design are defined in Table 3.2. The literature was initially screened by reading their abstracts, with those that met the inclusion criteria then having their full text obtained and screened. This ensured that only literature using a hybrid approach to conceptual modelling (i.e. used two or more methods/analytical

Table 3.1 Search terms for the literature search, which represent a list of synonyms for the search terms associated with the method and context of our method-focused hybrid approach

Method	Context
"Hybrid conceptual model"	Organisation
"Conceptual model"	Workplace
"Domain model"	Firm
"Concept model"	Company
"Functional model"	Business
	Corporation
	Institution

The concept of social and socio-technical systems is filtered during the documentation review

approaches for conceptual model development) progressed to full-text evaluation. A backwards search was then performed to identify any relevant literature that had not been returned using our search criteria. To ensure quality, we only included literature that had been peer-reviewed and also performed internal validity of the methods in the literature as per Petticrew and Roberts [12]. In addition, duplicate entries were removed, which resulted in the final dataset of published literature. This dataset was analysed using thematic coding analysis [13] and development of an adjacency matrix to investigate the pairwise relationships between individual techniques used to develop the hybrid conceptual models in the literature, visualised using the Gephi graph visualisation and manipulation software [14].

Table 3.2 Inclusion and exclusion selection criteria for population, interventions, setting, and study design

	Inclusion criteria	Exclusion criteria
Population	Social groups (incorporating employees and project teams) within organisations that informs about behaviours/dynamics of the group	Studies reporting about proxy accounts of group members' experiences. Research that does not include information about behaviours/dynamics of the group
Intervention	Conceptual model that uses two or more methods and is focused on work-based social or socio-technical systems working on normal day-to-day operations or involved in a project	Studies not involving conceptual model of work-based social or socio-technical systems and not having two or more methods used for conceptual modelling
Setting	Public sector or commercial/industry-based organisations	Sports/recreational groups, third-sector organisations
Study design	Primary research, secondary research, meta-analysis. Qualitative and quantitative studies. Literature review on conceptual modelling	Non-empirical research, literature reviews of domain, case reports, standards, opinion pieces, and editorials

3.3 Results

The results of the final searches and selection of literature are shown in Fig. 3.1, which presents our PRISMA flow diagram. We coded the literature using the following six parent categories: type of study, scope of study, type of research, analytical techniques used for conceptual modelling, publication type, and publication rank. With respect to the scope of study parent category, we followed Freixanet and Federo [15] in identifying the subcategories, as we considered the empirical information on the reviewed articles. Based on our reading of the articles, four general themes emerged: industry/sector being studied, the organisational setting, the specific research focus, and the country of the study. Figure 3.2 provides the final coding scheme, whilst Fig. 3.3 provides an overview of the literature reviewed. The reviewed articles consist of 21 that provide background context to conceptual modelling, 32 empirical and 6 theoretical. Furthermore, 24 used qualitative methods, 9 used quantitative methods, 11 used mixed methods, and 15 represented tutorials, conference panel discussions, and perspective/opinion pieces that did not analyse data.

3.3.1 Background Literature

The 21 articles that represented background literature on conceptual modelling comprised of 11 conference papers, 7 journal articles, and 3 book chapters from the 2011 edited book that focused on conceptual modelling for discrete-event simulation. Of these, 8 were reviews, 5 presented conceptual modelling frameworks, 3 summarised panel discussions at leading international conferences, 3 studied the activities and/or processes involved in conceptual modelling by experienced modellers, and 2 provided tutorials to conceptual modelling.

Both tutorials were written by Robinson [16, 17]. The former focuses on choosing what to model within a conceptual model that will subsequently be translated into a computational model for simulation-based experiments, using the case studies of an outpatients building model within a healthcare setting and a daycare nursery. The latter tutorial provides a definition of a conceptual model through situating it within the lifecycle of a modelling and simulation project, providing an overview of artefacts of conceptual modelling along with the purpose and benefits, followed by a discussion of the requirements of a conceptual model and how these can be documented. The 3 panel discussions focus on education in conceptual modelling for simulation [18], providing a definition to conceptual modelling and discussing the purpose and benefits [19], and setting the scene for conceptual modelling for hybrid simulations [20]. The latter is particularly pertinent to this chapter and clearly advises that the lack of standard modelling approaches over the different modelling methods used in OR (i.e. discrete-event simulation [DES], system dynamics [SD] or agent-based modelling and simulation [ABMS]), means that there is not currently a shared understanding of how to develop a hybrid conceptual model. Importantly, the

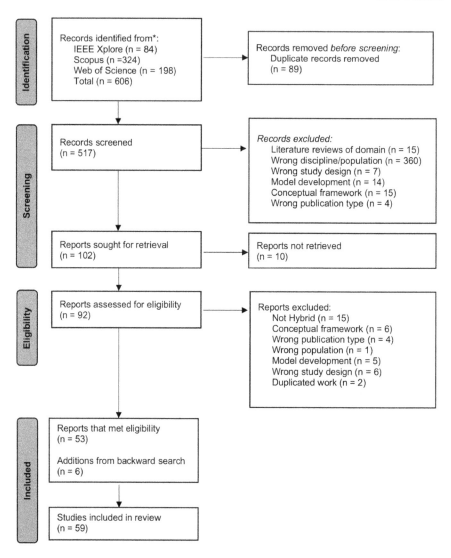

Fig. 3.1 Identification of studies from databases. The PRISMA flow diagram depiction of the identification, screening, and final set of studies from the systematic search process [9]

panel discussion notes that conceptual modelling is one of the most important and least understood stages within the lifecycle of modelling and simulation projects, which unfortunately culminates in conceptual modelling being one of the project activities that has the least focus by modellers or modelling teams. This was reasoned through the fact that there was limited guidance at the time of the Winter Simulation Conferences in 2016 [19] and 2019 [20], on how to perform hybrid modelling and simulation and hybrid conceptual modelling due to the focus at that time being on

Parent Category	Subcategories	Description
Type of Study	Background Empirical Theoretical	The article is background, empirical or theoretical
Scope of Study	Industry/Sector Setting Research Theme Country	The industry/sector that the study focus on The organization(s) that the study focuses on and the research particpants recruited The specific research theme of interest The country that data relates to
Type of Research	Qualitative Quantitative Mixed N/A(Background)	The type of research is qualitative, quantitative or mixed
Analytical Techniques	Soft Systems Methodology Unified Modeling Language Business Process Modeling Notation Nomological Network Diagram Informal Cartoon Diagram Activity Diagram Flow Diagram Process Diagram Structured Equation Modeling Statistical Analysis Thematic Coding Analysis Factor Analysis Content Analysis Social Network Analysis Enterprise Modeling	The tools/techniques used for development of hybrid conceptual model
Publication Type	Book Chapter Conference paper Journal Article	The article is a book chapter, conference paper or journal article
Publication Quality	Publication Name Publisher SciMago Index	The book/conference/journal name, the publisher and the SciMago ranking (for journal)

Fig. 3.2 Coding scheme

overcoming technical challenges with the modelling approaches/platforms to develop the computational model and perform simulation-based experimentation. This gap has however recently been filled by Tolk et al. [5], who discuss the methods that can be used to enable the conceptual alignment between the models used by experts of different disciplines.

The 8 reviews focused on various aspects of conceptual modelling. Robinson's [21] conference paper focused on the issues and research requirements when developing conceptual models for simulation. His review provided high-level summaries into defining: the notion of conceptual modelling, conceptual model requirements, the development process (including principles, methods of simplification and modelling frameworks), the validation of conceptual models, a high-level summary of university-level teaching of conceptual modelling, and a summary of research requirements and suggested areas for future research in order to advance the field. Robinson subsequently built upon this [22] with his review on the definition and requirements of conceptual models for DES as applied to the modelling of operations systems or operating systems. It is noted that these systems primarily focus on manufacture, transport, supply, and service functions within a business setting, and

Author(s)	Year	Type	Setting	Method & Techniques	Findings
Robinson [21]	2006	Background	N/A	N/A	Presented some of the issue and requirements when developing conceptual models for simulation-based research, in particular when research is focused on organisational change and uses DES
Wang, & Brooks [28]	2007	Background	N/A	Quantitative	The aim of the study was to improve understanding of the modeling process followed in practice by modelers, with a particular focus on conceptual modeling
Korunka, Hoonakker, & Carayon [43]	2008	Empirical	Employees in multiple organizations in USA and Austria	Qualitative, Quantitative, Nomological network diagram, SEM, Cartoon diagram	Developed a conceptual model of the quality of working life and turnover intention of IT workers
Kotiadis, & Robinson [29]	2008	Background	N/A	N/A	Discussion on SSM as a knowledge acquisition and model abstraction technique for conceptual modelling
Robinson [22]	2008	Background	Individual organization	Qualitative, Nomological network diagram	Provides an introduction to conceptual modeling for simulation, focused on defining conceptual modeling and the requirements of a conceptual model. Specific focus is applied to DES for modeling operations systems or operating systems
Robinson [31]	2008	Background	Individual organization	Qualitative, Nomological network diagram, Flow diagram	Provides a framework for conceptual modeling using the Ford Motor Company Engine Assembly Plant in South Wales as case study
Meng [44]	2009	Empirical	Multiple organizations	Nomological network diagram, Quantitative	Developed a conceptual model of the relationship between perceived organizational support, job satisfaction and employees intention to remain with their organization after industry restructuring
Onggo [36]	2009	Theoretical	N/A	Qualitative, Nomological network diagram, SSM, UML, BPMN, Stock and flow diagram	Developed a conceptual model of hospital Accident and Emergency
Vasilakis, Lecznarowicz, & Lee [45]	2009	Empirical	Individual employees in single hospital	Qualitative, SSM, UML	Developed a conceptual model of patient flow using SSM and UML in a UK hospital
Ou, Davison, Liang, & Zhong [46]	2010	Empirical	Individual employees who are part-time students at Universities in China	Qualitative, Quantitative, Nomological network diagram, PLS, ANOVA	Developed a conceptual model of the significance of Instant Messaging at Work
Ou, Davison, Zhong, & Liang [47]	2010	Empirical	Individual employees who are part-time students at Universities in China	Qualitative, Nomological network diagram, Quantitative, PLS, ANOVA	Developed a conceptual model on how instant messaging can empower teams at work

Fig. 3.3 Literature focused on development of hybrid conceptual models and background to conceptual modelling. The 59 entries are sorted by publication date (primary sorting) and then author surname (secondary sorting)

Author(s)	Year	Type	Setting	Method & Techniques	Findings
Tako, Kotiadis, & Vasilakis [39]	2010	Empirical	Healthcare Professionals in UK Hospital	Workshops, SSM, Qualitative, Nomological network diagram, Process diagram	Implemented a new modelling framework to develop conceptual models for simulation studies
Tako, Kotiadis, & Vasilakis [33]	2010	Background	N/A	N/A	Developed a modelling framework for simulation studies to develop conceptual models for simulation studies using Group Model Building (GMB) approach for System Dynamics models.
van der Zee, Brooks, Robinson, & Kotiadis [48]	2010	Background	N/A	N/A	Assessment of conceptual modeling research and highlighting opportunities for future research
van der Zee, Kotiadis, Tako, Pidd, Balci, Tolk et al. [18]	2010	Background	N/A	N/A	Panel discussion around the education/teaching of conceptual modeling for simulation with a view to bringing about improvements
Wang, & Brooks [30]	2010	Background	Individual organization	Quantitative	The study aimed to improve understanding of the modeling process followed in practice by modelers, with a particular focus on conceptual modeling.
Robinson [32]	2011	Background	Individual organization	N/A	Discussion around validation of conceptual models to ensure they represent the right model
Siddiqui, & Tripathi [49]	2011	Empirical	Multiple organizations	Quantitative, SSM, Nomological network diagram, SCA	Developed a conceptual model of the business models used by retail stores within a shopping mall in Lucknow, India.
Li, & Wu [50]	2011	Empirical	Individual employees in multiple organizations	Nomological network diagram, Quantitative	Developed a conceptual model of the relationship between work values and work performance in communication enterprises
Woo [51]	2011	Theoretical	N/A	Nomological network diagram, Flow diagram, Actor diagram, Venn diagram	Proposed that conceptual modelling of IS development focused on technology, but needed to be extended to include more business context (e.g. mission of the organization), which would help manage the business change that the IS Project would facilitate.
Yali, & Tuozhen [52]	2011	Theoretical	N/A	Qualitative, Nomological network diagram, Cartoon diagram	Developed a conceptual model of tacit knowledge transfer within organizations
Basl, Buchalcevova, & Gala [53]	2012	Theoretical	N/A	Nomological network diagram, Cartoon diagram, UML	Developed a conceptual model of the impact of enterprise information system innovation on sustainability
Duggirala, Mehta, Kambhatla, & Arya [54]	2012	Empirical	Individual employees in a large organization	Qualitative, Quantitative, Cartoon Diagrams	Developed a conceptual model of employee engagement to enable personalised actions for each employee to improve their engagement.

Fig. 3.3 (continued)

Author(s)	Year	Type	Setting	Method & Techniques	Findings
Hao, Ma, Zhao, & Fan [55]	2012	Empirical	Individual employees in multiple organizations	Quantitative, Qualitative, Nomological network diagram, SEM	Developed a conceptual model of back-office employee satisfaction and how it affects customer satisfaction
Heumuller, Richter, & Lechner [56]	2012	Empirical	Individual staff members at disaster response organizations in Germany	Action Research, Qualitative, Nomological network diagram, Cartoon diagram	Developed a conceptual model of staff in disaster response organizations
Robinson [16]	2012	Background	N/A	N/A	Tutorial on choices needed to be made before conceptual modeling starts
Rolland, & Prakash [34]	2012	Background	N/A	Nomological network diagram, Cartoon diagram	Discussed link between conceptual modeling and requirements engineering
van der Zee, Holkenborg, & Robinson [37]	2012	Theoretical	N/A	Nomological network diagram, Qualitative, Activity diagram, Cartoon diagram	Developed a conceptual model for simulation-based serious gaming, using a theoretical example of inventory control in retail management
Ahmed, Robinson, &Tako [23]	2014	Background	N/A	N/A	Presented lessons that can be learned from computer science for conceptual modeling in OR/MS
Ciby, & Baya [57]	2014	Empirical	Individual employees in multiple multi-national organizations in India	Qualitative, Nomological network diagram, Quantitative	Developed a conceptual model of victim experiences of workplace bullying
Kotiadis, Tako, & Vasilakis [40]	2014	Empirical	Individual employees in single hospital	Qualitative, SSM, Nomological network diagram, Flow diagram	Developed a participative and facilitative conceptual modeling framework for DES studies in healthcare
Furian, O'Sullivan, Walker, Vossner, & Neubacher [35]	2015	Background	N/A	Qualitative, Nomological network diagram, UML, Flow diagram	Developed a conceptual modeling framework for DES using hierarchical control structures
Li [58]	2015	Empirical	Individual employees at multiple Government Departments	Qualitative, Quantitative, Nomological network diagram, Factor analysis	Developed a conceptual model of relations between organizational justice, employees affection and organization performance
Robinson, Arbez, Birta, Tolk, & Wagner [19]	2015	Background	N/A	N/A	Provided a definition of conceptual modeling, its purpose and benefits
Roca, Pace, Robinson, Tolk, & Yilmaz [24]	2015	Background	N/A	N/A	Provided an overview of the Defense Handbook on Conceptual Models for Simulation Systems developed by the Department of Defense Modeling and Simulation Coordination Office
Rumanti, Hidayat, & Saputro [59]	2015	Empirical	Individual employees in an SME	Qualitative, Quantitative, Nomological network diagram, SEM	Developed a conceptual model of tacit knowledge transfer to empower SMEs
Robinson [17]	2017	Background	N/A	N/A	Tutorial on simulation conceptual modeling

Fig. 3.3 (continued)

Author(s)	Year	Type	Setting	Method & Techniques	Findings
Sepehrirad, Rajabzadeh, Azar, & Zarei [60]	2017	Empirical	Stakeholders in multiple organizations	Qualitative, SSM, Nomological Network Diagrams	Developed a conceptual model of occupational cancer control problem in the petroleum industries in Iran
Callaghan, Jackson, Dunnett, & Tako [42]	2018	Empirical	Regional UK Police Force	Qualitative, Flow Diagram, Nomological network diagram, Process diagram	Developed a conceptual model for police custody in the UK
Oliveira, & Martins [61]	2018	Empirical	Individual employees in a large organization	Quantitative, SEM	Develop a conceptual model of Project Management Office performance, focused on implementation strategies, capacitation and personnel training, and control of the operations environment in projects
AlKalbani, Deng, & Kam [62]	2019	Empirical	Individual employees in multiple public sector organizations	Quantitative, Nomological network diagram, SEM	Developed a conceptual model on the influence of organizational enforcement on the attidutes of employees towards information security compliance
Gelfert [25]	2019	Background	N/A	N/A	Discussed and assessed the credibility of conceptual models
Tako, Eldabi, Fishwick, Krejci, & Kunc [20]	2019	Background	N/A	N/A	Panel discussion from a conference, setting the scene towards conceptual modeling for hybrid simulation
Williams [63]	2019	Empirical	Individual employees in multiple organizations	Qualitative, UML, SNA	Developed a hybrid conceptual model of conflict propagation within large technology and software engineering implementations.
Fathian, Sharifi, & Nasirzadeh [64]	2020	Empirical	Individual employees in a large banking organization	Qualitative, Quantitative, Nomological network diagram, Process diagram	Developed a conceptual model of employee engagement when organizations use gamification strategies
Robinson [27]	2020	Background	N/A	N/A	Reviews progress made in the field of conceptual modeling for simulation since the 2006 meeting. Also sets a number of grand challenges for future conceptual modeling research
Samma, Zhao, Rasool, Han, & Ali [65]	2020	Empirical	Individual employees across a number of SMEs	Quantitative, Nomological network diagram, SEM	Developed a conceptual model of the relationships between workplace ostracism, incivility and innovative work behaviour
Sinha, Singh, Gupta, & Singh [66]	2020	Empirical	Social Media Platform	Qualitative, Quantitative, SEM, Nomological network diagram	Developed a conceptual model of Twitter data using various analytics to integrate findings
Williams [67]	2020	Empirical	Individual employees in multiple organizations	Qualitative, UML, Nomological network diagram	Used a Cybernetics perspective to develop a hybrid conceptual model of conflict propagation within large technology and software engineering implementations

Fig. 3.3 (continued)

Author(s)	Year	Type	Setting	Method & Techniques	Findings
Ding, Yang, Wang, & Xu [68]	2021	Empirical	Individual employees in multiple organizations	Nomological network diagram, Quantitative	Developed a conceptual model of the effects of employee volunteering on organizational loyalty
Fayoumi & Williams [38]	2021	Theoretical	N/A	Qualitative, Nomological network diagram, UML, Enterprise Modelling, Cartoon diagram, Process diagram	Developed a conceptual model of the complex socio-technical system associated with healthcare information systems
Prasiwi, & Ardi [69]	2021	Empirical	Individual employees in a large organization	Qualitative, Process diagrams, Nomological network diagram	Developed a conceptual model of employee behaviors and their resulting performance due to their fear of Covid-19
Slatten, Mutonyi, & Lien [70]	2021	Empirical	Individual employees in a large organization	Nomological network diagram, Quantitative, SEM	Developed a conceptual model of factors associated with organizational vision integration by hospital employees and tested this on a sample of employees
Engen, Muller, & Falk [41]	2022	Empirical	Individual employees in a large organization	Qualitative, Nomological network diagram, Cartoon diagram, Process diagram, Workflow diagram, Quantitative	Developed a conceptual model to support system-level decision-making in the Norwegian energy industry
Gabriel, Campos, Leal, & Montevechi [26]	2022	Background	N/A	N/A	Discussed good practices and deficiencies in conceptual modeling
Harper, & Mustafee [71]	2022	Empirical	ONS data and anonymized hospital demand data from the patient administration system	Qualitative, Quantitative, Nomological network diagram, Flow diagram	Developed a hybrid conceptual model of strategic resource planning of endoscopy services in a Clinical Commissioning Group in a major health service in South England
Kiely, Butler, & Finnegan [72]	2022	Empirical	Individual employees in a large organization	Qualitative, Nomological network diagram	Developed a conceptual model of global virtual teams coordination mechanisms in software development
Williams [1]	2022	Empirical	Individual employees in multiple organizations	Qualitative, SNA, UML, SSM	Developed a hybrid conceptual model for simulation, based on a case study of conflict development between team members in a large enterprise system implementation.
Cheng, Hsu, Li, & Brading [73]	2023	Empirical	Alumni of part-time MBA with IS specialism from a University in southern Taiwan	Quantitative, Nomological network diagram, PLS	Developed a conceptual model of intellectual capital and team resilience capability in information system project teams

Fig. 3.3 (continued)

the review uses the case study of modelling the new engine assembly plant in South Wales (UK) for the Ford Motor Company. The main concern of the manufacturing engineers who were involved in the case study was that of scheduling, specifically around the impact that scheduling has on the inventory of key components and how much space is required to hold sufficient stock. A simple schematic was developed showing the layout of the engine assembly plant and the various assembly lines in the production of engines. Through detailed analysis of the sequence of activities in assembling an engine, a simplified conceptual model was developed that grouped sections of the line and presented them as a queue with a delay. The review continues by discussing the requirements of conceptual models, with specific reference to validity, credibility, utility, and feasibility, along with the overarching requirement that the model is kept as simple as possible to meet the objectives of the simulation study.

The review by van der Zee, Brooks et al. [18] discusses the past, present and future of conceptual modelling for DES, providing very useful discussions around: conceptual modelling frameworks, the use of soft systems methodology (SSM) for conceptual modelling, software engineering for conceptual modelling, and importantly the need for domain-specific conceptual modelling due to the differing requirements across fields, e.g. the requirements for organisations involved in military applications will be very different to those involved in more traditional business/management operations. Ahmed et al. [23] complement these discussions in their review, which focuses on the lessons that can be learned from computer science, due to the long history of research into the pre-development phases of systems design that involve extensive analysis of requirements and objectives. The main message derived from this review is the need for a well-defined process for conceptual modelling within modelling and simulation (as applied to OR). Roca et al. [24] build upon this by highlighting the need for a simulation conceptual model paradigm, which positions the conceptual model as a core component of simulation validation. The edited book chapter by Gelfert [25] extends this line of research by reviewing how we can assess the credibility of conceptual models through verification, validation and testing of simulation output after the conceptual model has been translated into a computational model. More recently, the journal article by Gabriel et al. [26] provides a review of good practices and deficiencies in conceptual modelling, with specific focus on modelling language, the conceptual modelling process, syntactic quality, semantic quality, and pragmatic quality. In addition, they discuss the need to facilitate ease of audience understanding through appropriate visual modelling techniques, the use of multiple sources of data/information, and the need to ensure independence from any chosen simulation software. Finally, Robinson [27] provides a review of the progress made in conceptual modelling for simulation and sets a number of grand challenges for the OR community around: agreeing the definition of conceptual modelling, developing conceptual modelling frameworks, developing a standardised approach to representing conceptual models, and linking research to practice.

There were 3 articles that studied conceptual modelling processes. The first, by Wang and Brooks [28], aims to improve the understanding of conceptual modelling by comparing the processes followed by an expert in conceptual modelling and nine student groups who were studying for either a Master (MSc) or Bachelor (BSc) degree in a UK Management School. They discovered that the expert spent significantly more time on verification and validation of the conceptual model than the students who spent less time on conceptual modelling processes in order to maximise time on performing simulations with the resultant computational models. This was attributed to lack of experience and understanding by the novices of the need for robust and thorough validation and verification of the conceptual model before moving onto the development phase of the project where it was translated into the computational model. The second, by Kotiadis and Robinson [29], further contributes to improving the understanding of conceptual modelling, with specific focus on the processes of: knowledge acquisition, which is the process of identifying and analysing the problem situation and developing a system description, along with model abstraction, which relates to the simplifications required to translate the system description into a conceptual model. They used SSM to develop a conceptual model of a health and social care system, which would form the basis for a subsequent DES computational model. Finally, the book chapter by Wang and Brooks [30] builds upon their previous journal article [28] through additional analysis on the differences in the conceptual modelling processes between an expert and novice (student) modellers. Perhaps the most striking finding is that novices moved sequentially through the various activities associated with conceptual modelling, whereas the expert used an iterative approach that switched between activities and returned to previous artefacts of the conceptual modelling process as new knowledge and insights were gained from subsequent activities.

The final category of background literature relates to the 5 articles that present conceptual modelling frameworks. The first, by Robinson [31], presents a framework of five iterative activities: understanding the problem situation, determining the modelling and general project objectives, identifying the model outputs (e.g. responses), identifying the model inputs (e.g. experimental factors), and determining the model content (e.g. scope and level of detail), using the Ford Motor Company case study from the previous paper [22]. The later conference paper by Robinson [32] uses the case study of an outpatient clinic to build upon this framework by providing templates to describe each element of the conceptual model: modelling and general project objectives (e.g. organisational aim, modelling objectives, general project objectives); model outputs/responses (e.g. outputs to determine achievement of objectives, outputs to determine reasons for failure to meet objectives); experimental factors; model scope; model level of detail; modelling assumptions; and model simplifications. The third, a conference paper by Tako et al. [33], provides a framework for stakeholder participation in the development of conceptual models for simulation studies. Specifically, they used SSM as the problem structuring method, along with the group model building (GMB) approach that structures the participation of stakeholders throughout the conceptual modelling process. The fourth, by Rolland and Prakash [34], suggests that conceptual modelling for simulation is situated in the broader context of requirements engineering for computational systems. As such, they advocate the use of requirements engineering processes that relate to systems engineering (w.r.t. computer science and engineering) to augment the processes used by the OR community for conceptual modelling. Finally, the journal article by Furian, O'Sullivan et al. [35] proposes the use of hierarchical control structures when developing conceptual models for DES, which focuses on the identification of a models' system behaviour, and replacing queuing networks with more sophisticated mechanisms for entity behaviour that enables explicit, centralised control policies (such as dispatching routines).

3.3.2 Empirical and Theoretical Literature

There were 38 articles that developed conceptual models using either empirical data or theoretical case studies. The 6 articles that represent the theoretical literature to hybrid conceptual modelling all utilised a qualitative approach to analysis and consisted of 3 journal articles, 2 conference papers, and 1 book chapter from the 2010 edited book that focused on conceptual modelling for discrete-event simulation. Two of the articles used theoretical case studies specific to the UK, whilst the remaining 4 were generic and not localised to a particular country. Regarding the specific modelling techniques used to develop the hybrid conceptual models, all used nomological network diagrams (to represent constructs of interest) along with various mixtures of informal cartoon diagrams, unified modelling language (UML), SSM, business process modelling notation (BPMN), activity diagrams, process diagrams, actor diagrams, flow diagrams, enterprise modelling, and Venn diagram. The most important of these regarding hybrid conceptual modelling for simulation include Onggo [36], van der Zee et al. [37], along with Fayoumi and Williams [38]. The first of these used nomological network diagram, UML, BPMN, stock and flow diagram to develop a hybrid conceptual model of an Accident and Emergency Department for a District General Hospital within the UK. These different diagrammatic notations were found to complement each other in order to model multiple perspectives of the system and were also found to be highly expressive, simple to learn, scalable through the hierarchical layers of the organisational system, and intuitive so that domain experts can easily assist in validation. The second theoretical article used nomological network diagram, activity diagram, and cartoon diagram (as per informal diagrams developed in scientific disciplines) to develop a hybrid conceptual model of simulation-based serious gaming for decision-making by managers in large organisations. Whilst the third theoretical article used nomological network diagram, UML, enterprise modelling, process diagram, and cartoon diagram to develop a hybrid conceptual model of the organisational structure, people and technology associated with the back-office administrative functions (e.g. HR, recruitment, procurement, CRM) of a large hospital, and how a number of emerging technology solutions, such as AI, chatbots, and avatars might help make these processes more efficient.

The 32 articles that represent the empirical literature to hybrid conceptual modelling were published over 15 years (2008–2023). The majority used a qualitative approach ($n = 14$), with the fewest using purely quantitative approaches (n

= 7), and the remainder utilising a mixed-method approach of qualitative and quantitative techniques ($n = 11$). There was a wide range of techniques used to analyse the data and develop hybrid conceptual models, which came from a wide range of countries including the UK ($n = 9$), China ($n = 7$), India ($n = 2$), Indonesia ($n = 2$), Norway ($n = 2$), global studies ($n = 3$), and others (i.e. Austria, Brazil, Germany, Iran, Oman, Pakistan, and Taiwan). They also crossed multiple sectors, including healthcare ($n = 5$), information technology/systems ($n = 8$), manufacturing ($n = 2$), energy ($n = 2$), and others.

Figure 3.4 presents the range of techniques that were used to develop the conceptual models within the 38 articles that correspond to the empirical and theoretical literature. The literature overwhelming uses nomological network diagrams as one of the techniques used to develop the hybrid conceptual models ($n = 33$). Other techniques to note are informal cartoon diagrams ($n = 8$), structural equation modelling (SEM) to augment the diagrammatic models ($n = 8$), UML ($n = 7$), process diagrams ($n = 6$), SSM ($n = 6$), and flow diagrams ($n = 5$), with the other techniques only being utilised once or twice each. The combinations of techniques have been further analysed through development of a weighted adjacency matrix that defines the pairs of techniques used for hybrid conceptual model development (Fig. 3.5), along with the Gephi network visualiser software to develop the network graph that shows the network of topology of these pairs (Fig. 3.6). The adjacency matrix illustrates that the most commonly used pairs of techniques were nomological network diagram and SEM ($n = 7$) and nomological network diagram and cartoon diagram ($n = 7$). These were followed by nomological network diagrams and flow diagram ($n = 5$), nomological network diagram and process diagram ($n = 5$), nomological network diagram and SSM ($n = 4$), nomological network diagram and UML ($n = 4$), and nomological network diagram and PLS ($n = 3$). The remaining pairs of techniques were utilised once or twice. Regarding the number of techniques that literature used, these were grouped into those that utilised 2 different techniques ($n = 24$), 3 different techniques ($n = 9$), 4 different techniques ($n = 4$), and 5 different techniques ($n = 1$).

The article by Fayoumi and Williams [38] utilised the most techniques to develop the hybrid conceptual model, consisting of nomological network diagram, UML, enterprise modelling, cartoon diagram, and process diagram. This hybrid conceptual model used a theoretical case study to investigate the complex socio-technical system that arises within digitally implemented healthcare. Specifically, a systems perspective was used to analyse the case study, which consisted of a complex social system comprising healthcare professionals, patients, healthcare equipment suppliers and healthcare information technology and information system suppliers, along with a complex technical system that comprised the various technical architecture and software applications. The hybrid conceptual model was primarily underpinned by enterprise modelling and class, communication, and sequence diagrammatic notations (UML) to develop various views of the socio-technical system, such as the goal model, organisational model, agent model, resource and capability model, rule and decision model, and technology model. These diagrammatic representations of the system were augmented with a nomological network diagram, cartoon diagrams, and

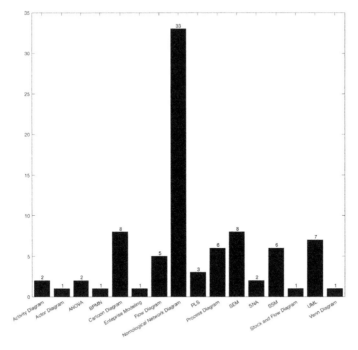

Fig. 3.4 Bar chart showing the frequency of conceptual modelling techniques used within the empirical and theoretical literature

	Activity Diagram	Actor Diagram	ANOVA	BPMN	Cartoon Diagram	Enterprise Modeling	Flow Diagram	Nomological Network Diagram	PLS	Process Diagram	SEM	SNA	SSM	Stock and Flow Diagram	UML	Venn Diagram
Activity Diagram																
Actor Diagram																
ANOVA																
BPMN																
Cartoon Diagram	1															
Enterprise Modeling					1											
Flow Diagram		1			1											
Nomological Network Diagram	2	1	2	1	7	1	5									
PLS			2					3								
Process Diagram	1				2	1	2	5								
SEM					1			7								
SNA																
SSM	1						1	4		1		1				
Stock and Flow Diagram				1			1									
UML				1	2	1		4				2	2	1		
Venn Diagram		1					1	1								

Fig. 3.5 Adjacency matrix showing the number of times each pair of techniques was used for development of hybrid conceptual models in the empirical and theoretical literature

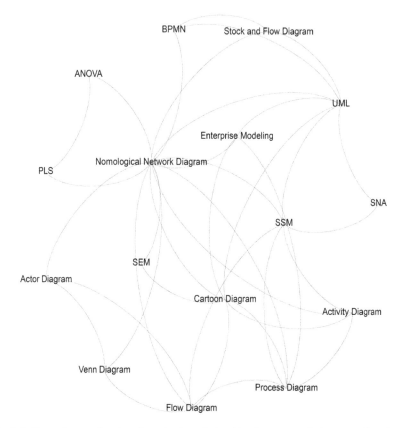

Fig. 3.6 Network map showing the pairwise relationships between techniques used to develop hybrid conceptual models within the empirical and theoretical literature

process models to develop the comprehensive conceptual model. The authors believe their approach of using enterprise modelling augmented with 4 other techniques has provided a powerful "hybrid" approach for conceptual model development.

A good example of literature that utilises 4 different techniques is a conference paper by Tako et al. [39], where their previously published participative framework [33] is used to develop a hybrid conceptual model of the obesity care system within the UK healthcare setting. They used SSM to develop the initial iteration of the conceptual model and as a framework to engage 12 healthcare professionals, which included consultants, fellows, and nurses from a wide range of specialties, such as general surgery, clinical biochemistry, anaesthetics, endocrinology, and the medical director for the trust. This initial conceptual model was then augmented by using nomological network diagram, activity diagram, and process diagram, before translating into a DES computational model. This work was built upon in their later journal article [40].

Another example of literature that utilised 4 techniques is Engen et al. [41] who developed a hybrid conceptual model to support system-level decision-making (via simulation) in the Norwegian energy industry. They utilise process diagram to define the field development process, along with informal cartoon diagrams to provide a visual overview of the field development infrastructure. Nomological network diagram and flow diagrams were then used to model the detailed interactions between components in the complex physical and spatiotemporal system. They found the hybrid conceptual model to be extremely useful to the system engineers and wider project team as a mechanism to unambiguously communicate concepts, the system context, and support a holistic mindset.

Additional examples of note include Williams [1] who developed a hybrid conceptual model of relationship conflict between team members of large information system projects, using SNA, SSM, and UML, for subsequent development of an ABMS; Callaghan et al. [42] who developed a hybrid conceptual model of the police custody process, with a view to subsequent translation into a DES computational model. Focus was applied to the resources (people and rooms) and system phases (e.g. booking, holding, Interviewing, charging, holding/release). They used nomological network diagram, flow diagram, and process diagram to model secondary data from previously published work, which was validated through observation at a police custody suite and discussions with police officers before updating with these new requirements.

3.4 Discussion

Our focus was on the methods and techniques used for developing hybrid conceptual models of social and socio-technical systems, with a view to subsequent computational modelling and simulation. Importantly, our use of the term socio-technical system here is consistent with the conceptual framework presented by Ropohl [43]. The systematic search process, which followed PRISMA guidelines, resulted in 59 publications meeting our inclusion criteria from the total items retrieved ($n = 606$) across the IEEE Xplore, Scopus, and Web of Science databases. This final set of publications that met our inclusion criteria followed the removal of duplicates ($n = 89$), removal of those that could not be retrieved ($n = 10$) or did not meet the inclusion criteria ($n = 454$), and addition of a small set from backward search ($n = 6$). This represented background literature to conceptual modelling ($n = 21$), hybrid conceptual models that utilised theoretical case studies ($n = 6$), and hybrid conceptual models that utilised empirical cases ($n = 32$).

The empirical studies were focused on a wide range of countries, indicating that conceptual modelling of social and socio-technical systems is widespread across the globe and not just a Western activity. Interestingly, the UK ($n = 9$) and China ($n = 7$) represent the most frequent, with other studies coming from countries in Europe, South America, Asia, and the Middle East, along with a wide range of industry sectors, including information technology/systems ($n = 8$), healthcare ($n = 5$), public sector ($n = 2$), manufacturing ($n = 2$), energy ($n = 2$), and others (including 6 cross-sector studies). The majority of these empirical studies used a qualitative approach ($n = 14$), closely followed by a mixed-method approach of qualitative and quantitative techniques ($n = 11$), and the remainder utilising quantitative approaches ($n = 7$). These analytical approaches comprised a range of diagrammatic, descriptive, and statistical techniques (defined in the network map of Fig. 3.6), a range of which were utilised in the development of the various hybrid conceptual models, as defined by the pairs of techniques within the adjacency matrix (Fig. 3.5).

3.4.1 Implications for Practice

Regardless of the organisational research setting, industry/sector or country, the evidence from this semi-systematic literature review reveals that the overwhelming majority of hybrid conceptual modelling publications that used empirical data utilised either a qualitative approach or a mixed-method approach (i.e. both qualitative and quantitative). Of note is that nomological network diagrams are nearly universal throughout (all of the theoretical [$n = 6$] and most of the empirical [$n = 27$]) due to its flexibility in representing constructs of interest (and their interactions) within the system. Additional diagrammatic notations are used related to UML, activity, flow and process diagrams, and informal cartoon diagrams, which like the nomological network diagram provide the researcher with considerable flexibility in modelling the system of interest. Importantly, the UML diagrammatic language contains 14 diagrammatic notations, and a number of these, along with flow (including stock and flow) and process diagrams, were often combined in order to develop complementary perspectives/views of the system. Indeed, the pairwise combinations of techniques are presented as a weighted adjacency matrix (Fig. 3.5), along with the corresponding network graph (Fig. 3.6) that shows the network of topology of these pairs. The most commonly used pairs of techniques were nomological network diagram and SEM ($n = 7$), along with nomological network diagram and cartoon diagram ($n = 7$).We believe that the pairing of a qualitative approach (such as nomological network diagram) with SEM and other statistical techniques (e.g. PLS, ANOVA, etc.), highlights the power of a mixed-method approach when developing hybrid conceptual models of social systems within an organisation. Overall, most publications were conservative in the number of different analytical techniques utilised, with the overwhelming majority using only 2 or 3 different techniques. In fact, there was only 1 publication that utilised 5 different techniques, consisting of cartoon diagrams, nomological network diagram, enterprise modelling, process diagram, and UML, to develop a hybrid conceptual model [38] of a theoretical case in healthcare. Of the 4 publications that utilised 4 different techniques, the most comprehensive was the hybrid conceptual model by Williams [1], which used SSM, UML, SNA, and nomological network diagrams. We believe that this publication provides a comprehensive example of how to structure, document and visually represent a hybrid conceptual model of a complex socio-technical system within an organisational setting, due to the utilisation of multiple analytical techniques and its leverage of proven conceptual modelling frameworks, such as those discussed by Kotiadis and Robinson [29] and Robinson [22, 31] from the OR community, along with Stepney and Polack [44] from the computer science community, who strongly advocate comprehensive and robustly validated conceptual models (which they term *domain models*) as a precursor to the engineering of simulations as scientific instruments. A very recent publication [45] has used a similar approach for resilience planning of water dams, by utilising a data collection and analysis framework that is analogous to SSM and complementing this with SNA to develop the conceptual model. The main difference in this paper however is that instead of using the computational model for simulation, it is actually intended to be used as a decision support tool for project leadership that deepens their awareness of the effects of their decisions and the impacts on the general population and specific communities in the geographical vicinity.

3.4.2 Limitations and Directions for Future Research

This study has several limitations that should be mentioned. First, despite the considerable effort involved with using a systematic approach to the literature search and selection in order to avoid any bias within the semi-systematic literature review, it is not possible to state that the final sample is entirely bias-free. Second, as indicated in the results section, publications in the final sample came from a diverse range of industry/sectors and countries; however, it is unclear whether certain techniques are used uniformly across the globe, or whether certain regions of the world favour particular techniques, for example, SSM in the UK, SNA, and UML in the Western hemisphere and statistical techniques in other parts of the world. Third, in order to maximise the quality of publications, we targeted peer-reviewed conference papers, peer-reviewed book chapters, and peer-reviewed journal articles through using the IEEE Xplore, Scopus, and Web of Science databases, which means that grey literature, such as reports, white papers, PhD Theses, or MSc Dissertations are not included in the sample. Fourth, the term *hybrid conceptual model* is relatively recent and is primarily focused within the OR community at present, so the majority of the publications within the sample do not explicitly use this term within the searched fields (i.e. keywords, abstract, or title), even though their models conform to the definition of a hybrid conceptual model.

For future research, several suggestions can be made. First, as the focus of this review was on hybrid conceptual models of social and socio-technical systems within an organisational setting, future research by others could widen up the scope and consider hybrid conceptual models of other complex systems or settings, such as manufacturing, transportation, and logistics, etc. Second, we have not focused on a specific computational modelling paradigm, but hybrid modelling for simulation in general, meaning that future research could target ABMS, DES, or SD, which are the most commonly used simulation approaches in the OR community. Third, a detailed investigation on the extent to which techniques have been used across the globe could be performed to ascertain whether there are preferences by global region, i.e. SSM in the UK, SNA, and UML in the Western hemisphere and statistical techniques in other parts of the world.

3.5 Conclusions

The current study contributes to knowledge by reviewing the extant literature around hybrid conceptual modelling for simulation of social and socio-technical systems within organisations. The evidence shows that hybrid conceptual models have predominantly taken qualitative or mixed-method approaches, with the majority using nomological network diagram to convey system constructs of interest, along with one or two additional techniques. For the purely qualitative approach, the most common combination is nomological network diagram along with either cartoon diagram, activity diagram, process diagram, or flow diagram. When mixed-method approaches are used, the most common combination was nomological network diagram and a statistical technique to further analyse the empirical quantitative data that had been collected, such as ANOVA, PLS, or SEM. It is evident that very few hybrid conceptual models use 4 or more techniques, but we believe that these publications with the highest number of techniques are of benefit to the OR community due to them providing multiple views of the complex social or socio-technical system, thus providing a more comprehensive conceptual model which can then be translated into the subsequent computational model for simulation-based experimentation. We accept that there could be a counterargument made here in that the incorporation of more techniques could create a higher barrier for comprehension and engagement by subject matter experts and industry professionals but believe that this can be managed by the modelling team performing a familiarisation activity around the tools and techniques that will be used. This study further makes contributions through analysis of the composition of extant hybrid conceptual models, which used thematic coding along with a number of visual techniques, such as pairwise technique adjacency matrix, technique network map, and simple bar chart. Finally, we have identified a number of areas for future research, whereby interested members of the OR community might wish to extend this work around the specific techniques used within hybrid conceptual modelling.

References

1. Williams RA (2022) Towards an agent-based model using a hybrid conceptual modelling approach: a case study of relationship conflict within large enterprise system implementations. J Simul. https://doi.org/10.1080/17477778.2022.2122741 Accessed 2023-06-10
2. Alam SJ, Geller A (2012) Networks in agent-based social simulation. In: Heppenstall AJ, Crooks AT, See LM, Batty M (eds) Agent-Based models of geographical systems. Springer, Dordrecht, Netherlands, pp 199–216

3. Mustafee N, Powell JH (2018) From hybrid simulation to hybrid systems modelling. In: Rabe M, Juan AA, Mustafee N, Skoogh A, Jain S, Johansson B (eds) Proceedings of the winter simulation conference. IEEE, Gothenberg, Sweden, pp 1430–1439
4. Powell JH, Mustafee N (2017) Widening requirements capture with soft methods: an investigation of hybrid M&S studies in health care. J Oper Res Soc 68(10):1211–1222
5. Tolk A, Harper A, Mustafee N (2021) Hybrid models as transdisciplinary research enablers. Eur J Oper Res 291(3):1075–1090
6. Xiao Y, Watson M (2019) Guidance on conducting a systematic literature review. J Plan Educ Res 39(1):93–112
7. Snyder H (2019) Literature review as a research methodology: an overview and guidelines. J Bus Res: 333–339
8. Kraus S, Breier M, Lim WM, Dabic M, Kumar S, Kanbach D et al (2022) Literature review as independent studies: guidelines for academic practice. Rev Manag Sci 16:2577–2595
9. Page JP, Boutron I, Shamseer L, Brennan SE, Grimshaw JM, Li T et al (2021) The PRISMA 2020 statement: an updated guideline for reporting systematic reviews. BMJ 372:n71
10. Rowley J, Slack F (2004) Conducting a literature review. Manage Res News. 27(6):31–39
11. Ouzzani M, Hammady H, Fedorowicz Z, Elmagarmid A (2016) Rayyan: a web and mobile app for systematic reviews. Sys Rev. 5:210
12. Petticrew M, Roberts H (2006) Systematic reviews in the social sciences: a practical guide. Blackwell, Oxford, UK, p p336
13. Braun V, Clarke V (2006) Using thematic analysis in psychology. Qual Res Psychol 3:77–101
14. Bastian M, Heymann S, Jacomy M (2009) Gephi: an open source software for exploring and manipulating networks. In: Proceedings of the international AAAI conference on weblogs and social media. AAAI, San Jose, CA, USA
15. Freixanet J, Federo R (2023) Learning by exporting: a system-based review and research agenda. Int J Manag Rev. https://doi.org/10.1111/ijmr.12336Accessed2023-06-10
16. Robinson S (2012) Tutorial: choosing what to model—conceptual modeling for simulation. In: Proceedings of the winter simulation conference. IEEE, Berlin, Germany, pp 1–12
17. Robinson S (2017) A tutorial on simulation conceptual modeling. In: Proceedings of the winter simulation conference. IEEE, Las Vegas, Nevada, USA, pp 565–579
18. van der Zee DJ, Kotiadis K, Tako AA, Pidd M, Balci O, Tolk A et al (2010) Panel discussion: education on conceptual modeling for simulation—challenging the art. In: Proceedings of the winter simulation conference. IEEE, Baltimore, Maryland, USA, pp 290–304
19. Robinson S, Arbez G, Birta LG, Tolk A, Wagner G (2016) Conceptual modeling: definition, purpose and benefits. In: Proceedings of the winter simulation conference. IEEE, Arlington, Virginia, USA, pp 2812–2826
20. Tako AA, Eldabi T, Fishwick P, Krejci CC, Kunc M (2019) Towards conceptual modeling for hybrid simulation: setting the scene. In: Proceedings of the winter simulation conference. IEEE, National Harbor, Maryland, USA, pp 1267–1279
21. Robinson S (2006) Conceptual modeling for simulation: issues and research requirements. In: Proceedings of the winter simulation conference. IEEE, Monterey, California, USA, pp 792–800
22. Robinson S (2008) Conceptual modelling for simulation part I: definition and requirements. J Oper Res Soc 59(3):278–290
23. Ahmed F, Robinson S, Tako A (2014) Conceptual modelling: lessons from computer science. In: Onggo BS, Heavey C, Monks T, Tjahjono B, van der Zee D-J (eds) Proceedings of the operational research society simulation workshop. OR Society, Birmingham, UK, pp 154–166
24. Roca R, Pace D, Robinson S, Tolk A, Yilmaz L (2015) Paradigms for conceptual modeling. In: Proceedings of the spring simulation conference, 47(2). SCS, Alexandria, Virginia, USA, pp 202–209
25. Gelfert A (2019) Assessing the credibility of conceptual models. In: Beisbart C, Saam NJ (eds) Computer simulation validation: fundamental concepts, methodological frameworks, and philosophical perspectives. Springer Nature, Cham, Switzerland, pp 249–269

26. Gabriel GT, Campos AT, Leal F, Montevechi JAB (2022) Good practices and deficiencies in conceptual modelling: a systematic literature review. J Simul 16(1):84–100
27. Robinson S (2020) Conceptual modelling for simulation: progress and grand challenges. J Simul 14(1):1–20
28. Wang W, Brooks RJ (2007) Improving the understanding of conceptual modeling. J Simul 1(3):153–158
29. Kotiadis K, Robinson S (2008) Conceptual modelling: knowledge acquisition and model abstraction. In: Proceedings of the winter simulation conference. IEEE, Miami, Florida, USA, pp 951–958
30. Wang W, Brooks RJ (2010) Improving the understanding of conceptual modeling. In: Robinson S, Brooks R, Kotiadis K, van der Zee DJ (eds) Conceptual modeling for discrete-event simulation. CRC Press, Boca Raton, Florida, USA, pp 57–70
31. Robinson S (2008) Conceptual modelling for simulation part II: a framework for conceptual modelling. J Oper Res Soc 59(3):291–304
32. Robinson S (2011) Choosing the right model: conceptual modeling for simulation. In: Jain S, Creasey R, Himmelspach J, White KP, Fu MC (eds) Proceedings of the winter simulation conference. IEEE, Phoenix, Arizona, USA, pp 1423–1435
33. Tako AA, Kotiadis K, Vasilakis C (2010) A conceptual modelling framework for stakeholder participation in simulation studies. In: Proceedings of the operational research society simulation workshop. OR Society, Worcestershire, UK, pp 76–85
34. Rolland C, Prakash N (2000) From conceptual modelling to requirements engineering. Ann Softw Eng 10(1):151–176
35. Furian N, O'Sullivan M, Walker C, Vossner S, Neubacher D (2015) A conceptual modeling framework for discrete event simulation using hierarchical control structures. Simul Model Pract Th 56:82–96
36. Onggo BSS (2009) Towards a unified conceptual model representation: a case study in healthcare. J Simul 3(1):40–49
37. van der Zee DJ, Holkenberg B, Robinson S (2012) Conceptual modeling for simulation-based serious gaming. Decis Support Syst 54(1):33–45
38. Fayoumi A, Williams R (2021) An integrated socio-technical enterprise modelling: a scenario of healthcare system analysis and design. J Ind Inf Integr 23:100221
39. Tako AA, Kotiadis K, Vasilakis C (2010) A participative modelling framework for developing conceptual models in healthcare simulation studies. In: Proceedings of the winter simulation conference. IEEE, Baltimore, Maryland, USA, pp 500–512
40. Kotiadis K, Tako AA, Vasilakis C (2014) A participative and facilitative conceptual modelling framework for discrete event simulation studies in healthcare. J Oper Res Soc 65(2):197–213
41. Engen S, Muller G, Falk K (2023) Conceptual modeling to support system-level decision-making: an industrial case study from the Norwegian energy domain. Syst Eng 26(2):177–198
42. Callaghan H, Jackson L, Dunnett S, Tako AA (2018) Developing a conceptual model for police custody in the UK. In: Proceedings of the winter simulation conference. IEEE, Gothenberg, Sweden, pp 2781–2791
43. Ropohl G (1999) Philosophy of socio-technical systems. Phil Tech 4(3):186–194
44. Stepney S, Polack FAC (2018) Engineering simulations as scientific instruments: a pattern language. Springer Nature, Cham, Switzerland
45. Tolk A, Richkus JA, Shults FL, Wildman WJ (2023) Computational decision support for socio-technical awareness of land-use planning under complexity: a dam resilience planning case study. Land 12:952
46. Korunka C, Hoonakker P, Carayon P (2008) Quality of working life and turnover intention in information technology work. Hum Factors Ergon Manuf 18(4):409–423
47. Meng X (2009) Perceived organizational support, job satisfaction, and the retention of employees after industry restructuring. In: Proceedings of the international conference on innovation management. IEEE, Wuhan, China, pp 22–25
48. Vasilakis C, Lecznarowicz D, Lee C (2009) Developing model requirements for patient flow simulation studies using the unified modelling language. J Simul 3(3):141–149

49. Ou CXJ, Davison R, Liang X, Zhong X (2010) The significance of instant messaging at work. In: Proceedings of the 5th international conference on internet and web applications and sciences. IEEE, Barcelona, Spain, pp 102–109
50. Ou CXJ, Zhong X, Davison R, Liang Y (2010) Can instant messaging empower teams at work? In: Proceedings of the 4th international conference on research challenges in information systems. IEEE, Nice, France, pp 589–598
51. van der Zee DJ, Brooks RJ, Robinson S, Kotiadis K (2010) Conceptual modeling: past, present, and future. In: Robinson S, Brooks R, Kotiadis K, van der Zee DJ (eds) Conceptual modeling for discrete-event simulation. CRC Press, Boca Raton, Florida, USA, pp 473–490
52. Siddiqui MH, Tripathi SN. Application of soft operations research for enhancing servicescape as a facilitator. Vikalpa 36(1):33–49
53. Li W, Wu K (2011) The relationship between work values and work performance in communication enterprises. In: Proceedings of the 2nd international conference on artificial intelligence, management science and electronic commerce. IEEE, Zhengzhou, China, pp 1733–1737
54. Woo C (2011) The role of conceptual modeling in managing and changing the business. In: Proceedings of the 30th international conference on conceptual modeling. Lecture notes in computer science, vol 6998. Springer, Brussels, Belgium, pp 1–12
55. Yali C, Taozhen H (2011) Conceptual model of tacit knowledge transfer within organizations. In: Proceedings of the international conference on product innovation management. IEEE, Wuhan, China, pp 151–154
56. Basl J, Buchalcevova A, Gala L (2012) Conceptual model of the impact of enterprise information systems innovation on sustainability. In: Advances in enterprise information systems II. Taylor & Francis, London, pp 27–34
57. Duggirala M, Mehta S, Kambhatia N, Arya P (2012) Employee engagement: conceptual model and computation framework. In: Annual SRII global conference. IEEE, San Jose, California, USA, pp 850–858
58. Hao J, Ma Q, Zhao X, Fan G (2012) Does back-office employee satisfaction affect customer satisfaction? In: International joint conference on service sciences. IEEE, Shanghai, China, pp 115–119
59. Heumuller E, Richter S, Lechner U (2012) Towards a conceptual model of staffs in disaster response organizations. In: Proceedings of the 25th Bled eConference–eDependability. Bled eCommerce conference, Bled, Slovenia, pp 250–264
60. Ciby M, Raya RP (2014) Exploring victims' experiences of workplace bullying: a grounded theory approach. Vikalpa 39(2):69–81
61. Li MS (2015) Empirical study on relations between organizational justice, employees affection and organization performance. In: Zhang Y, McAnally E, Hylind M, Solovjeva I (eds) Proceedings of the international conference on economics, management, law and education. Atlantis Press, Kaifeng, China, pp 174–178
62. Rumanti AA, Hidayat TP, Saputro YD (2015) Tacit knowledge transfer and its implementation on small and medium enterprises. In: Proceedings of the IEEE international conference on industrial engineering and engineering management. IEEE, Singapore, Singapore, pp 587–590
63. Sepehrirad R, Rajabzadeh A, Azar A, Zarei B (2017) A soft systems methodology approach to occupational cancer control problem: a case study of the ministry of petroleum of Iran. Syst Pract Action Res 30(6):609–626
64. Oliveira RR, Martins HC (2018) Strategy, people and operations as influencing agents of the project management office performance: an analysis through structural equation modeling. Gest Prod 25(2):410–429
65. AlKalbani A, Deng H, Kam B (2019) The influence of organizational enforcement on the attitudes of employees towards information security compliance. In: Proceedings of the 10th international conference on information and communication systems. IEEE, Irbid, Jordan, pp 152–159
66. Williams RA (2019) Conflict propagation within large technology and software engineering programmes: a multi-partner enterprise system implementation as case study. IEEE Access 7(1):167696–167713

67. Fathian M, Sharifi H, Nasirzadeh E (2020) Conceptualizing the role of gamification in contemporary enterprises. IEEE Access 8:220188–220204
68. Samma M, Zhao Y, Rasool SF, Han X, Ali S (2020) Exploring the relationship between innovative work behavior, job anxiety, workplace ostracism, and workplace incivility. Healthcare 8(4):508
69. Sinha N, Singh P, Gupta M, Singh P (2020) Robotics at workplace: an integrated Twitter analytics—SEM based approach for behavioral intention to accept. Int J Inform Manage 55:102210
70. Williams RA (2020) Cybernetics of conflict within multi-partner technology and software engineering programmes. IEEE Access 8:94994–95018
71. Ding Z, Yang H, Wang J, Xu J (2021) The effects of employee volunteering on organizational loyalty: the moderating effects of perceived organization support. In: Proceedings of the IEEE international conference on industrial engineering and engineering management. IEEE, Singapore, Singapore, pp 1289–1293
72. Prasiwi NF, Ardi R (2021) Fear of covid-19: a conceptual model for behavior and employee performance in service and manufacturing industries. In: 4th Asia Pacific conference on research in industrial and systems engineering. ACM, Depok, Indonesia, pp 121–127
73. Slatten T, Mutonyi BR, Lien G (2021) Does organizational vision really matter? An empirical examination of factors related to organizational vision integration among hospital employees. BMC Health Serv Res 21(1):483
74. Harper A, Mustafee N (2022) Strategic resource planning of endoscopy services using hybrid modelling for future demographic and policy change. J Oper Res Soc. https://doi.org/10.1080/01605682.2022.2078675Accessed2023-06-10
75. Kiely G, Butler T, Finnegan P (2022) Global virtual teams coordination mechanisms: building theory from research in software development. Behav Inf Technol 41(9):1952–1972
76. Cheng KT, Hsu JSC, Li YZ, Brading R (2023) Intellectual capital and team resilience capability of information system development project teams. Inf Manage 60(1):103722

Chapter 4
Towards Hybrid Modelling and Simulation Concepts for Complex Socio-technical Systems

Andreas Tolk, Jennifer A. Richkus, and Yahya Shaikh

Abstract Complex socio-technical systems have many challenging facets, comprising multiple, highly interconnected, and highly diverse technical and social components. Each entity represented by these components comes with many perspectives, methods, and tools developed in the supporting communities to address the subset of overall challenges the entity is concerned about. Various works on cross-disciplinarity have addressed coping with such diverse entity viewpoints within a community. While technologies have been developed to address multiple views technically, new methods of participatory modelling are needed to capture the inputs of social groups that are usually not familiar with computational support devices but whose lived experience and views are pivotal for understanding the socio-technical systems, such as underserved community and groups in socially unfavourable situations. These approaches may help cross "the final mile" between these groups and researchers by serving two purposes: getting the necessary research input and explaining the research results to the community. This chapter provides an overview of the conceptual framework, some useful tools and methods, and open research questions that still need to be addressed by the hybrid modelling and simulation community.

Keywords Hybrid M&S · Socio-technical systems · Participatory modelling · Hard and soft OR · Cross-disciplinary research · Cross-cultural research

---

A. Tolk (✉)
The MITRE Corporation, 1001 Research Park Blvd #220, Charlottesville, VA 22911, USA
e-mail: atolk@mitre.org

J. A. Richkus
The MITRE Corporation, 7515 Colshire Drive, McLean, VA 22102, USA
e-mail: jasrichkus@mitre.org

Y. Shaikh
The MITRE Corporation, 2275 Rolling Run Drive, Windsor Mill, MD 21244, USA
e-mail: yshaikh@mitre.org

4.1 Introduction

The modelling and simulation (M&S) community of scholars and practitioners has been interested in hybrid approaches since its beginning. Scholars were interested in the applicability of mixed approaches that used different methods to find better solutions faster [1]. Practitioners often face existing solutions based on different methods that must be integrated to support a common task in their application domain, such as among many other fields in cyber-physical systems [2]. The combination of physical and virtual components to study the dynamic behaviour of complex engineering systems is of interest, as they also require representing the physical components in the form of models to the virtual ones, resulting in a high variety of interconnected components [3]. They result in the need for integration of many solutions independently developed contributions enabling intelligent, adaptive, and autonomous behaviour of such systems [4]. Overall, the authors perceive hybrid M&S has matured significantly in recent years.

The focus of a majority of hybrid M&S work, however, has been predominantly on technical challenges in support of engineered systems. Within this chapter, understanding socio-technical systems is the main objective. The application domain provides decision support in complex socio-technical systems requiring technical and social competence. Gilbert et al. argue in [5] that *"where the costs or risks associated with a policy change are high, and the context is complex, it is not only common sense to carry out policy modelling, but it would be unethical not to."*

There is a strong argument for using models to gain insights into the thinking and value processes and to communicate valuable insights. Gelfert observes in the summary of his philosophical primer on models:

> Whereas the heterogeneity of models in science and the diversity of their uses and functions are nowadays widely acknowledged, what has perhaps been overlooked is that not only do models come in various forms and shapes and may be used for all sorts of purposes, but they also give unity to this diversity by mediating not just between theory and data, but also between the different kinds of relations into which we enter with the world. Models, then, are not simply neutral tools that we use at will to represent aspects of the world; they both constrain and enable our knowledge and experience of the world around us: models are mediators, contributors, and enablers of scientific knowledge, all at the same time. [6, p. 127]

In his introduction to the wide variety of models, Page [7] mentions seven uses of models: *Reason* (to identify conditions and deduce logical implications), *Explain* (to provide testable explanations for empirical phenomena), *Design* (to choose features of institutions, policies, and rules), *Communicate* (to relate knowledge and understanding), *Act* (to guide policy choice and strategic actions), *Predict* (to make numerical and categorical predictions of future and unknown phenomena), and *Explore* (to investigate possibilities and hypotheticals). In his introduction to these uses, Page states, *"these form the acronym REDCAPE, a not-so-subtle reminder that many-model thinking endows us with superpowers."* [7, p. 13]. These are powerful arguments to focus on the modelling part of hybrid M&S methods at least as much as on the simulation part.

What does this mean for hybrid M&S? How can we benefit from the technically increasing maturity of hybrid modelling and simulation to enable decision support that is not only technically competent, but that allows us to consider a full complement of ways to explore different ways of thinking, understanding, and response? How can we expand our efforts to address diversity, equity, and inclusion for social aspects, engaging relevant stakeholders, providing tools to increase contributions to the process, and ensuring that all voices are represented accordingly in the decision support system? Furthermore, how do we communicate important, helpful, and applicable research results to the community, preferably in their language and models?

In the social context, this challenge is sometimes called bridging "the final mile" between the researching organisations, often supporting the government or social institutions, and the underserved and underrepresented groups. As just discussed, this gap exists in both directions. The two central questions are: First, how do we capture the value and belief systems, the lived experience, and the other important aspects of value to the underrepresented groups in our models to consider them for any optimisation processes, and second, how do we communicate recommended changes and their expected benefits in the context of their social and cultural background? For both directions, participatory modelling can provide the means to solicit the required information and provide information in the context needed to fit into the value and belief systems of the targeted group if models meaningful to the group are used. If we don't elicit the necessary information, we are likely to solve the wrong problem and optimise something that is not valued. Suppose we don't provide the recommended changes and expected benefits in a way that addresses the true needs. In that case, the help is likely to be ignored, or recommended changes are ineffective in solving the right problem if implemented. That is why the principle of "no model without the modelled," as discussed in [8], is important when working with recommendations within complex socio-technical systems.

In this chapter, we propose to conduct research enabling the better capture of views and value systems of underrepresented groups via participatory modelling. This approach requires addressing indigenous methods and epistemologies. These concepts are closely related but currently do not necessarily entail each other. We make the case to solve common challenges in complex socio-technical systems. We need the cognitive diversity of all groups, which requires participatory modelling that enables equal participation in the process, so that all members are fully integrated into the team. To this end, we evaluate the state of the art of participatory modelling, including motivating the need to extend those for broader social applications. We then extend the categories of research methods for hybrid M&S research to include the comprehensive view of participatory modelling, and we close with a selected set of research topics.

4.2 Participatory Modelling

The need to integrate stakeholders into the model development process has been recognised for decades. Voinov and Bousquet [9] provide examples from the seventies. However, this early work focuses on supporting and eliciting information from managers and technical experts on technical excellence. These traditional stakeholders in the process, modelled to understand it better and very often to optimise it, were usually professionals with managerial or engineering education. As such, these stakeholders were used to formal models to think about their processes. The challenge was identifying tools and methods, as discussed in [10]. The various efforts to map these tools and methods to provide a common understanding of the process are foundational to ensuring semantic and conceptual consistency of the resulting common model. While such groundbreaking work was necessary, a new group of stakeholders requires rethinking and extending participatory modelling approaches.

Muthukrishna [11, pp. 26–27] observed that *"without a common understanding, common goals, and a common language, the flow of ideas is stymied, preventing recombination and reducing innovation."* If there is no way to express views and ideas commonly, diversity is more likely to stifle than to support finding common solutions. While diversity is the key to understanding complex socio-technical systems, adapting the modelling and methods to reflect the views of all stakeholders is critical to building a shared sense of value in the modelling approach. The multitude of views and this shared sense of value is pivotal to capturing all aspects of a socio-technical challenge, ensuring that we are not solving the wrong problem, and building confidence in the model and modelling outputs. By ensuring equity in a modelling process that is understandable and applicable to all stakeholders, they can be truly integrated into the team.

4.2.1 Participatory Modelling in the Context of Research Methods

Modelling, including participatory modelling, requires establishing a certain common understanding. In turn, the model represents the common understanding that helps capture entities, relations, behaviours, assumptions, and constraints to be shared within the group [11]. Several research methods are necessary to achieve successful participatory modelling, as described in [12], and must be informed and orchestrated by an overall process ensuring that all stakeholders in the complex socio-technical systems are heard, including those of underserved and underrepresented groups. In [12], we adapted the process recommended originally by [13] to ensure to reach this objective. All research methods captured here support this activity but are neither complete nor exclusive.

Research methods contribute to understanding the actors, entities and attributes of interest, the components and relations, and possible boundaries of the system that will guide all other methods to evaluate the socio-technical system. Initial scoping guides the application of the methods, and the results of application may provide additional insights that will change the scope and follow-on methods, e.g. if additional stakeholders are identified, or new relationships are discovered. Furthermore, even without discoveries of follow-on methods, complex adaptive systems continuously adapt themselves to new developments, so that the method of scoping becomes an integrated and continual task.

As Padilla et al. describe in [14] regarding data collection, the viewpoints of researchers of the humanities and those of engineering disciplines may be surprisingly diverse, but they can be mutually supportive. Humanities generally employ interviews and storytelling, while engineers utilise data engineering methods and utilisation of big data methods. Interviewing is a collective term that includes role-plays, life histories, and art-based knowledge elicitation that, in many cases, result in a collection of descriptive data that provide facets of various socio-ecological views of participants. Interviews may also elicit information from secondary sources, especially in the initial collection stages. For example, Kumm used narratives from YouTube videos to generate stories to initialise life histories. The insights in this first set of interviews are often evaluated in more detail by facilitating dialogs between actors to appreciate the values and decision-making processes better.

Another interview research method category often used to collect stakeholder information is role-playing and serious games. Particularly when the stakeholders have little to no engineering background or have no formal decision support practice, like is the case for many underrepresented groups whose life experience is needed to get a complete picture to understand the socio-technical system of interest, these methods can be very helpful.

All these research methods contribute to and prepare the participatory modelling approach. As the hybrid M&S community, we may consider expanding our view on Modelling and how to use it to support better decision-making in socio-technical settings. So far, traditional scientific approaches still dominate our ideas about hybrid M&S support, which must be expanded to create a more integrative understanding. The following chapter expands on this concept and gives an overview of its introduction into the operation research communities of the western world.

4.2.2 Motivating Extended Participatory Modelling

Many current participatory modelling approaches concerning underrepresented communities, ontologies, and epistemologies fall short in many ways. The idea of representation points to a problem within the approach: who is being represented and to/for whom? While community members' participation may happen during the effort, the originators of the project and the ultimate audience of the modelling output may not be the community(s) being represented. This concern is captured very cogently in the chapter entitled "Anthropologists and Other Friends" in "Custer Died for Your Sins: An Indian Manifesto" by Vine Deloria Jr., a member of the Standing

Rock Sioux Tribe, a lawyer, theologian, and activist. In this work, he observes the observers, in which the observed are neither the originators of the project nor the audience for the knowledge produced:

> An anthropologist comes out to Indian reservations to make observations. During the winter these observations will become books by which future anthropologists will be trained, so that they can come out to reservations years from now and verify the observations they have studied. After the books are written, summaries of the books appear in scholarly journals in the guise of articles. These articles "tell it like it is" and serve as a catalyst to inspire other anthropologists to make the great pilgrimage next summer. [15]

The activity of "modelling" in "participatory modelling" is a foundational challenge to the representation of communities. Frequently, modelling takes the shape of simulating a system using software and other tools, which often requires quantifying aspects of the system so outputs can be characterised given a certain set of inputs. Underlying these approaches are beliefs related to the reason for modelling, philosophical underpinnings, ontological assumptions, place of values in the modelling process, assumptions about the nature of knowledge, considerations about what counts as truth, and more [16]. Existing Western paradigms, however, may be ineffective in modelling underrepresented communities. Attempting to model underrepresented epistemologies may be determined by relationships between the seen and the unseen, a primordial past that is alive in the present, and a present that shapes a transcendent future; the animate living, the inanimate living, the seen and the unseen living, and the non-living; the network of dependencies between the human as a steward of their environment, the environment, and the cosmos; and a way of living where sensory experience in addition to generational wisdom and revealed knowledge shape thoughts, actions, and communities [16], although these considerations are less common in the conceptualisation of participatory modelling despite the observed need [17–21].

Not starting with the ontological realities and epistemologies of the underrepresented means the potential for even further marginalisation—by design or by accident. Among a countless number of examples include the genocidal impacts of the Kinzua dam on the Seneca Tribe [17], the diversion of the Gila River through the Florence Casa Grande Project as an attempt by the U.S. federal government to benefit Pima Indians [18, 19], but which resulted in several severely negative impacts including one of the highest rates of diabetes in the world among a formerly healthy population [20–22]; and the displacement of and detriment to farmers and communities by the Peligre Dam in Haiti [23].

In addition, once recommendations from insufficiently comprehensive modelling approaches are implemented, communicating and implementing the proposed solutions can be hindered by a lack of awareness or understanding of lived experiences and beliefs. In these cases, cross-epistemic communication of results emerging is at risk of entrenching marginalising structures and reinforcing epistemic injustice [24, 25].

The above and other challenges to current participatory modelling approaches mean substantial opportunities exist to innovate and reinvigorate our approach to underrepresented communities for truly participatory decision-making. It means an opportunity for the space to centre and learn from knowledge systems that have been marginalised and to examine how those systems may enable innovation that benefits holistic representation. This issue is of particular urgency with challenges of existential relevance to humanity in general, such as climate change [26–28], where there is a need for making societal demands and aspirations the basis for real solutions [29].

4.2.3 Examples of Participatory Modelling

Public participation is not a new concept. It has been applied particularly in land use and water resources planning for over four decades through the passage of the National Environmental Policy Act of 1969 (42 U.S.C. §§ 4321 et seq.), among others reported by the Institute for Water Resources [30]. Although methods described by the U.S. Army Corps of Engineers (USACE) are mainly document-driven and focus on public comments, the authors repeatedly emphasise the importance of understanding the values held by the public that drive public participation and conclude that "*information about values held by the public was the most important information this planner could receive*" [30, p. 52].

The International Association for Public Participation (IAP2) was founded in 1990 to recognise the importance of understanding public values, to respond to the rising global interest in public participation. While the initial focus was to support the practitioners to make better public decisions, the scope quickly widened to allow for more active public representation in decision-making with the ultimate goal of public empowerment. Table 4.1 shows their public participation spectrum, in which the public's impact on the decision increases from left to right.

Although the full empowerment of the public is still a future objective, public participation has already been implemented and demonstrated to improve public perception of environmental management policies and practices and increase the speed and cost of implementation [31]. Early examples of increased public participation include translating USACE's scientific and technical materials into more accessible language and formats for public consumption in the '70s, and the first Public Involvement training course was conducted by the Institute of Water Resources [30, 32]. USACE continued to evolve from inform and consult to involve and collaborate over decades, culminating in a successful environmental permitting process in Sanibel Island, FL, approved by citizens, environmental groups, state and local agencies, and USACE. USACE notes the further success of the process through the lack of disputes throughout the permit, as public concerns were addressed early in the process and solutions comprised inputs by the stakeholders. The handbook for Corps engineers captured the lessons learned and recommended good practice [33].

As implemented in the form of community stewardship of local environmental resources across the IAP2 spectrum, public participation has been significantly correlated with higher species diversity and biomass [34]. The authors note that participants were empowered to provide educational and restoration-related support through

Table 4.1 How the IAP2 spectrum supports public participation

	Limited impact → Full control				
	Inform	Consult	Involve	Collaborate	Empower
Public participation goal	To provide the public with balanced and objective information to assist them in understanding the problem, alternatives, opportunities and/or solutions	To obtain public feedback on analysis, alternatives and/or decisions	To work directly with the public throughout the process to ensure that public concerns and aspirations are consistently understood and considered	To partner with the public in each aspect of the decision including the development of alternatives and the identification of the preferred solution	To place final decision-making in the hands of the public
Promise to the public	We will keep you informed	We will keep you informed, listen to, and acknowledge concerns and aspirations, and provide feedback on how public input influenced the decision	We will work with you to ensure that your concerns and aspirations are directly reflected in the alternatives developed and provide feedback on how public input influenced the decision	We will look to you for advice and innovation in formulating solutions and incorporate your advice and recommendations into the decisions to the maximum extent possible	We will implement what you decide

a strong connection between people and policy at the local level. Those connections and stewardship reinforce a social norm that significant societal value is placed on local ecological resources and sustainability, reinforcing participation and ecological protection.

Although the inclusion of public participation in environmental management has come a long way in the past century, challenges persist in establishing connections for participation and incorporating the many and often conflicting perspectives that arise from multi-stakeholder participation [35]. The 1964 treaty has been hailed as a prime example of international cooperation. However, public participation and underserved populations were not considered in the terms of the treaty. Beginning in 2024, the

Columbia River Treaty, which establishes the international cooperative use of the Columbia River, can be nullified, or modified [36]. Both the U.S. and Canada have been engaged in public participation for over a decade to gather public insights and opinions, and both Canadian and U.S. governments have recognised, considered, and begun to incorporate public comment requesting ecosystem functions be explicitly considered in the Treaty [37]. However, the U.S. and Canada are still in session to determine the extent of the public's input, thus falling short of empowerment.

One recent example of integrated modelling is the application of serious gaming with an integrated watershed management decision support system in the Cedar River Watershed, Iowa [38]. 60 participants representing local communities, government, organisations, academia, and state and federal entities were convened for a multi-hazard watershed management tournament. Enabling access to social, environmental, and economic impact modelling and opportunities to work with entities with varied perspectives and goals increased collaboration among different stakeholders and resulted in the changes to existing hazard mitigation plans to be more inclusive. Bakhanova et al. [39] observe that some forms of serious games "are good for learning about the perspectives of others and building communication skills," which addresses a need identified in this chapter. Their journal article "suggests possible extensions of game design used for each stage of the participatory modelling process, aiming at better learning, communication among stakeholders, and overall engagement." Overall, the use of serious games to be considered for the toolbox of hybrid M&S methods seems promising to increase participatory modelling opportunities.

Castilla-Rho [40] applies serious game methods to identify and represent competing stakeholder perspectives in his approach, which uses an agent-based model to understand the complexity of the decision-making process and project stakeholders to alternative futures. He notes that participatory modelling transforms study outcomes from a written report into a living document. In his article, he refers to this as follows:

> … participatory groundwater agent-based models (ABMs) as 'management flight simulators.' Just like pilots use flight simulators to hone their judgment and make sound decisions in the face of uncertain flight conditions, a management flight simulator allows stakeholders to identify robust policy options under uncertainty. The management flight simulator becomes a laboratory in which a wide range of policies, decisions, and models can be tried and tested, without the risk of making mistakes on the real groundwater system. [40, p. 622]

An example that addressed many of the needs and proposed support discussed in this chapter is given in [12]. The research described in this article copes with the need for computational support in a complex decision space under deep uncertainty to address dam resilience planning to ensure technical excellence and increase the decision-makers' social competence. First, all relevant stakeholders in this socio-technical system are identified, and their various viewpoints and value systems are captured. Creating the models is done by subject matter experts, but the need to improve the participation of the stakeholders using better participatory modelling methods is recognised. The resulting model is then used to drive an agent-based simulation that allows the computation and visualisation of the many objectives of the stakeholders and the effects and consequences of decisions. For example, while engineers may model the safety and a dam, solutions to increase resilience, the resilience of the dam and the dam system may be informed by accessibility to points of interest important for tourism, environmental protection, and protection of places of cultural interest. Better-informed decision-making provides computational insights into the socio-economic impacts of decisions in the model's constraints. It provides ethical insights for project leadership, ensuring that all the relevant viewpoints, values, and objectives of the members of the society are considered. The more the public can be integrated into developing and using this computational repository of models, the more informed and socially competent the decisions.

Within the hybrid modelling and simulation community, Harper and Mustafee [41] provide a first example of participatory modeling for healthcare applications. In their paper they describe how subject matter experts can integrated into the modeling and calibration processes of simulation-based research.

4.3 The Role of Hybrid Modelling and Simulation

The need to address social challenges with models is not new, as evidenced in examples in [42]. However, over recent years, societal change has put more emphasis on social responsibilities. Participatory modelling using various models and simulations is increasingly a topic of published research [5, 43–45]. The role of hybrid M&S methods in this context, however, is still in its beginning. Mustafee et al. [46] provide one of the most inclusive definitions for the various types of hybrid M&S and give classification examples. They follow the example of the operations research community, which integrates hard operations research methods, predominantly working with quantitative and soft operations research methods. Mustafee, Harper, and Onggo predominantly apply qualitative methods to support many application domains. However, social sciences [47] and healthcare [48] are often used as examples of the need to bring experience into projects that can only be accessed by soft methods. A general overview of recent applications of these mixed operations research methods combining quantitative and qualitative aspects is provided in [49]. Simulation methods are usually categorised as hard methods, as discrete and continuous methods are quantitative.

Focussing on hybrid M&S, Mustafee et al. [46] distinguish between quantitative, qualitative, and cross-disciplinary research methods. Under quantitative methods, they enumerate discrete simulation methods, continuous simulation methods, and hard operations research methods. The various soft operations methods, such as interviews and focus groups, make up the qualitative side as described by Mustafee et al., and cross-disciplinary research methods incorporate methods that are not part of the operations research realm, such as ethnographic methods, cloud computing, cyber-physical systems, etc., calling them cross-disciplinary research methods.

Using this taxonomy, researchers can differentiate between various types of hybrid M&S mixing methods. The first three types belong to the realm of mixed simulation methods. Multi-technique hybrid simulation combines multiple simulation techniques of the same method group, either discrete or continuous. Multi-methodology hybrid simulation combines discrete and continuous methods, such as agent-based models and system dynamics. They use both methods and multiple techniques resulting in multi-technique multi-methodology hybrid simulation. For this chapter, the following types are more interesting as they address multiple modelling methods. Hybrid operations research M&S (OR/MS) models add operations research methods to the simulation mix. If soft methods are applied, they are referred to as multi-paradigm hybrid OR/MS models. Finally, cross-disciplinary hybrid models utilise research methods from other disciplines. For such support, cross-disciplinary hybrid models can use all hybrid modelling categories identified before. The hybrid use is not limited to choosing just two research methods, but it can combine as many research methods as needed to support the research.

Motivated by the examples in the previous section, it seems beneficial to add another category to this spectrum comprising methods rooted in underrepresented epistemologies and methodologies [50]. As these methods may incorporate contradictory value and belief systems, it is worth capturing them in their category to extend the taxonomy. Table 4.2 gives selected examples for the categories. Note that some research method examples may belong to additional research method categories.

For studying complex socio-technical systems and providing computational decision support, multi-paradigm hybrid OR/MS models are particularly interesting to advance the field and empower underrepresented communities. It becomes even more interesting when these methods are augmented by cross-disciplinary methods addressing the social aspects of the problem to be understood and solved, such as those described in [51]. Using the perspective of Mustafee et al. [46] and the research methods addressed for social–ecological systems, Fig. 4.1 captures the resulting hybrid M&S categorisation. Only the blocks of discrete and continuous M&S address hybrid simulation, while hybrid M&S focuses on the modelling part and uses all research methods to support cross-disciplinary research. As discussed, many research methods can be selected, but at least one must be a hybrid simulation method. The topic of cross-disciplinary research support has been addressed in detail in [52].

The types of hybrid M&S mixing methods identified in [46] can be generalised to facilitate discussions with the research communities. They already show the gradual extension from mixing simulation methods to mixing simulation and hard operations research methods, extending this to soft operation research methods, and finally including methods from disciplines other than OR/MS. Figure 4.1 shows the methods' growing inclusion into the hybrid M&S methods realm. The categorisation is important, as with every extension, additional steps are needed to align the tools, underlying methods, and foundational research views characteristic of cross-disciplinary research on various maturity levels [52]. Accordingly, the integration effort must meet different challenges.

To ensure that methods are aligned and conceptually consistent, we need a framework helping to reach a consensus on *what is modelled* and *how it is modelled*. The need to address a referent's referential and methodological aspects in building

Table 4.2 Research method categories and examples

Category	Research method examples
Discrete simulation	Discrete event simulation (DEVS) Finite element methods Agent-based modelling
Continuous simulation	System dynamics Continuous simulation Computational fluid dynamics
Quantitative operations research	Linear programming Network analysis Dynamic optimisation Game theory Queueing theory Markov processes Decision theory
Qualitative operations research	Data collection Interviews Observational field notes Focus groups Documents and archival data Data recording Data analysis and interpretation Data coding
Socio-ecological research (cross-disciplinary)	Data generation Systems scoping Knowledge co-production Serious games Resilience assessment Qualitative content analysis Comparative case study analysis Historical assessment
Underrepresented communities and cultures research (cross-cultural)	Relational accountability Meaningful involvement Story as a method Decolonizing theory Integrative knowledge systems Participatory action research Metaphysical and holistic understanding Kinesthetic and spiritual Intelligence Trans-rational knowing

compositions has been addressed from the theoretic side in detail by Hofmann et al. [53]. If the mixed methods stay in the simulation methods domain, the experts belong to the same discipline and understand conceptual foundations, applicable methods, and useful tools, such as those captured in the M&S Body of Knowledge [54]. Some additional alignment is needed when integrating hard operations research methods, but at least all experts follow the quantitative paradigm. The views must be extended, and underlying theories must be aligned when allowing qualitative research methods

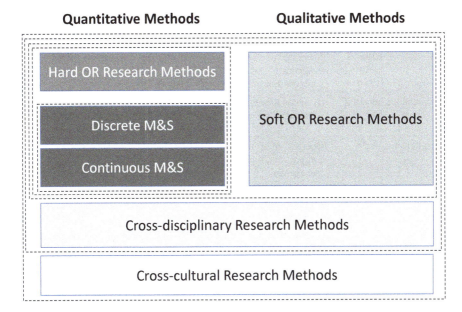

Fig. 4.1 Categories of research methods for hybrid M&S socio-ecological research

from the soft operations research community. However, the experts are all rooted in the operations research disciplines. A common understanding between the disciplines is also needed when cross-disciplinary methods are applied.

Traoré gives an example of how to meet these integration challenges [55]. The author provides a framework to capture the multiple perspectives on the conceptual level underlying the methods to allow a holistic hybrid simulation approach on the implementation level. To do so, he distinguishes between *concepts* captured in *formalisms*, such as Discrete Event Specification (DEVS), Petri Nets, cellular automata, ordinary differential equations, and operations research and artificial intelligence methods, *specifications* captured in *models*, such as discrete or continuous simulation models, or algorithms, and *operations* provided by *engines*, such as executable simulations, solvers. The author observes that while hybrid compositions execute on the operations level, their sound analysis also requires alignment on the specifications and concepts level. Nonetheless, most hybrid M&S research contributes to the operations and composing of different engines. Table 4.3 slightly contradicts his recommendations on addressing heterogeneous composition challenges.

This framework aligns well with the recommendations given in [52], and supports the alignment of tools (operations), methods (specifications), and theory of the discipline (concepts). The degree of alignment and the maturity of the cross-disciplinary research co-developed across a spectrum of multidisciplinary, interdisciplinary, and trans-disciplinary collaboration. Experts from multiple disciplines come together in multidisciplinary research to solve a common problem. Often, ad-hoc groups need

Table 4.3 Composition challenges in hybrid M&S studies

Level	Challenge	Example situations
Concepts (formalisms)	Integrated perspectives (semantical consistency and compositional validity)	Aligning a disease-spreading perspective and a population dynamics perspective
Specifications (models)	Syntactic model (syntactic composability)	Aligning the perspectives of an agent-based approach with the results of an ethnographic study
Operations (engines)	Operational semantics (hybrid composition, time management, ...)	Aligning discrete event simulator with a solver
	Data management (data exchange format and protocols)	Aligning Linux-Java based code and Windows-C++ based code
	Code/event synchronisation (parallel and distributed Computation)	Applying parallel and distributed computing methods to align executables

to be able to conduct commons operations. They may be dissolved once the problem is solved, but they may also realise that a whole class of problems can be solved with their approach. They may develop common methods foundational to tools and solutions, resulting in interdisciplinary teams. They still belong to different disciplines, but they share common ground regarding the classes of problems they can address. This may further lead to the development of a common theoretical understanding so that the experts no longer approach the challenges from the viewpoint of their original discipline. However, a new perspective evolves, making the research transdisciplinary [52]. Although transdisciplinary research approaches developed and implemented do not necessarily incorporate underrepresented and underserved viewpoints, the approaches are likely to be more extensible and positioned for holistic, participatory modelling.

Our objective is to support the evaluation of complex socio-technical systems using hybrid M&S methods that allow the equitable contribution of minority communities via participatory modelling. All social groups shall understand the diverse challenges, participate in considering options, and understand the proposed solutions. Figure 4.2 extends the earlier mentioned process first introduced by Shults and Wildman [13] for interdisciplinary teams, and later modified for enabling computational decision support by Tolk et al. [12]. The five process steps can be iterative and have feedback loops, e.g., when the identification of objectives and belief systems leads to the discovery of additional members, or when conducting the experiments leads to the need for refining the various metrics.

The first step "identification of stakeholders" must be conducted without any hybrid M&S method support but is important to establish credibility with the participating groups. To address all aspects of interest under the various viewpoints, selecting a diverse team and integrate the members as equal member is the first step. The "identification of objectives and belief systems, social values, etc." follows in the

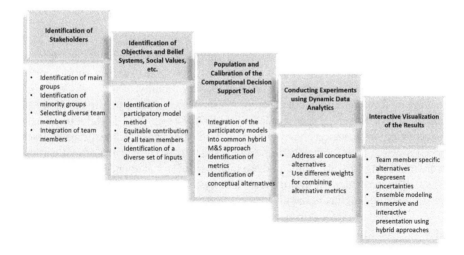

Fig. 4.2 A process for evaluating complex socio-technical systems using hybrid M&S methods

second steps. To elicit those often deeply rooted beliefs from the team members, identifying the various and best applicable methods supporting participatory modelling is essential. This will lead to the diverse set of inputs needed to ensure the equitable contribution of team members, no matter of their ability to support traditional hybrid M&S methods. For the next step, the "population and calibration of the computational decision tool," the integration of the participatory models is required. Due to the high variance of belief systems, it is likely that no all of them can be integrated into a conceptually coherent set of hybrid M&S methods, as they represent conceptual alternatives. As shown by Tolk et al. [56], it is possible to derive an ontological reference model that captures all inputs while allowing for conceptual diversity and use the reference model to derive a set of conceptually consistent conceptual models from it. Using an ensemble approach, the next step becomes now feasible, "conducting experiments using dynamic data analytics" that addresses the various alternative and furthermore allows to use different weights for combining the underlying metrics. The last step is the "interactive visualisation of results." Again, this will require multiple, member specific methods to ensure that the importance of the results and the applicability to address the challenges identified as urgent for the team members are communicated back using the best participatory methods to do so. Being immersive and interactive will create the empathy needed for all team members to add to the technical competence of engineering solutions and enable to understand the socio-technical system and the belief systems of all members.

To be able to realise this vision requires to improve various methods identified so far and drive the research to enable the integrated approach and reaching a fuller potential. Some of the most urgent topics are captured in the following section.

4.4 Research Topics

Section 4.2 established the need to include participatory modelling into our concepts supporting hybrid modelling and simulation. We gave some examples of successful proofs of feasibility to extend our reach to the community of socio-ecology and integrate their research methods. Section 4.3 showed that this extension is a logical continuation of the work spearheaded by Mustafee et al. [46], introducing a new interpretation of the various types of hybrid M&S categories. This section addresses several research topics that need to be addressed to make this vision of real participatory modelling a reality. Some of these topics are essential to understanding previously underrepresented viewpoints and value systems, which can be quite different from those of our engineers and the hybrid M&S community is currently accustomed to. We also need to extend our viewpoints for real integration of these diverse views. Other topics are more custodian in style and beneficial for other domains of hybrid M&S support in a more general sense. However, still, they are necessary to support our view of full integration, such as visualisation and representation of classes of uncertainties and many objectives.

Capturing the epistemological differences in the various research methods may be the most challenging task, far out of our scholastic comfort zone of expertise, but captured in this section as a request for further community consideration and research. They all contribute to the objective of the proposed computational decision support based on the extended view of hybrid M&S methods, to elicit knowledge from all social groups to understand the diversity challenges, to provide equality in participation by proper participatory models, and to communicate the proposed solutions in understandable form back to the diverse groups.

4.4.1 Immersive Visualisation

The importance of providing the results of a decision support system tailored to the needs of the decision-makers has been discussed by Castilla-Rho [40] and Rouse [57]. Both authors envision creating a "flight simulator for the decision maker," though the immersive visualisation is also critical for communication with other stakeholders. This vision includes a meaningful visualisation of the results and the full immersion of the deciding group into a system that immediately reacts to collected information and decisions. Haberman and Page also describe such an interactive, immersive visualisation system [58]. New technologies in virtual reality technology are opening additional opportunities to provide a truly immersive experience that allows a team to experience the effects of decisions. City and urban planning teams already take advantage of such technologies to increase participation in their planning [59]. Creating immersive and interactive virtual places to elicit information collection, decision-making, and results communication across all stakeholders should therefore be among the research topics of priority.

4.4.2 Using Real-Time Data

As with many other analytic methods, big data analytics methods are often categorised as descriptive, prescriptive, or predictive [60]. This observation is often interpreted as "how things are," "how things should be," and "how things will be." Predictive analysis is particularly interesting in combination with simulation methods, often used to conduct what-if analyses by parametric variations of the input parameters. There is a close relationship between the fields of dynamic data-driven simulation [61] and dynamic data analytics [62].

Mustafee et al. [63] introduce ideas for applying hybrid M&S methods and real-time data to improve digital twins. Suppose the research recommended in this chapter is successful. In that case, it leads to improved digital twins via real-time data utilisation. However, it also ensures that the digital twins look and behave like previously underrepresented groups are characterised in the models. The serious game applications discussed earlier can be used to create this behaviour, and once immersive capabilities are established, behaviour can be extracted by observation and reflection with the groups. Again, we recommend that urban design groups use this method to increase participation [64].

4.4.3 Using Artificial Intelligence Support

The creation of a set of conceptually consistent simulation specifications derived from a common reference [56] can be supported by artificial intelligence (AI) methods. Ontologies have been chosen because they are used to capture knowledge for AI applications. Nonetheless, their use in support hybrid M&S to capture what is modelled in referential ontologies and how it is modelled in methodological ontologies, as described by Hofmann et al. [53], is not yet common practice.

As within many other domains, using AI methods is generally perceived to help create models [65, 66]. Social sciences are already actively driving research in this direction forward, emphasising the need for a clear understanding of addressing biases as well as data fidelity [67]. Two interesting use cases for AI in participatory modelling are eliciting knowledge and observing and mimicking behaviour.

To elicit knowledge, interviews, and questions are often used. Shen et al. [68] evaluate the positive and negative aspects of using large language models (LLM), such as OpenAI's chatGPT [69], to communicate medical results with patients as well as supporting medical experts in their work by providing information used within their training. LLMs usually go through a two-stage training process. In the first phase, large amounts of not annotated data are used to learn the statistical likelihood of word combinations used, correlations of terms and phrases, and other language contexts. This part is usually conducted as unsupervised learning. In the second phase, the LLM is trained in the language of a specific domain. This training results in some understanding. However, the debate is still ongoing if this is pure language understanding or if the LLM understands the underlying concepts, including the social situations language encodes [70]. In any case, LLMs create a model of the underlying concepts and relations, which may help create machine-readable and executable representations of problem situations comprising many different viewpoints. A known shortcoming of current solutions is the tendency to close gaps in the underlying knowledge by statistical language constructs that fit into the gap instead of looking for clarifications. However, research has been conducted to improve this behaviour. To what degree such an LLM model can support hybrid M&S methods and how they can and should be integrated must become a subject of research of our community in this domain. First results on related efforts recently have been published by Giabbanelli [71].

Observing and mimicking behaviour is a more traditional use of AI in model development. Machine learning and rule sets that can be calibrated have been used for a significant time to represent behaviour. In particular, the defence industry provided some pioneering work in human and behavioural representation [72]. These research efforts have recently been broadened to non-defence-related modelling [73, 74]. Like in many computational social science publications, using agent-based models to represent human and social behaviour is often recommended. An overview of current research on using serious games, agent-based models, and AI methods is presented in [75]. Applying these ideas to integrate the behaviour of stakeholders with no systems engineering background is an interesting research topic for the hybrid M&S community.

4.4.4 Dealing with Deep Uncertainty

Quantitative operations research methods work with numbers and numerical expressions. It assumes that the entities, attributes, behaviour, processes, and relations are known. They also apply probabilistic methods to cope with uncertainties in these parameters. However, as the examples in this chapter show, uncertainties go beyond probability distributions for parameter values or uncertainty intervals of outcomes. The operations research community refers to this challenge as deep uncertainty. The term and its definition were introduced in a report by Lempert as conditions:

> …where analysts do not know, or the parties to a decision cannot agree on, (1) the appropriate conceptual models that describe the relationships among the key driving forces that will shape the long-term future, (2) the probability distributions used to represent uncertainty about key variables and parameters in the mathematical representations of these conceptual models, and/or (3) how to value the desirability of alternative outcomes. [76]

Decision-making under deep uncertainty has become a research topic involving members from the soft and hard operations research communities. The Society for Decision Making under Deep Uncertainty (https://www.deepuncertainty.org/) is a multidisciplinary association of practitioners, scholars, and students working to improve processes, methods, and decision-making tools under deep uncertainty.

This society has published a summary of supporting methods in an open-access compendium [77].

Hybrid simulation methods are already part of the recommended toolbox to cope with deep uncertainty, among others described by Kwakkel and Pruyt [78], who also provide open-access to their exploratory modelling working bench. An alternative to exploratory modelling is the ensemble simulation approach, such as described in [79]. It must be expected that in highly diverse teams, such deeply uncertain situations will more likely be the rule than the exception, so extensions of these methods should also become part of hybrid M&S methods.

An important aspect is also to make stakeholders aware of such uncertainties in the current situation and accompany all options to address the issues of concern. Accordingly, the immersive visualisation discussed in Sect. 4.1 must provide visual representations of uncertainty, as discussed in [80], for the situation and the effect of options.

4.4.5 Many-Objective Challenges

Multi-objective optimisation is a well-known topic in quantitative operations research and deals with optimisation problems with two or three objectives [81]. Terminology and methods are consolidated. The effect of preferences on multi-objectives is well understood in this community. However, the complexity grows rapidly in size with the number of objectives. To cope with such multi-objective optimisation challenges with many objectives, in practical terms, more than four operations research organised contributions under the domain of many-objective optimisation. Many methods aim to mathematically reduce the dimension, as described by Copado-Mendz et al. [82]. Others try to incorporate user preferences more [83], which leads to the question of which users get such privileged treatment. So far, no standard methods have emerged, and using heuristics is the norm.

As participatory modelling aims to increase the participation of stakeholders with multiple views and value systems, and this diversity can lead to an increase of objectives, most challenges in the socio-technical realm are highly likely to become many-objective challenges. As such, the participation of an inclusive and equitable group of stakeholders is critical.

4.4.6 Human Simulation and Artificial Societies

The hybrid M&S community can reach back to a growing body of knowledge on supporting taking the human aspects into account when using M&S to understand social and economic factors better, as several examples in this chapter have demonstrated. Diallo et al. [84] provide several examples of how to apply M&S

to research subjects in the humanities disciplines. Human simulation is a transdisciplinary approach to studying societal problems and is a pivotal means to study the human condition, generating insights through valid simulation. While this work inspired scholars of computational social sciences to use M&S more frequently, a broader, deeper, more inclusive analysis is needed to form the foundation for simulation that better serves society.

In particular, the studies to better understand the spreading of the SARS-CoV-2 Coronavirus in 2020 showed the need to integrate artificial societies and social science simulation more into the toolbox of researchers of complex socio-technical systems. Compartment models, a long-time public health tool, led the early phase. In these models, differential equations describe the flow between compartments of the population of Susceptible, Exposed, Infectious, and Recovered, leading to the SEIR model. System dynamics provides a well-supported set of methods and tools to allow the rapid development of such models, which then could be configured by empirically observed data on contact rates, infection rates and incubation time, and recovery/death rates and recovery time. However, SEIR can only show trends and effects as a standalone model. The differential equations approach to these models requires assumptions of even mixing heterogeneous populations or cumbersome workarounds. The alternative is agent-based models that cover the behaviour of various types of individuals in their situated environment and consider the networks between these agents as Epstein envisioned decades before COVID-19 hit [85]. Artificial society applications to better understand the pandemic, its spread, and its effects are described, among others, in [86, 87]. Within those artificial societies, the SEIR model was still useful in modelling the status of individuals going through the circle.

It seems to be a valuable research topic to bring human simulation closer into the hybrid M&S community to allow for better use of computational social science methods. This closer relationship enriches artificial societies with such human simulation solutions to represent a community as a complex socio-technical system. Furthermore, the resulting approach can be powerful if these models are better informed and calibrated by participatory models.

4.4.7 Capturing Epistemological Differences

How participation is conceptualised as part of participatory modelling can itself be problematic. Even efforts that include communities from the inception of an effort need to recognise that participatory modelling doesn't stop with the participation of community *members*. Participation entails the participation of ontological realities and epistemological approaches. It entails considering who is describing and theorising about whom and whether the other is naming, knowing, and generating knowledge from their frame of reference. A large body of literature describing epistemic violence stresses the criticality of this consideration as part of any description of the term "participation," such as [88–92].

These observations motivate that one of the most challenging tasks is to capture underrepresented epistemological methods allowing us to address these ontological realities and epistemological approaches across cultures. Examples of these research methodologies are captured in books such as [17, 93], which can be used as a starting point. Although there is a growing awareness of alternative value systems and world views, which are captured in their epistemologies, the nature of these epistemological methods often remains unclear, and they are often perceived as strange to the Western-educated engineer. What is the philosophical basis from which they derive? Can they be validated, and is it comparable to Western traditions? What is their contribution to scholarship and academic insights? What must we know to integrate them into the cross-disciplinary framework discussed in Sect. 4.3 in this chapter? Can we even capture spiritual elements and transrational knowledge to make it accessible to the quantitative M&S methods still at the core of hybrid M&S methods?

Acknowledging the differences between the Western tradition and underrepresented views is essential to understand post-colonial epistemology and methodology. Some examples of participatory approaches in geographical research can be found in [94]. Botha recommends mixing traditional and indigenous methods [95], our goal when integrating hybrid M&S methods with indigenous methods. Similarly, Levac et al. provide examples of learning across indigenous and Western knowledge systems, such as it is envisioned to be supported by participatory modelling [96]. Most of these contributions come from the social science domain, and it requires additional research to make such ideas and concepts applicable to engineers. For the hybrid M&S community, this research topic may remain the most challenging for the foreseeable future.

Figure 4.3 depicts the research topics and their contribution to the overall vision and solution proposed in this chapter.

The research topics enumerated in this section are neither complete nor exclusive but were selected to start the discussion on what is needed to make the objective of hybrid M&S support for complex socio-technical systems feasible and put them onto the agenda for future conferences and symposia.

4.5 Discussion and Conclusion

In his book "the difference," Scott Page makes a strong case for cognitive diversity [97]. Cognitive diversity refers to the differences in how different stakeholders approach a challenge, including how they think, their work and learning styles, and the experience they have gained in related areas. Over recent years, the need to include the various viewpoints of a challenge can be as important as various ways to find a solution, as solutions are always tied to the viewpoint used to understand the problem to be solved. As Padilla observes in [98], we often define understanding as matching our knowledge, worldview, and view of the problem to drive towards the solution. The outcome of this process is assigning a truth value to a problem, the generation of knowledge, and the generation of worldview. We posit that socio-ecological support

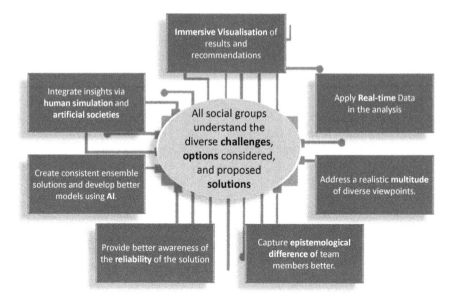

Fig. 4.3 Contributions of the identified research topics

must extend these insights into multiple viewpoints, as these multiple viewpoints best reflect the cognitive diversity needed to address such problems.

However, a cognitively diverse team alone is not sufficient. Underserved and underrepresented voices must be heard, and all possible solutions can be expressed. To allow for this integration, however, the various members need equal opportunities to express their views, which requires different forms of participatory modelling to provide the equity needed to allow for a true democratising of the model-building process. As discussed by [12], diversity, integration, and equity are necessary but insufficient. The team needs to be united by a common objective and metrics defining success, which may be a many-objective challenge. However, ultimately, we are interested in building a team with the most valuable merits to address the challenges by allowing us to apply diverse cognitive views and solutions. We support this team's integration by giving all team members equal participation opportunities. Participatory modelling in hybrid M&S plays a pivotal role in this process.

The results of participatory modelling have the potential to address several other social challenge domains as well, such as helping to create a better understanding of communities and their needs that are marginalised by current practices. For example, various recent publications posit that within our Western healthcare systems, more algorithms than expected are built on faulty assumptions due to discrimination or misunderstanding of the target groups that are embedded in medical and science training and practice, as well as confounding identities that are socially constructed with the biological phenomenon [99, 100]. Allowing these groups to express their views and concerns via participatory models should help address them effectively and provide better communication between social groups.

There are also many promising research efforts in hard and soft operations research, and the opportunities to create immersion by visualisation are continuously increasing. However, only integrating challenge-relevant cognitive diverse teams and incorporating underrepresented viewpoints can ensure that we solve the right problems faster, not just our assumed problems. We increasingly realise that we not only need to strive for technical excellence but social responsibility as well. We submit that participatory modelling, as envisioned in this chapter, provides some support to ensure we can meet this responsibility.

Acknowledgements and Disclaimer Part of the underlying research leading to this chapter was supported by the MITRE Innovation Program. The content was discussed with many colleagues at meetings and conferences, and the authors are thankful for the contributions and recommendations for improvement. Among these colleagues, we particularly thank Dr. Jeffrey Arnold, Dr. Uba Backonja, Dr. Navonil Mustafee, Dr. Jose Padilla, and Mr. Joshua Stadlan.

The authors' affiliation with The MITRE Corporation is provided for identification purposes only and is not intended to convey or imply MITRE's concurrence with, or support for, the positions, opinions, or viewpoints expressed by the author. This paper has been approved for Public Release; Distribution Unlimited; Case Number 23-02030-02.

References

1. Brailsford SC, Eldabi T, Kunc M, Mustafee N, Osorio AF (2019) Hybrid simulation modelling in operational research: a state-of-the-art review. Eur J Oper Res 278(3):721–737
2. Tolk A, Page EH, Mittal S (2018) Hybrid simulation for cyber physical systems: state of the art and a literature review. In: Proceedings of the annual simulation symposium, Baltimore, MD
3. Saouma V, Sivaselvan M (eds) (2008) Hybrid simulation—theory, implementation and applications. Taylor & Francis, London
4. Mittal S, Tolk A (eds) (2020) Complexity challenges in cyber physical systems—using M&S to support intelligence, adaption and autonomy. Wiley, Hoboken, NJ
5. Gilbert N, Ahrweiler P, Barbrook-Johnson P, Narasimhan K, Wilkinson H (2018) Computational modelling of public policy: reflections on practice. J Artif Soc Soc Simul 21:3669
6. Gelfert A (2016) How to do science with models: a philosophical primer. Springer, Cham, Switzerland
7. Page SE (2018) The model thinker: what you need to know to make data work for you. Basic Books, New York, NY
8. Tolk A, Gilbert TCN, Macal CM (2022) How can we provide better simulation-based policy support? In: Annual modeling and simulation conference (ANNSIM), San Diego, CA
9. Voinov A, Bousquet F (2010) Modelling with stakeholders. Environ Model Softw 25:1268–1281
10. Voinov A, Jenni K, Gray S, Kolagani N, Glynn PD, Bommel P, Prell C, Zellner M, Paolisso M, Jordan R, Sterling E, Olabisi LS, Giabbanelli PJ, Sun Z, Le Page C, Elsawah S, BenDor TK, Hubacek K, Laursen BK, Jetter A, Basco-Carrera L, Singer A, Young L, Brunacini J, Smajgl A (2018) Tools and methods in participatory modeling: selecting the right tool for the job. Environ Model Softw 109:232–255
11. Muthukrishna M (2020) Cultural evolution and the paradox of diversity. The Bridge, pp 26–28
12. Tolk A, Richkus JA, Shults FL, Wildman WJ (2023) Computational decision support for socio-technical awareness of land-use planning under complexity—a dam resilience planning case study. Land 12(5):952

13. Shults FL, Wildman WJ (2020) Human simulation and sustainability: ontological, epistemological, and ethical reflections. Sustainability 12:10039
14. Padilla JJ, Frydenlund E, Wallewik H, Haaland H (2018) Model co-creation from a modeler's perspective: lessons learned from the collaboration between ethnographers and modelers. In: Social, cultural, and behavioral modeling: 11th international conference, Washington, DC
15. Deloria Jr V (1969) Custer died for your sins: an Indian Manifesto. University of Oklahoma Press, Norman, OK
16. Chilisa B (2019) Indigenous research methodologies. Sage Publications, Washington, DC
17. Bilharz JA (2002) The Allegany Senecas and Kinzua Dam: forced relocation through two generations. University of Nebraska Press, Lincoln, NE
18. Martínez D (2011) Stealing the Gila: the Pima agricultural economy and water deprivation, 1848–1921. Am Indian Q 35(1):143–145
19. DeJong DH (2004) An equal chance? The Pima Indians and the 1916 Florence-Casa Grande irrigation project. J Ariz Hist 45(1):63–102
20. Bennett PH, Miller NBRM, LeCompte PM (1976) Epidemiologic studies of diabetes in the Pima Indians. In: Proceedings of the 1975 Laurentian hormone conference. Academic Press, pp 333–371
21. Knowler WC, Bennett PH, Hamman RF, Miller M (1978) Diabetes incidence and prevalence in Pima Indians: a 19-fold greater incidence than in Rochester, Minnesota. Am J Epidemiol 108(6):497–505
22. Krosnick A (2000) The diabetes and obesity epidemic among the Pima Indians. N J Med J Med Soc N J 97(8):31–37
23. Farmer P (2012) The water refugees (from AIDS and accusation: Haiti and the geography of blame). In: Haitian history: new perspectives. Routledge, New York, NY, pp 295–303
24. Matheson J, Chock VJ (2019) Science communication and epistemic injustice. Soc Epistemol Rev Reply Collective 8(1):1–9
25. Medvecky F (2018) Fairness in knowing: science communication and epistemic justice. Sci Eng Ethics 24(5):1393–1408
26. Berry LH, Koski J, Verkuijl C, Piggot CSG (2019) Making space: how public participation shapes environmental decision-making. Stockholm Environment Institute, Stockholm, Sweden
27. Devine-Wright P (2017) Environment, democracy, and public participation. In: Richardson D, Castree N, Goodchild M, Kobayashi A, Liu W, Marston R (eds) International Encyclopedia of geography: people, the earth, environment and technology
28. Dietz T (2013) Bringing values and deliberation to science communication. PNAS 110(supplement_3):14081–14087
29. Jasanoff S (2018) Just transitions: a humble approach to global energy futures. Energy Res Soc Sci 35:11–14
30. Creighton JL, Priscoli JD, Dunning CM (1983) Public involvement techniques: a reader of ten years experience at the institute for water resources. U.S. Corps of Engineers, Alexandria, VA
31. Langsdale SM, Cardwell HE (2022) Stakeholder engagement for sustainable water supply management: what does the future hold? AQUA Water Infrastruct Ecosyst Soc 71(10):1095–1104
32. Creighton JL, Dunning CM, Priscoli, Delli J, Ayres DB (1998) Public involvement and dispute resolution: a reader on the second decade of experience at the institute for water resources. U.S. Army Corps of Engineers, Alexandria, VA
33. Pamphlet #5 (1996) Overview of alternative dispute resolution (ADR): a handbook for corps managers. U.S. Army Corps of Engineers, Alexandria, VA
34. Turnbull JW, Clark GF Johnston EL (2021) Conceptualising sustainability through environmental stewardship and virtuous cycles—a new empirically-grounded model. Sustain Sci 16:1475–1487
35. Falconi SM, Palmer RN (2017) An interdisciplinary framework for participatory modeling design and evaluation—what makes models effective participatory decision tools? Water Resour Res 53:1625–1645

36. Shrestha A, Souza FAA, Park S, Cherry C, Garcia M, Yu DJ, Mendiondo EM (2022) Socio-hydrological modeling of the tradeoff between flood control and hydropower provided by the Columbia River Treaty. Hydrol Earth Syst Sci 26(19):4893–4917
37. Baltutis WJ, Moore M-L, Tyler S (2018) Getting to ecosystem-based function: exploring the power to influence Columbia River Treaty modernization towards ecosystem considerations. Int J Water Gov 6(3):43–64
38. Bathke DJ, Haigh T, Bernadt T, Wall N, Hill H, Carson A (2019) Using serious games to facilitate collaborative water management planning under climate extremes. J Contemp Water Res Educ 167(1):50–67
39. Bakhanova E, Garcia JA, Raffe WL, Voinov A (2020) Targeting social learning and engagement: what serious games and gamification can offer to participatory modeling. Environ Model Softw 134:104846
40. Castilla-Rho JC (2017) Groundwater modeling with stakeholders: finding the complexity that matters. Groundwater 55(5):620–625
41. Harper A, Mustafee N (2023) Participatory design research for the development of real-time simulation models in healthcare. Health Syst 12(4):375–386
42. Stroh DP (2015) Systems thinking for social change: a practical guide to solving complex problems, avoiding unintended consequences, and achieving lasting results. Chelsea Green Publishing, White River Junction, VT
43. Goldingay S, Epstein S, Taylor D (2018) Simulating social work practice online with digital storytelling: challenges and opportunities. Soc Work Educ 37(6):790–803
44. Đula I, Größler A (2021) Inequity aversion in dynamically complex supply chains. Eur J Oper Res 291(1):309–322
45. Shaikh Y, Jeelani M, Gibbons M, Livingston D, Williams D, Wijesinghe S, Patterson J, Russell S (2023) Centering and collaborating with community knowledge systems: piloting a novel participatory modeling approach. Equity Health 22:45
46. Mustafee N, Harper A, Onggo BS (2020) Hybrid modelling and simulation (M&S): driving innovation in the theory and practice of M&S. In: Winter simulation conference, virtual conference
47. Tashakkori A, Johnson RB, Teddlie C (2020) Foundations of mixed methods research: integrating quantitative and qualitative approaches in the social and behavioral sciences. SAGE Publications, Los Angeles, CA
48. Brailsford S, Vissers J (2011) OR in healthcare: a European perspective. Eur J Oper Res 212(2):223–234
49. Howick S, Ackermann F (2011) Mixing OR methods in practice: past, present and future directions. Eur J Oper Res 215(3):503–511
50. Wilson S (2001) What is an indigenous research methodology? Can J Nativ Educ 25(2):175–179
51. Biggs R, de Vos A, Schlüter M, Preiser R, Maciejewski K, Clements H (2021) The Routledge handbook of research methods for social-ecological systems. Routledge, New York, NY
52. Tolk A, Harper A, Mustafee N (2021) Hybrid models as transdisciplinary research enablers. Eur J Oper Res 291(3):1075–1090
53. Hofmann M, Palii J, Mihelcic G (2011) Epistemic and normative aspects of ontologies in modelling and simulation. J Simul 5(3):135–146
54. Ören T, Zeigler BP, Tolk A (2023) Body of knowledge for modeling and simulation: a handbook by the society for modeling and simulation international. Springer Nature, Cham, Switzerland
55. Traoré MK (2019) Multi-perspective modeling and holistic simulation. In: Complexity challenges in cyber physical systems: using modeling and simulation (M&S) to support intelligence, adaptation and autonomy. Wiley, Hoboken, NJ, pp 83–110
56. Tolk A, Diallo SY, Padilla JJ, Herencia-Zapana H (2013) Reference modelling in support of M&S—foundations and applications. J Simul 7:69–82
57. Rouse WB (2021) Understanding the complexity of health. Syst Res Behav Sci 38(2):197–203

58. Haberlin RJ, Page EH (2021) Visualization support to strategic decision-making. In: Simulation and Wargaming. Wiley, Hoboken, NJ, pp 317–334
59. Hanzl M (2007) Information technology as a tool for public participation in urban planning: a review of experiments and potentials. Des Stud 28(3):289–307
60. Hazen BT, Skipper JB, Boone CA, Hill RR (2018) Back in business: operations research in support of big data analytics for operations and supply chain management. Ann Oper Res 270:201–211
61. Hu X (2023) Dynamic data-driven simulation: real-time data for dynamic system analysis and prediction. World Scientific, Hackensack, NJ
62. Tolk A (2022) Dynamic data analytics support for multi-criteria and multi-objective decision making in complex environments under deep uncertainty. In: IISE annual conference & expo, Seattle, WA
63. Mustafee N, Harper A, Viana J (2023) Hybrid models with real-time data: characterising real-time simulations and digital twins. In: e Operational research society simulation workshop, Southampton, United Kingdom
64. Yang L, Zhang L, Philippopoulos-Mihalopoulos A, Chappin EJ, van Dam KH (2021) Integrating agent-based modeling, serious gaming, and co-design for planning transport infrastructure and public spaces. Urban Des Int 26:67–81
65. Shuttleworth D, Padilla JJ (2021) Towards semi-automatic model specification. In: Winter simulation conference, virtual
66. Shuttleworth D, Padilla JJ (2022) From narratives to conceptual models via natural language processing. In: Winter simulation conference, Phoenix, AZ
67. Grossmann I, Feinberg M, Parker DC, Christaki NA, Tetlock PE, Cunningham WA (2023) AI and the transformation of social science research. Science 380(6650):1108–1109
68. Shen Y, Heacock L, Elias J, Hentel KD, Reig B, Shih G, Moy L (2023) ChatGPT and other large language models are double-edged swords. Radiology 307(2):239163
69. OpenAI 2023 (2023) ChatGPT: optimizing language models for dialogue. OpenAI [Online]. Available: https://openai.com/blog/chatgpt/. Accessed 7 June 2023
70. Mitchell M, Krakauer DC (2023) The debate over understanding in AI's large language models. Proc Natl Acad Sci 120(13):e2215907120
71. Giabbanelli PJ (2023) GPT-based models meet simulation: how to efficiently use large-scale pre-trained language models across simulation tasks. In: Winter simulation conference, San Antonio, TX
72. Numrich SK, Picucci PM (2012) New challenges: human, social, cultural, and behavioral modelling. In: Engineering principles of combat modeling and distributed simulation. Wiley, Hoboken, NJ, pp 641–667
73. Davis PK, O'Mahony A, Gulden TR, Osoba OA, Sieck K (2018) Priority challenges for social and behavioral research and its modelling. RAND Corporation, Santa Monica, CA
74. Davis PK, O'Mahony A, Pfautz J (2019) Social-behavioral modeling for complex systems. Wiley, Hoboken, NJ
75. Dyulicheva YY, Glazieva AO (2022) Game based learning with artificial intelligence and immersive technologies: an overview. In: CEUR workshop proceedings
76. Lempert RJ, Popper SW, Bankes SC (2003) Shaping the next one hundred years: new methods for quantitative long-term policy analysis. RAND Report MR-1626, Santa Monica, CA
77. Marchau VAWJ, Walker WE, Bloemen PJTM, Popper SW (2019) Decision making under deep uncertainty: from theory to practice. Springer Nature, Cham, Switzerland
78. Kwakkel JH, Pruyt E (2013) Exploratory modeling and analysis, an approach for model-based foresight under deep uncertainty. Technol Forecast Soc Chang 80(3):419–431
79. Kovalchuk SV, Boukhanovsky AV (2015) Towards ensemble simulation of complex systems. Procedia Comput Sci 51:532–541
80. Chen H, Zhang S, Chen W, Mei H, Zhang J, Mercer A, Liang R, Qu H (2015) Uncertainty-aware multidimensional ensemble data visualization and exploration. IEEE Trans Visual Comput Graphics 21:1072–1086

81. Marler RT, Arora JS (2004) Survey of multi-objective optimization methods for engineering. Struct Multidiscip Optim 26:369–395
82. Copado-Méndez PJ, Guillén-Gosálbez G, Jiménez L (2012) Rigorous computational methods for dimensionality reduction in multi-objective optimization. Comput Aided Chem Eng 30:1292–1296
83. Ferreira TN, Vergilio SR, de Souza JT (2017) Incorporating user preferences in search-based software engineering: a systematic mapping study. Inf Softw Technol 90:55–69
84. Diallo SY, Wildman WJ, Shults FL, Tolk A (2019) Human simulation: perspectives, insights, and applications. Springer Nature, Cham, Switzerland
85. Epstein JM (2009) Modelling to contain pandemics. Nature 460(7256):687–687
86. Keskinocak P, Oruc BE, Baxter A, Asplund J, Serban N (2020) The impact of social distancing on COVID19 spread: state of Georgia case study. PLoS ONE 15(10):e0239798
87. Ozik J, Wozniak JM, Collier N, Macal CM, Binois M (2021) A population data-driven workflow for COVID-19 modeling and learning. Int J High Perform Comput Appl 35(5):483–499
88. Alatas SH (1974) Captive mind and creative development. Int Soc Sci J 26(4):691–700
89. Brunner C (2021) Conceptualizing epistemic violence: an interdisciplinary assemblage for IR. Int Polit Rev 9(1):193–212
90. Rajack-Talley TA (2018) Ethics, epistemology and community-based research on African Americans. In: World conference on qualitative research
91. Reardon J, TallBear K (2012) "Your DNA Is Our History": genomics, anthropology, and the construction of whiteness as property. Curr Anthropol 53(S5):S233–S245
92. Santos BdS (2014) Epistemologies of the south: justice against epistemicide. Routledge, New York, NY
93. Denzin NK, Lincoln YS, Smith LT (2008) Handbook of critical and indigenous methodologies. Sage
94. Louis RP (2007) Can you hear us now? Voices from the margin: using indigenous methodologies in geographic research. Geogr Res 45(2):130–139
95. Botha L (2011) Mixing methods as a process towards indigenous methodologies. Int J Soc Res Methodol 14(4):313–325
96. Levac L, McMurtry L, Stienstra D, Baikie G, Hanson C, Mucina D (2018) Learning across indigenous and western knowledge systems and intersectionality: reconciling social science research approaches. University of Guelph, Guelph, ON
97. Page S (2008) The difference. Princeton University Press, Princeton, NJ
98. Padilla JJ (2010) Towards a theory of understanding within problem situations. PhD thesis, Old Dominion University, Norfolk, VA
99. Obermeyer Z, Powers B, Vogeli C, Mullainathan S (2019) Dissecting racial bias in an algorithm used to manage the health of populations. Science 366(6464):447–453
100. Greaves L, Ritz SA (2022) Sex, gender and health: mapping the landscape of research and policy. Int J Environ Res Public Health 19(5):2563

Part II
Formalisms and Methods

Chapter 5
Defining Families of Hybrid Models with the πHyFlow++ Modeling and Simulation Integrative Framework

Fernando J. Barros

Abstract The area of hybrid systems has been subjected to intense research leading to different definitions of the concept of "hybrid model". For example, these models have been defined as combinations of discrete events, system dynamics, and/or agent-based models. Some modeling formalisms support an alternative and more basic view considering that hybrid models can read and produce continuous and discrete flows. This paper considers the latter perspective and presents πHyFlow++, an implementation of the πHyFlow integrative formalism, which supports the definition of modular and hierarchical hybrid components. πHyFlow supersedes conventional multi-paradigm representations, by providing a unifying framework for describing the basic constructs required by several modeling and simulation (M&S) approaches. These operators include the support for generalized sampling and the exact representation of continuous signals. πHyFlow is based on the process interaction worldview, enabling both the transaction and the network perspectives. In this paper, we exploit the potential of πHyFlow++ hybrid components for creating a set of related models, or model product lines (MPLs). We use πHyFlow++ inheritance of topology (IT) for creating new models from existing ones, by defining the differences between model versions. Additionally, we also consider the ability to add/remove processes as an extension to IT. We present results from the development of a family of microwave oven models.

Keywords Hybrid models · Model product lines · Inheritance of topology · Simulation

F. J. Barros (✉)
Department of Informatics Engineering, University of Coimbra, 3030 Coimbra, Portugal
e-mail: barros@dei.uc.pt

5.1 Introduction

The area of hybrid systems has been subjected to intense research leading to different definitions of the concept of "hybrid model". Hybrid models have also been defined as combinations of discrete event (DES), system dynamics (SD), and/or agent-based models (ABM) [11, 21, 22]. However, sound semantics for integrating DES, SD, and ABM has not been developed. In defining hybrid models, in this paper we have considered the perspective underlying the πHyFlow formalism [7], where hybrid models are able to read and produce continuous and discrete flows.

πHyFlow is an integrative formalism intended to represent other modeling approaches like Fluid Stochastic Petri Nets [6], or numerical integrators for solving ODEs. An advantage of integrative formalisms is that they guarantee, by design, the composition and interoperability of all models. πHyFlow supersedes conventional multi-paradigm representations, by providing a unifying framework for describing the basic constructs required by several modeling and simulation (M&S) approaches. These operators include the support for generalized sampling and the exact representation of continuous signals. πHyFlow is based on the process interaction worldview, enabling both the transaction and the network perspectives. The πHyFlow^{++} is a modeling and simulation framework based on the πHyFlow formalism supporting the definition of modular and hierarchical hybrid components. In this paper, we exploit the potential of πHyFlow^{++} hybrid components for creating a set of related models, or model product lines (MPLs).

MPLs enable the development of a family of related models. Since MPLs are usually constrained to a limited design space, the benefits of reuse can be exploited, i.e., both model code and design (topology) can be reused. In fact, by exploiting locality in the design space, one can develop new models from existing ones. New models can be created by applying a small set of operations that perform the adaptation of an existing model to new requirements. This kind of approach can contribute to lower cost of software development, short time-to-market, and higher quality products [9, 12]. These advantages can be leveraged by software reuse, since new products can be created based on existing ones, being the effort made on the *small* changes to the original applications. Reuse also contributes to software quality, since components that are used and in different applications are exposed to a large variety of testing and operational conditions, leveraging their correctness.

In the area of software engineering, the development of families of software products has been subjected to intense research that includes model-driven engineering [16, 24], UML [29] and software architectures [10]. These approaches are, however, in general very abstract and may be difficult to apply in practice. A limitation is the lack of a formal semantics for converting high-level architecture diagrams into code, requiring commonly an error-prone manual mapping process. Given, for example, a set of UML diagrams, it may prove complex and error prone to create executable models [15]. Feature diagrams although provide a description of application functional requirements [29], they do not offer the operators required to implement those features. As shown in Sect. 5.3, components representing features are only a part of the solution, being the other part the links among components that cannot be fully described in feature diagrams.

In this paper, we exploit the use of modular modeling components as a technique for the implementation of MPLs. Modular models enable the creation of new models through composition, making it possible to create simulation models through model reuse.

We consider here that effective support for MPLs requires not only the ability to reuse components but also the capacity to reuse the overall design of complete models. The latter kind of reuse can be accomplished through the concept of inheritance of topology (IT). Since components enable composition, we exploit the ability to incrementally modify these network topologies through IT. It becomes possible to reuse a whole model and derive a new one through simple and incremental operations over the topology/design. Operations include the ability to add/remove components and their interconnections, and the possibility to add/remove simulation processes, enabling new models to be developed using simple transformations over existing ones.

In this paper, we describe the results of using components and inheritance of topology in the development of a microwave oven model product line that has been previously studied using an UML-based approach [15].

The realization of the microwave oven product line was achieved using πHyFlow^{++} [7], a C++20 implementation of the πHyFlow formalism [8]. This formalism supersedes conventional multi-paradigm representations, by providing a unifying framework for describing the basic constructs required by several modeling approaches. The πHyFlow formalism inherits HyFlow [4, 5] support for continuous flows, discrete flows, generalized sampling, dense outputs—for the exact representation of continuous signals, and modular and hierarchical dynamic topology models. The πHyFlow formalism also supports the transition and network process interaction worldviews, being the former enabled by πHyFlow ability to create/destroy processes at simulation runtime.

πHyFlow^{++} provides a flexible interface based on ports, enabling the detailed discrimination of the coupling relationship. This contrasts with the interface of UML components that are based on the concept of UML-ports [25] that aggregate method signatures offering a smaller flexibility in component coupling and reuse.

The paper is organized as follows: Sect. 5.2 introduces the $_\pi$HyFlow^{++} modeling and simulation (M&S) framework, a C++ implementation of the πHyFlow formalism based on the C++ language. Section 5.3 describes the microwave oven MPL. Section 5.4 gives simulation results for the MPL. Related work is discussed in Sect. 5.5. Future research is proposed in Sect. 5.6. Conclusion is provided in Sect. 5.7.

5.2 πHyFlow++ M&S Framework

πHyFlow++ is a modeling and simulation framework based on the πHyFlow formalism [7] and implemented in the C++ language. πHyFlow++ components (πHFCs) provide a realization of independent models. We describe next the two kinds of πHFCs: base and network.

5.2.1 Base Component

Base components provide the basic constructs for model development, supporting processes and port communication. Each component is defined as a modular entity. Fully independence is achieved by limiting πHFC communication to be exclusively made through its own interface. A component provides a realization of the πHyFlow base model concept, represented in Fig. 5.1.

A base model defines a modular interface with an input set X, and an output set Y. Inputs and output sets enable the representation of hybrid signals that combine continuous flows and discrete events. Input values are handled by the input function ξ that modifies the state p of the model. This state is shared by a dynamic set of processes $\pi_1, \ldots \pi_n$ that can also define their own private state, minimizing the dependency among processes. Model output is described by the output-function Λ_p that is based on the output of all processes. A base model supports a modular unified process interaction worldview that combines both transaction and network perspectives.

In the πHyFlow++ implementation, inputs and outputs are structured by ports. There is a buffer associated with every discrete input port. The continuous flows associated with a continuous output port are available through the sampling operation. A continuous flow can describe any function, enabling the representation of numerical

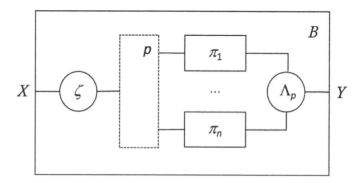

Fig. 5.1 πHyFlow base model

integrators for solving ordinary differential equations [7]. The next sections describe πHyFlow++ base components.

5.2.2 Holder

As an example of a base component with communication ports, we consider the holder depicted in Fig. 5.2. This component is used to store/retrieve a value. It has the discrete input port "set" to set the current value. The holder has the continuous output port "value" where the current value can be sampled, and the discrete output port "event" that produces a discrete signal when it is set to a new value.

πHyFlow++ holder implementation is represented in Listing 1. The stored value is defined in line 4, and it is shown in the widget of line 5, where class "wx::label" is based on the wxWidgets GUI library [23]. Holder ports are defined in lines 7–9. The component defines the process "manager" that is created in line 15 and defined in lines 17–29.

Holder continuous output flow is defined in line 19, being available at the continuous output port "value". This flow is piecewise constant, and it provides the current value of variable "value" that is available through sampling. The manager process is responsible for updating the stored value. Since new values are received at input port "set", the manager defines variable "values" to reference the buffer associated with the discrete input port "set" (line 21). The manager performs a loop where it waits for an incoming value (line 23), removes the value (line 24), presents the value in the GUI (line 25), and sends a discrete signal through port "event", signaling that the stored value has changed (line 26).

A typical holder input/output trajectory is depicted in Fig. 5.3. The holder receives values v_0, v_1, and v_2 at times t_0, t_1, and t_2, respectively. After receiving a discrete value, the holder sends the value through the discrete output port "event". The continuous output port "value" holds the current value so it can be sampled.

The holder can be reused in arbitrary contexts. The adaptation to a specific model needs only to be made when a composition is established. In the next section, we describe a πHyFlow++ timed model.

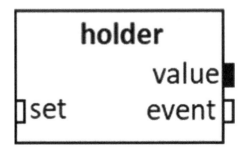

Fig. 5.2 Holder component input/output interface

```
template<class T = int>
class holder: public sim::component {
private:
    T value;
    wx::label* label;
public:
    std::vector<sim::port> in_ports_d() {return {"set"};}
    std::vector<sim::port> out_ports_c() {return {"value"};}
    std::vector<sim::port> out_ports_d() {return {"event"};}
    holder(std::string_view name, T value, wx::point p, wx::size s):
    sim::component(name), value{value}
    {
        init();
        label = new wx::label(aFrame, "", p, s);
        manager("holder");
    }
    sim_void manager(std::string_view name) {
        sim_start(name);
        output_c("value", [this] (const double&) {return value;});
        label.setValue(value);
        auto& values = buffers_d("set");
        while (true) {
            sim_wait until_cond([&values] {return values.any();});
            value = values.remove<T>();
            label.setValue(value);
            sim_wait out("event", value);
        }
        sim_end;
    }
};
```

Listing 1 Holder component

5.2.3 Exponential Time Integrator

We show πHyFlow[++] timed behavior by representing the exponential time differencing (ETD) method [13]. This integrator is used in the next section to model the microwave oven grill temperature. The ETD is an accurate integration method for solving 1st-order ordinary differential equations (ODEs). Given an ODE in the form: $y' = ky + F(x(t), y(t))$, $y(0) = y_0$. A numerical solution based on the 0th-order approximation $F(x(t), y(t)) \approx F(x(0), y(0))$, is given by:

$$y_{k+1} = -\frac{F(x_n, y_n)}{k} + \left(y_n + \frac{F(x_n, y_n)}{k}\right)e^{kT}, \tag{5.1}$$

where T is a fixed time step. This equation is *exact* if F is constant in the interval [0, T]. πHyFlow[++] ETD implementation is described in Listing 2. The component has the continuous input port "value" (line 3) to sample the input value. Discontinuities

Fig. 5.3 Holder trajectories

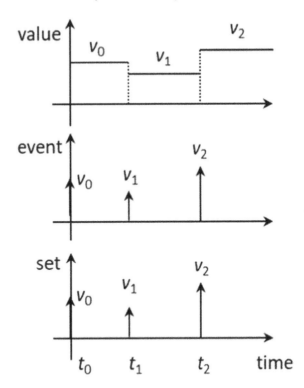

in the input function are received at the discrete input port "event", while changes in the ODE coefficient k are received at the discrete input port "k" (line 4). The update of k is used in the next section for representing changes in oven door status: open ↔ close. ODE solution is available at continuous output port "value" (line 5). Discontinuities are signaled through discrete output port "event" (line 6), along with the current value (line 25).

ODE numerical solution is performed by the "integrator" process (line 12). ODE continuous output value is set, in line 15, to Eq. (5.1). The ETD uses a default constant sampling interval "T" that can be interrupted by an event or by a change in the ODE coefficient k (line 20). When interrupted, the ODE produces an event (line 25). The sampling of the input function value "F" is made in line 28. External input events are accumulated in buffers (lines 17–18), and they can be used in conditional expressions (line 20). The next section describes πHyFlow^{++} network components.

```
class ETD: public component {
public:
    std::vector<sim::port> in_ports_c() {return {"value"};}
    std::vector<sim::port> in_ports_d() {return {"event", "k"};}
    std::vector<sim::port> out_ports_c() {return {"value"};}
    std::vector<sim::port> out_ports_d() {return {"event"};}
public:
    ETD(const std::string_view name, double k, double y, double h = 5.e-1):
    sim::component(name) {
        init();
        integrator("ETD", k, y, h);
    }
    sim_void integrator(const std::string_view name, double k, double y, double T) {
        sim_start(name);
        double F = 0.;
        output_c("value", [&] (const double& dt) {return -F / k + (y + F / k) * exp(k * dt);});
        sim_wait sample("value", F);
        auto& events = buffers_d("event");
        auto& k_buf = buffers_d("k");
        while (true) {
            sim_wait duration(T, [&] {return events.any() || k_buf.any();});
            if (events.any())
                events.clear();
            y = sample_out<double>("value");
            if (events.any() || k_buf.any())
                sim_wait out("event", y);
            if (k_buf.any())
                k = k_buf.remove<double>();
            sim_wait sample("value", F);
        }
        sim_end;
    }
};
```

Listing 2 ETD component

5.2.4 Network Component

Base components can be composed to define networks. Network components have the same kind of interface of base components. This property makes both types of components equivalent enabling components to be hierarchically defined. We illustrate network components through the definition of a microwave oven grill subsystem. This component is used in Sect. 5.3 to define the overall oven model. The grill network component is depicted in Fig. 5.4, where network topology is controlled by the executive component. The grill is also composed by the ETD defined in the last section. This component is responsible for computing the grill temperature.

Fig. 5.4 Grill network component

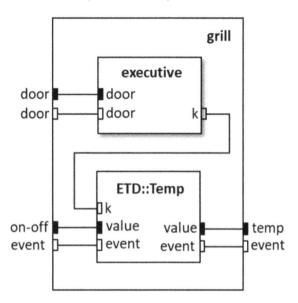

The grill has two input ports to sense the oven door, and two input ports to sense the digital controller that switches the grill heater on and off. The grill temperature is available through the continuous output port "temp". Discontinuities in temperature derivative are signaled through the discrete output port "event". The grill network is described in Listing 3. Grill ports are defined in lines 3–6. Network executive ports are defined in lines 7–9. Components and links are defined in lines 15–31. Links connecting discrete ports are commonly represented by identity functions, where values are transmitted unchanged. This is the case of all discrete links like, for example, the link in line 18 that connects network input discrete port "door" to executive input port "door". Continuous ports require merger functions for their definition. This is the case of continuous input port "value" of the ETD component "Temp" (line 23) that sets the input value of the ODE describing the grill temperature. More details on the grill model are given in Sect. 5.3.

The executive also creates the "door" process (line 32) that is responsible to modify grill network topology to represent whether the door is open or closed. This process is described in Listing 6.

We have described the development of new components from composition of existing simpler ones. However, model reuse, and particularly the development of variations of existing products, also requires the ability to reuse model design. We will describe this type of reuse in the next section.

```
class oven_grill: public sim::executive {
public:
    std::vector<sim::port> net_in_ports_c() {return {"on_off", "door"};}
    std::vector<sim::port> net_in_ports_d() {return {"event", "door"};}
    std::vector<sim::port> net_out_ports_c() {return {"value"};}
    std::vector<sim::port> net_out_ports_d() {return {"event"};}
    std::vector<sim::port> in_ports_c() {return {"door"};}
    std::vector<sim::port> in_ports_d() {return {"door"};}
    std::vector<sim::port> out_ports_d() {return {"k"};}
public:
    oven_grill(const std::string_view name, double k_opn, double k_cld, double t_ext, double stime):
    sim::executive(name)
    {
        init();
        link_c("network", "door", "executive", "door");
        merger_c("executive", "door", [] (const sim::vector<sim::cvalue>& v) {
            return v[0];
        });
        link_d("network", "door", "executive", "door");
        add(new sim::ETD("Temp", k_opn, t_ext, stime));
        link_c("network", "on_off", "Temp", "value");
        merger_c("Temp", "value", [k_opn, t_ext] (const sim::vector<sim::cvalue>& v) {
            return v.get<double>(0) - k_opn * t_ext;
        });
        link_d("network", "event", "Temp", "event");
        link_d("executive", "k", "Temp", "k");
        link_c("Temp", "value", "network", "value");
        merger_c("network", "value", [] (const sim::vector<sim::cvalue>& v) {
            return v[0];
        });
        link_d("Temp", "event", "network", "event");
        door("door", k_opn, k_cld, t_ext);
    }
    ...
};
```

Listing 3 Grill network definition

5.3 The Microwave Oven Model Product Line

The microwave oven MPL used in this paper is adapted from [15]. We provide here a πHFC-based design to the original approach that is based on UML models. The aim of this study is to exploit the potential of inheritance of topology and πHFC in the development of MPLs. In this work, we have developed an oven simulation model. An alternative approach based on software components was developed by [1]. This solution, however, did not use simulation processes requiring more components. It was also more complex to develop. A comparison between this solution and the πHyFlow[++] process-based approach is presented in Sect. 5.5.

5.3.1 Oven Base Topology

The oven base topology is depicted in Fig. 5.5. The executive plays the role of coordinator, defining many of the processes that define the oven behavior. In the oven model, some components are used as surrogates for the real hardware.

This is the case of the "grill", "scale", "heater", "table", and "door" components that are represented by the "holder" component described in Sect. 5.2. For simplifying model development, several discrete output ports, like "heater_emul", "door_emul", "scale_emul", and "grill_emul", are used as hardware emulators for making tests easier. The "scale_emul" port, for example, is used to emulate the scale sensor that weighs the food placed inside the oven. The test driver described in Listing 7 uses port emulators to operate the oven under specific conditions that can be replicated, since they do not depend on external factors, like the actions of a human operator.

One of the oven key features is to ensure that cooking can only occur if the door is closed, there is some food in the oven, and the cooking time is larger than zero.

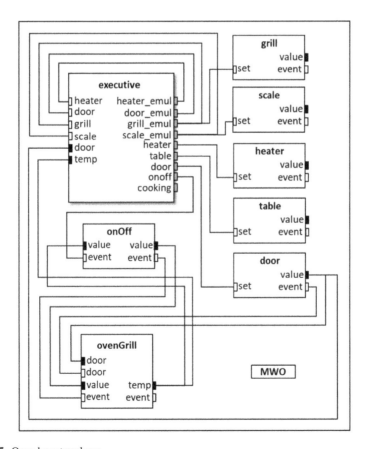

Fig. 5.5 Oven base topology

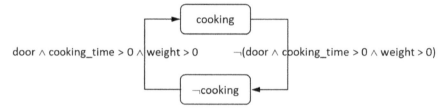

Fig. 5.6 Cooking state diagram

These constraints consider functional requirements and safety considerations. The cooking process has the state diagram shown in Fig. 5.6.

The oven executive has several processes that define the global device operation. In general, every discrete port is controlled by a process. We found it useful to split the oven state diagram among the several processes that run in the executive. The result was a set of state diagrams attached to each oven feature that were easy to develop, when compared to global state diagrams presented in [15, pp. 401, 403].

Listing 4 shows the turntable controller process. The control of physical elements, such as the electric motor, is achieved through the port "table". This port allows the use of the values -1, 0 and 1. Value 0 stops the motor, value − 1 sets motor rotation in the clockwise (CW) direction, and value 1 sets the motor in a counterclockwise (CCW) direction. While the oven is cooking, the controller defines the cycle stop-CW-stop-CCW, holding each command for an interval of 5 s. The loop is modified whenever the oven stops cooking. The conditions for cooking are set by the shared variable "cooking" that is managed by a different process already described by the state diagram of Fig. 5.6.

```
   sim_void turntable(const std::string_view name) {
2      sim_start(name);
       while (true) {
4          sim_wait out("table", 0);
           sim_wait duration(5.);
6          sim_wait until_cond([this] {return cooking;});
           sim_wait out("table", 1);
8          sim_wait duration_or_until(5., [this] {return ! cooking;});
           if (! cooking) continue;
10         sim_wait out("table", 0);
           sim_wait duration_or_until(5., [this] {return ! cooking;});
12         if (! cooking) continue;
           sim_wait out("table", -1);
14         sim_wait duration_or_until(5., [this] {return ! cooking;});
       }
16     sim_end;
   }
```

Listing 4 Over turntable process

```
class mw_oven_alt: public mw_oven {
public:
    mw_oven_ii(std::string_view const name): mw_oven(name) {
            remove_process("turntable");
            remove_component("Table");
    }
};
```

Listing 5 Topology of the simplified oven reusing the original model

The turntable process sends the command to stop the table (line 4) and waits for 5 s (line 5). It then waits until the oven is cooking (line 6). It then sends a command to rotate the turntable CCW during 5 s (lines 7–8). The turntable stops for 5 s (lines 10–11), and then it rotates CW during 5 s (lines 13–14). These steps are interrupted when the oven stops cooking (lines 8, 11, 14). After an interruption, the process restarts the cycle. In the next section, we describe πHyFlow++ support for creating a family of oven models.

5.3.2 Alternative Products

In the last section, we developed a specific oven model. We illustrate now how new ovens can be derived from an existing model. We consider here inheritance of topology (IT) as the key construct to support variability in MPLs. From the base oven model, we can use IT to create new ovens using add/removal operations. These operations can be made not only over the set of components [1] but also over the set of processes. A simplified oven without the turntable is defined by Listing 5, and the new topology is shown in Fig. 5.7.

Line 4 removes the "turntable" process, and line 5 removes the "Table" component. When removing a component, all links from and to the components are also removed. The modified oven is just one from a family of similar models. Other models can be easily developed through IT.

5.3.3 The Grill Model

πHyFlow++ supports network topologies that can be modified at runtime. This feature was used in the oven MPL to enable link reconfiguration within the grill component. The hybrid automaton [18] describing grill temperature is depicted in Fig. 5.8. Temperature depends on the door status: open or closed; and on the on–off digital controller output. When the temperature is below the reference value T_{ref}, the digital controller turns on (A), and when above T_{ref}, the controller turns off (B). The external temperature T_{ext} is considered constant. When the digital controller is on, grill power

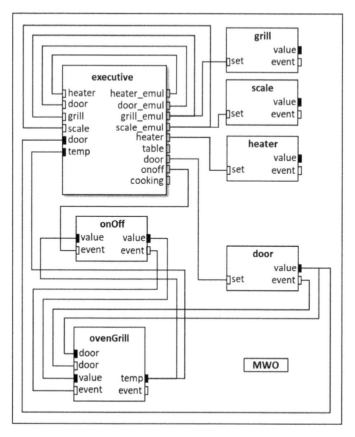

Fig. 5.7 Simplified oven topology without the turntable

is given by P_{grill}. For saving energy, the controller is turned off when the door is open. When open, the door is modeled by equation (C). The temperature coefficient is given by k_{cld}, for the closed door and k_{opn} for the open door. The sampling rate is set by the digital controller, and it is not represented in the automaton. The automaton state change is achieved by modifying model topology as described in Listing 6 that describes oven grill executive process "door".

The executive starts by sampling the continuous input gate "door" to check door status (line 6). If the door is closed, the "Temp" exponential time differentiating integrator (ETD) has its input described by line 9, corresponding to state (A) of the hybrid automaton. The grill executive sends the value "k_{cld}" through the discrete port "k" that is linked to ETD port "k", as shown in Fig. 5.4. Figure 5.1 to set the new integrator coefficient. The door is initially closed (Listing 3, line 23). The executive does not need to take any action if this condition holds.

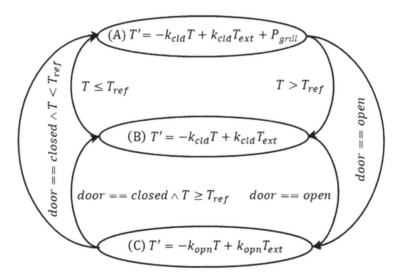

Fig. 5.8 Grill oven hybrid automaton

After the initial topology is set, the executive waits for a door event (line 15) and chooses the appropriate temperature coefficient "k" (line 17). ETD input is set in line 19 and the new value "k" is sent to the ETD component (line 21).

5.4 Simulation Results

The oven has been simulated using the driver given in Listing 7. The cooking time is set to 3 m 20 s (line 3). Initially, the door is open, the grill is turned on, and 250 g of food is placed in the oven (line 6). The door is closed at time 10 s (line 7) and opened at time 60 s (line8). The door is closed again at time 70 s (line 9) and opened at time 120 s (line 10). The door is closed at time 150 s, and the grill knob is turned off at time 270 s (line 12). Finally, the door is opened at time 310 s (line 13).

For obtaining simulation results, we have used the oven parameters listed in Table 5.1. External and reference temperatures are kept constant during the overall simulation.

Grill temperature is shown in Fig. 5.9. The temperature starts to increase when the door is closed. The temperature is kept around T_{ref}, where the oscillations are related to the on–off digital controller sampling rate sTime. During cooking time, the door is opened twice. In these intervals, the controller is turned off and the grill temperature drops to values close to the external temperature T_{ext}. Cooking finishes at time 250 s. Although grill now is turned off time 270 s, the controller is switched off at time 250 s, and the temperature starts to drop slowly according to coefficient k_{cld}, since

```
class oven_grill: public sim::component {
    ...
    sim_void door(std::string_view const name, double k_opn, double k_cld, double t_ext) {
        sim_start(name);
        int door = 0;
        sim_wait sample("door", door);
        if (door) {
            merger_c("Temp", "value", [k_cld, t_ext] (const sim::vector<sim::cvalue>& v) {
                return v.get<double>(0) - k_cld * t_ext;
            });
            sim_wait out("k", k_cld);
        }
        auto& door_buf = buffers_d("door");
        while (true) {
            sim_wait until_cond([&] {return door_buf.any();});
            int door = door_buf.remove<int>();
            double k = (door == 0)? k_opn: k_cld;
            merger_c("Temp", "value", [k, t_ext] (const sim::vector<sim::cvalue>&
v) {
                return v.get<double>(0) - k * t_ext;
            });
            sim_wait out("k", k);
        }
        sim_end;
    }
    ...
};
```

Listing 6 Grill component dynamic topology

```
sim_void test(const std::string_view name) {
    sim_start(name);
    cooking_time = hhmmss(0, 3, 20);          //cooking time = 3m20s
    sim_wait out("door_emul", 0);             //open door
    sim_wait out("grill_emul", 1);            //grill knob on
    sim_wait out("scale_emul", .25);          //place 250g of food in oven
    sim_wait out(10., "door_emul", 1);        //10 s
    sim_wait out(50., "door_emul", 0);        //60 s
    sim_wait out(20., "door_emul", 1);        //70 s
    sim_wait out(50., "door_emul", 0);        //120 s
    sim_wait out(30., "door_emul", 1);        //150 s
    sim_wait out(120., "grill_emul", 0);      //270 s
    sim_wait out(40., "door_emul", 0);        //310 s
    sim_end;
}
```

Listing 7 Test driver for the base oven

Table 5.1 Oven parameters

T_{ext}	T_{ref}	P_{grill}	sTime	k_{cld}	k_{opn}
20 °C	140 °C	7	1 s	− 0.02	− 0.1

the door is still closed. After the door is opened at time 210, the temperature depends on coefficient k_{opn}, and it drops more rapidly, reaching T_{ext} at time ~ 400 s.

Turntable rotation direction is shown in Fig. 5.10. The table is rotating only in intervals [10, 60], [70, 120], and [150, 250], corresponding to the periods when the door is closed, and the oven is cooking.

The table executes the cycle stop-CW-stop-CCW as described by Listing 4. The turntable stops at time 250 s when the cooking programs ends.

Results were taken in the piHyFlow3.0 modeling and simulation environment, a C++20 implementation of the piHyFlow formalism in the MS VisualStudio 2022 using the compiler version 17.7.5. GUI support is provided by the wx components library that wraps the wxWidgets object library. Simulation xy-plots are currently made using MS Excel and the text files produced by the modeling components. A snapshot of piHyFlow3.0 is depicted in Fig. 5.11.

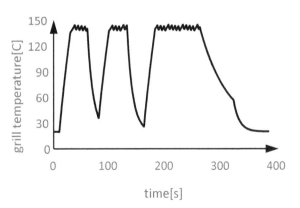

Fig. 5.9 Grill temperature

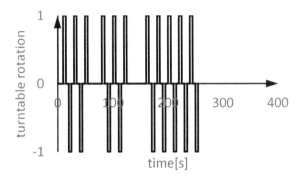

Fig. 5.10 Turntable rotation direction

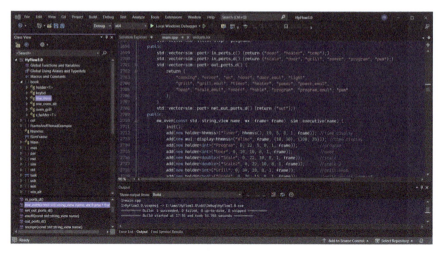

Fig. 5.18 piHyFlow3.0 M&S environment

5.5 Related Work

An earlier definition of model product line was provided in modeling and simulation, where the concept of system entity structure (SES) was developed to represent families of models [28]. Product lines have been currently specified by feature diagrams requiring their translations into an executable by code generation techniques [14]. In some cases, product features can be expressed as aspects [20]. However, some features may become difficult to capture in just one component, requiring the cooperation of several components and/or processes. In our work, the mapping of features into πHFCs is made manually. Manual mapping is common in the design of other kinds or systems, like computer hardware and automobiles, for example. Although the automatic mapping of features into code can be proven to be an attainable goal, we conjecture that it may become simpler to employ a different type of code generator. Instead of creating the complete code for a model, a generative procedure for choosing the required πHFCs and their connections may prove to be more effective. In this case, πHFC components would be reused making model generation simpler.

Reusable components, like πHFCs, provide an approach between general-purpose and domain-specific languages (DSLs), taking advantages from both. Components provide a general framework for modeling any type of system, since they are compatible with the features of a general-purpose language. However, since domain libraries can be easily developed and reused, πHFCs also enable the advantages of DSLs. Components can represent the high-level concepts related to a specific domain, surpassing the low-level constructs offered by programming languages [27]. Instead of developing specific compilers to represent concrete domains, we were able to keep a common language that can be extended with domain components, simplifying the creation of MPLs.

One can consider two basic forms of product line development: simple to complex (S2C) and complex to simple (C2S). We have found that the C2S strategy enables the creation of specific products by just changing the topology without introducing new code. From this study, we also found that C2S simplifies the derivation of new products while keeping traceability through the IT relationship.

In this work, we have used several heuristics that may prove effective in other product lines. Whenever possible, we have reused components from existing libraries. This was the case with the ETD integrator and the on–off digital controller that are general-purpose components. We also found that creating domain-specific processes in the executive was easier than creating domain-specific components that are likely not to be reusable. πHyFlow^{++} ability to support several cooperating processes in the same base model has simplified model design when compared to formalisms like HyFlow that can only represent one event in a base model. We have found that shared variable communication can be simpler and more effective than a solution relying only on modular communication. For example, having one process to compute cooking status and updating the corresponding shared variable can easily be achieved in πHyFlow^{++}. This approach makes process communication and synchronization quite simple to achieve. On the contrary, a modular solution based on HyFlow [1], for example, would require to re-organize each process into a modular component, making it more complex to provide communication and coordination among these base components. We have kept each process functionality to a minimum by creating a process for each discrete port. As a result, the state diagram of all the processes in the oven model was kept quite simple, like the one in Fig. 5.6. In contrast, and as mentioned before, oven original state diagrams tend to be more complex since they try to capture all interactions among oven elements.

There are, currently, different definitions of the concept of "Hybrid Model". We have considered here the perspective underlying the πHyFlow formalism [8], where hybrid models are able to read and produce continuous and discrete flows. πHyFlow is an integrative formalism intended to represent other modeling approaches like Fluid Stochastic Petri Nets [6], agents [7], and numerical integrators for solving ODEs, like the ETD of Sect. 5.2. The ability to explicitly represent integrators is crucial, for example, to model 2nd-order energy conservation systems through the use of geometric integrators that enable long-time simulations [17].

Hybrid models have also been defined as combinations of discrete event (DES), system dynamics (SD) and/or agent-based models (ABM) [11, 21, 22]. However, sound semantics for integrating DE, SD, and ABM has not been developed. Furthermore, these modeling approaches also provide a limited set of operators to describe hybrid models. A more general description should include, for example, dynamic topologies [3], generalized sampling, continuous flows [4, 5], ETD integrators, 2nd-order integrators [3], and chattering avoidance using sliding mode controllers [4, 5]. The representation of spatially moving entities can also benefit from publish/subscribe operators supported, for example, by the high-level architecture [19] and the heterogeneous flow system specification [2]. In the latter work, publish/subscribe communication was mapped into point-to-point interactions and dynamic topologies.

5.6 Future Research

Likewise SPLs, model product lines have been proposed to reduce the cost and time to create new applications. The concept of MPL was illustrated with the microwave oven domain. Additional models need to be developed to further demonstrate the effectiveness of the approach across other areas. A quantitative study is necessary to access the benefits of πHyFlow^{++} approach to MPLs when compared with other methods developed to support SPLs. The automatic mapping of features into models seems still an elusive goal. Although features can easily be associated with components and/or processes, the final application behavior still depends on how components are linked. This aspect needs to be considered when automatizing MPLs. We consider however that inheritance of topology can provide effective support for this problem. The microwave oven product line shows the integration of numerical ODE solvers to simulate the real-time environment. Future works will address the seamless composition of models with the hardware components used in the final product. Given πHyFlow^{++} ability to modify model topology at runtime, we will exploit this ability to support model adaption at runtime. This concept has been studied in the context of dynamic software product lines [26].

5.7 Conclusion

The πHyFlow formalism and the πHyFlow^{++} implementation provide a unifying framework for describing the basic constructs used in many multi-paradigms modeling approaches. Modeling constructs include generalized sampling, dense outputs, and the support for transaction and network process interaction worldviews. πHyFlow^{++} inheritance of topology (IT) leverages the development of model product lines. IT enables the incremental modification of a base model topology simplifying the creation of new models. New designs can be developed with the help of add/remove operations over the set of components and links that define a model. We have extended the concept of IT with the capability to add/remove simulation process from a base model. IT enables that both design and testing efforts from previous models can be reused enabling productivity gains. IT has shown to be a declarative construct that makes it easier to keep traceability between model versions.

References

1. Barros F (2013) On the representation of product lines using pluggable software units: results from an exploratory study. In: Symposium on theory of modeling and simulation
2. Barros F (2016) Modeling mobility through dynamic topologies. In: Simulation modelling practice and theory, 113–135

3. Barros F (2018) Handling overlapping collisions: a dynamic topology approach. In: Spring simulation conference
4. Barros F (2019a) A modular stabilization approach for chattering-free simulation. In: Spring simulation conference
5. Barros F (2019b) A unifying framework for the hierarchical co-simulation of cyber-physical systems. In: Mittal S, Tolk A (eds) Complexity challenges in cyber physical systems: using modeling and simulation (M&S) to support intelligence, adaptation and autonomy. Wiley, pp 111–130
6. Barros F (2023a) πHyFlow: a modular process interaction worldview. In: Winter simulation conference
7. Barros F (2023b) The unified process interaction worldview. J Simul
8. Barros F (2023c) Operational semantics of πHyFlow. Retrieved from http://arxiv.org/abs/2310.19818
9. Barros-Justo J, Pinciroli F, Matalonga S, Martínez-Araujo N (2018) What software reuse benefits have been transferred to the industry? A systematic mapping study. In: Information and software technology, 1–21
10. Bosch J (2000) Design and use of software architectures. Addison-Wesley
11. Brailsford S, Eldabib T, Kunca M, Mustafee N, Osorio A (2019) Hybrid simulation modelling in operational research: a state-of-the-art review. Euro J Oper Res
12. Chimalakonda S, Lee D (2021) A family of standards for software and systems product lines. In: Computer standards & interfaces
13. Cox S, Matthews P (2002) Exponential time differencing for stiff systems. J Comput Phys: 430–455
14. Czarnecki K, Antkiewicz M (2005) Mapping features to models: a template approach based on superimposed variants. In: Generative approaches to component enginnering. Springer, pp 422–437
15. Gomaa H (2004) Designing software product lines with UML. Addison-Wesley
16. Greenfield J, Short K (2004) Software factories. Wiley
17. Hairer E, Lubich C, Wanner G (2002) Geometric numerical integration: structure-preserving algorithms. Springer
18. Henzinger T (1996) The theory of hybrid automata. In: Eleventh annual IEEE symposium on logic in computer science, pp 278–292
19. Kuhl F, Weatherly R, Dahmann J (1999) Creating computer simulation systems: an introduction to the high level architecture. Prentice Hall
20. Liu J, Lutz R, Rajan H (2006) The role of aspects in modeling product line variabilities. In: Workshop on aspect oriented product line engineering, pp 69–76
21. Mykoniatis K, Angelopoulou A (2020) A modeling framework for the application of multi-paradigm simulation methods. In: Simulation, 55–73
22. Nguyen L, Howick S, Megiddo I (2022) Interfaces between SD and ABM modules in a hybrid model. In: Winter simulation conference
23. Smart J, Zeitlin V, Dunn R, Csomor S (2023) wxWidgets manual. Retrieved from https://docs.wxwidgets.org/3.2/
24. Stahl T, Volter M (2006) Model-driven software development. Wiley
25. Taylor R, Medvidovic N, Dashofy E (2010) Software architecture. Wiley
26. Valdezate A, Capilla R, Crespo J, Barber R (2022) RuVa: a runtime software variability algorithm. IEEE Access
27. Visser M, Voelter E (2011) Product line engineering using domain specific laguage. In: Software product line conference, pp 1530–1605
28. Zeigler B (1984) Multifacetted modelling and discrete event simulation. Academic Press
29. Ziadi T, Jezequel J-M (2006) Software product line engineering with UML: deriving products. In: Software product lines. Springer, pp 557–586

Chapter 6
CELL-DEVS Modelling of Individual Behaviour Towards Influencers in Social Media

Saptaparna Nath and Gabriel A. Wainer

Abstract With the exponentially rising popularity of social media, and the provided convenience of digital advertising across various platforms, individuals have found ways to sustain a lifestyle by providing companies with a personalised platform for advertising their products. These individuals are popularly labelled as 'influencers'. Determining the impact an influencer has on a product market is an important aspect in determining whether a company should invest in the influencer's platform. The model presented in this paper, namely, the influencer model employs the Cell-DEVS formalism, an extension of cellular automata that can be used to build discrete-event spaces, implemented through the Cadmium tool to simulate the reaction of individuals towards an influencer. The model simulates the rate of increase of follower count under various scenarios that exercise various reactions and employs an opinion-based approach that simulates an individual's evolving opinion state towards a subject.

Keywords CELL-DEVS modelling · Social media · Cellular automata · Behavioural modelling

6.1 Introduction

Social interactions of any kind play a role in the changing opinions of individuals and human behaviour itself. Social influence has been affected by the introduction of social media platforms. Social media has become an increasingly popular way to date, make friends, explore interests, share passions, and from certain points of view, improve our quality of life [1]. Bond et al. [2] study on Facebook message propagation determined that there was an influence of opinion in the recipients and

S. Nath · G. A. Wainer (✉)
Carleton University, 3216 V-Sim Building, 1125 Colonel By Dr, Ottawa, ON K1S 5B6, Canada
e-mail: Gabriel.Wainer@sce.carleton.ca

S. Nath
e-mail: saptanath@cmail.carleton.ca

their corresponding 'neighbours', which can be defined as close friends and or relatives. Such behaviour has led to the recognition of social influencers, people who are 'opinion leaders' and have the ability to affect buying habits or opinions of others by posting content (usually sponsored) on social media. From a marketing standpoint, influencers can directly reach more people than the average ad campaign and therefore have the potential to increase revenue. With the growing use of technology, more people spend time on their phones browsing social media, not only for entertainment purposes but also to connect to friends and family and explore various interests. As such, businesses have found it highly practical to employ social media and influencers to expand their global market reach, it is important to understand just how far that reach can go and how impactful the use of influencers will be to their product campaign.

The notion of evolving opinions cannot be verified or observed in real-time, as the product of the action founded by the evolved opinion would require constant supervision over unknown periods. Thus, modelling and simulation (M&S) presents itself as a useful tool that mitigates this drawback. The application of M&S allows for the prediction of future results based on events that took place in the past. Furthermore, M&S can be used to simulate certain specified events that have low probabilities of occurring in the natural environment.

This paper presents a method to simulate the evolution of opinions and the resulting events regarding following or not following an influencer as a conclusion of the influenced human behaviour. The model, referred to as the 'influencer' model, is based on the methodologies presented by Behl et al. [2], Wang et al. [3], and White et al. [4], which presented simulation strategies for opinion evolution, sentiment propagation, and the spread of COVID-19, respectively. This is a hybrid model [5] that uses multiple M&S techniques as well as different techniques to one or more stages in study [6]. The model presented in this paper combines network-based information diffusion based on a differential equation model, its definition as a cell space using the Cell-DEVS formalism, a discrete-event specification based on the DEVS formalism and spatial modelling with explicit delays [7]. The model is built using the DEVS Cadmium tool [8]. The 'influencer' model is experimented on with different scenarios to analyse the behaviour of the model and the subsequent effects of its incorporated variables/parameters. Further experiments were performed to determine the effects of changing the rate at which individuals tend to follow an influencer as well as differing personality groups.

6.2 Background and Related Work

Social interactions can be represented using a network diffusion process that allows users to understand the dynamics and propagation of the process in the network (which could represent information, infectious disease, etc.). The method of propagation can be either physical, verbal, communicated over the Internet, or planned group events, depending on the situation [9]. Network theory is part of graph theory and is a popular technique used to model the spread of infectious diseases and in the study of sociology. In a network, the individuals in a population are represented by nodes and the interactions among them as edges [10]. Network models aim to represent individuals within a population and the relationship between the identified individuals.

Diffusion is defined by Rogers [11] as a special type of communication that transmits innovation through channels consisting of individuals of a social system over a given period. Innovations can pertain to new ideas or technologies that are the result of these new ideas. Rogers first introduced such a process to explain the theory behind how new ideas and rates of innovation spread among the populous. The theory known as diffusion of innovation theory was first published in 1962 and is founded on five elements: the innovation itself, adopters, communication channels, time, and social systems, with another five stages to the adoption process: awareness, persuasion, decision, implementation, and confirmation. The theory further elaborates on techniques to achieve critical mass—the point at which enough people have adopted the innovation resulting in the adoption of the innovation to become self-sustaining with an important focus on 'top officials'.

Expanding on the idea of 'top officials' Rogers penned the term 'opinion leaders' defined as those individuals that create an informal leadership within society, and is a position earned by the individual's technical competence, social accessibility, and conformity to the systems norms—system norms are described to be the behaviour patterns present in members in a particular social system. The idea of opinion leaders is largely founded on the two-step flow of communication presented by Katz and Lazarsfeld [12], a theory of information diffusion that defines the propagation of information from *mass media* to the *opinion leaders* and finally to the *locals*. Additionally, opinion leaders are more exposed to external communications, have a higher social status, and are more innovative on average. Opinion leaders tend to be the centre of their social communication network and hold a unique influential position, while serving as a potential social model, whose behaviour can be imitated by individuals within its social communication network. However, according to Rogers, opinion leaders can lose the respect they hold in their communication network if they deviate too far from the social norm, or in retrospect, may lose credibility if they are unable to keep up with the latest trends. An opinion leader can be monomorphic—is an opinion leader in one topic—or polymorphic—is well-informed about a variety of topics.

Diffusion models are popularly used to simulate the propagation of information in social networks and the adoption of ideas by characterising the social interactions. Most diffusion models that implement social interactions as a process are extensions of the independent cascade and linear threshold models. The independent cascade model [13] focuses on the dissemination of information and interactions from individuals to their friends along a social network by considering the weak and strong ties between individuals. This is strongly correlated to the susceptible-infected-recovered (SIR) model for epidemic spread [14], which takes a similar approach to simulating disease spread through interactions between individuals based on strong and weak ties. The linear threshold model [15] describes a threshold-based perspective on influence propagation, that is, when enough of your peers have adopted a certain idea, you are more than likely to adopt it too.

AlFalahi et al. [16] evaluated different, used to measure influence probability, focusing on (1) static models, (2) dynamic models, (3) diffusion models, and (4) models based on users' behaviour. The research brought focus on how the study of social networks with the applications in graph theory while employing social network analysis to help trace sources and distribution of influence, can lead to a better understanding of the evolution of social networks, resulting in a better investigation of social structures and social influence in such networks. The research determines that the diffusion of influence can be modelled through probabilistic framework, with the probability of the individual embracing a new idea dependent on the neighbouring nodes within the network.

In an effort to model information diffusion with consideration to how information can be exchanged between individuals, various research studies have explored the similarities between epidemic and information diffusion. Goffman and Newill [17] introduced the similarities between epidemic and information diffusion processed by focusing on the aspect that the epidemic process can be defined as a more generalised abstract process of transitioning from one state to another due to exposure to an external phenomenon, with individuals susceptible to both diseases and ideas. Although epidemic and information processes are extensions of the general abstract process, there is one key difference. In the case of information diffusion, the phenomenon, in this case the information itself, is desired, whereas in the case of epidemic diffusion, the phenomenon is undesired.

Different research has explored methods of modelling social interaction and the resulting human behaviour using diffusion models as a foundation. Wang and Li [18] explored the similarities of the epidemic spread models to create the online social networks information spreading (OSIS) with the implementation of cellular automata. They discussed the nature of opinions fading over time, and the inherited aspects of the state change equations from the SIR model (susceptible-infected-recovered; a popular epidemic method). Similarly, Bouanan et al. [9] presented a method to simulate the spread of information across a network and the subsequent influence on their behaviour, with a focus on message propagation with trust factor dependencies and opinion characterisation using confidence bounds. Other social influence models focused on the strength of an individual's influence rather than the network propagation. Peng et al. [19] presented a model to evaluate social influence based on entropy, which assesses the influence of individuals based on various methods of social media interactions and designs an algorithm to characterise propagation dynamics of social influence based on the individual's entropy.

Social interaction modelling can be made more accurate by considering the opinions of the individuals. Opinions are an important aspect in predicting how an individual makes decisions, their behaviour, and how they react to information. An individual's opinions can evolve over time due to external social interactions or by new information diffused through the network. As such, it is important to define how to represent the evolution of opinion and its interactions with the social network formally. The resulting opinion calculation presented in this paper is founded on previous work done by Behl et al. [2] and Wang et al. [3].

Behl et al. [2] discussed the application of Cell-DEVS on modelling human behaviours based on social interactions, including social influence of human behaviour and its evolution considering the population size, the number of interactions, the degree of influence of each interaction, and the threshold of an individual to adopt an opinion or change in behaviour. The opinion update equation is defined as:

$$O(x) = O(x) + \sum_{y \text{ in neighbourhood } |O(x)-O(y)| \leq \text{Threshold}} \text{influence} \times (O(x) - O(y)). \tag{6.1}$$

$O(x)$ is the current opinion of the cell, $O(y)$ is the current opinion of a neighbouring cell, *influence* is the degree of influence of y on x and *Threshold* is the threshold of x that determines if x can be influenced.

An additional extension of how opinions can evolve over time was defined by Wang et al. [3], who presented a method to visualise public sentiments by analysing online posts and predicting future trends on a topic. As not all individuals contribute to the 'comments' section, a sentiment parameter was introduced. The evolution of sentiment offset is defined to be affected by:

- Social emergencies and external stimulus—an individual's interest in a topic can be influenced by an external stimulus other than neighbouring cells.
- A topic can fade over time and the sentiment offset can therefore reduce. This is represented by Wang et al. as:

$$m_i(t)' = m_i(t) \times \left(1 - \alpha^{\frac{m_i(t)}{20}}\right), \tag{6.2}$$

where α is the fading rate and m_i represents the sentiment offset.
- Influence between neighbours.

Social interactions can be seen as similar to the spread of infectious diseases and thus can inherit model structures used by epidemiology studies. We designed the influencer model using an approach similar to Wang and Li [18], Bouanan et al. [9], and Peng et al. [19], which used infectious disease models to define features in social interactions. As in the case of Behl et al. [2], social interactions can be affected by differing personality traits, a subject that is applied to the influencer model. In an effort to simulate certain behaviour patterns in an attempt to assess the implications of different personality traits, research on how to create differing neighbourhoods and definitions of personality traits is needed. Research into the creation of various neighbourhoods to fulfil the requirement of simulating social divisions and different personality traits in the real world has been explored also in Khalil and Wainer [20] which included case studies on the spread of avian flu, interactions affecting the well-being of organisms, and drug usage involving individuals with different personality traits. The case studies explored boundary conditions that manipulate a cell space and various personality definitions within the cell space. The research defined various dynamic states the agents could possess as well as the transition functions.

An important foundation of the influencer model is thus the definitions of the transition functions. As previously mentioned, the design of social interaction equations can be adapted from physical interaction equations defined in infectious disease models. The influencer model was predominantly based on the state transition equations presented by various COVID-19 cellular automata models, in particular those defined by White et al. [4]. The local transition functions used to inspire the transition functions in this model are shown in Eqs. (6.3), (6.4), and (6.5).

$$I_{ij}^t = (1 - \varepsilon) \cdot \upsilon \cdot S_{ij}^{t-1} \cdot I_{ij}^{t-1} + S_{ij}^{t-1} \cdot \sum_{(\alpha,\beta) \in V^*} \frac{N_{i+\alpha,j+\beta}}{N_{ij}} \cdot \mu_{\alpha\beta}^{i,j} \cdot I_{i+\alpha,j+\beta}^{t-1} \quad (6.3)$$

$$S_{ij}^t = S_{ij}^{t-1} - \upsilon \cdot S_{ij}^{t-1} \cdot I_{ij}^{t-1} - S_{ij}^{t-1} \cdot \sum_{(,\beta) \in V^*} \frac{N_{i+\alpha,j+\beta}}{N_{ij}} \cdot \mu_{\alpha\beta}^{i,j} \cdot I_{i+\alpha,j+\beta}^{t-1} \quad (6.4)$$

$$R_{ij}^t = R_{ij}^{t-1} + \varepsilon \cdot I_{ij}^{t-1} \quad (6.5)$$

The key points to take note of are as follows and are used in the influencer model.

- $\mu_{\alpha\beta}^{i,j}$ is the product of $\mu_{\alpha\beta}^{i,j} = m_{\alpha\beta}^{i,j} \cdot c_{\alpha\beta}^{i,j} \cdot \upsilon$, where $m_{\alpha\beta}^{i,j}$ and $c_{\alpha\beta}^{i,j}$ are the movement and connection factors, respectively, between the main cell and the neighbouring cell $(i + \alpha, j + \beta)$. The parameters υ and ε are the virulence factor and the recovery factor, respectively.
- The number of infected individuals I is the summation of infected individuals that have not recovered, susceptible individuals S affected by the infected individuals in the cell, and the susceptible individuals with the probability of being infected by neighbouring cells that have travelled to the current cell.
- The number of recovered individuals R is the addition of currently recovered individuals from the previous time step and infected individuals with probability to recover in the current time step.
- The connection factor considers various methods of transportation available to the individual that allows it to travel to other cells, whereas the movement factor is the probability of the individual moving to a neighbouring cell.

The influencer model was defined using Cell-DEVS and was implemented using the Cadmium tool [8]. A brief explanation of Cell-DEVS is defined in the following section.

6.3 Cellular Automata and Cell-DEVS

Cellular automata (CA) is the usual form of cellular modelling that uses a regular uniform n-dimensional lattice structure, with discrete variables at each cell. The values of the variables for each cell are synchronously calculated every time stamp and is determined by the values of the variables of a finite state of neighbouring cells from the previous time stamp. Since the calculations for the new variable values occur in a synchronous manner, CA evolves in discrete time stamps. The neighbuorhood of a cellular model can be defined using various approaches, and the most popular configurations are the Moore neighbourhood or Von Neumann neighbourhood. The Moore neighbourhood comprises of the central cell and its eight closest neighbouring cells, while the Von Neumann neighbourhood comprises of the central cell and its nearest four neighbuoring cells. Both the neighbourhoods can be extended using their respective radius, and the calculations for the number of cells based on the radius are $r^2 + (r + 1)^2$, $(2r + 1)^2$ for the Von Neumann and Moore neighbourhood, respectively. There are advantages and disadvantages of cellular automata, CA is simple enough to allow for detailed mathematical analysis while also allowing for simple mathematical calculations to be applied to the cells to create a complex mathematical system which make CA a popular method to model complex mathematical systems. However, CA does have its drawbacks, (1) performance and precision is lacking due to the discrete time-based calculations, (2) since CA is asynchronous in nature it requires the use of a synchronous digital computers, and (3) it is difficult to create time triggered events for the cells in the CA model.

The discrete-event system specification (DEVS) formalism is a discrete-event system that employs modular hierarchal formalism for modelling and analysing systems which can be described by a set of states. With regard to hybrid simulations, the DEVS formalism is a popular M&S tool since it allows defining multiple models coupled to work together in a singular model by connecting their input and output messages [21]. The DEVS formalism can be defined as either an atomic model; which defines the behaviour of a system as transition between states as a result of external events or coupled model; which defines how subcomponents of the system interconnect.

The Cell-DEVS formalism overcomes most of the limitations introduced by the CA model and is an extension if the DEVS formalism. Cell-DEVS models can be best described as an n-dimensional lattice of cells where every cell is an atomic model that is interconnected using the DEVS coupled formalism. Additionally, the cells within the cell space can not only interact with each other following the DEVS coupled model conventions but can also interact with DEVS models outside the cell space. When a cell receives an input, a local computation function is triggered, which calculates the future states of the cells. The output of the computation (the new states of the cell) is transmitted from the cells output port to other coupled cells after a defined delay elapses. The formalism includes a delay function and dictates when a change can occur once an external event is received from a neighbouring coupled cell, thus preventing any scheduled changes from occurring before the predefined time.

When an output is received at the cell, the external transition function is triggered. An additional duration function controls the lifetime of the cell, once the lifetime has expired an event is triggered to invoke the internal transition function to update the states of the cell. According to the formalism, before the internal transition function is triggered the output function and the output events are generated.

6.3.1 Cadmium Tool

The Cadmium library provides the necessary libraries to translate conceptual Cell-DEVS models to a computational model [22]. Cadmium is a header only C++, with the cells being defined in C++ and the cell space defined using JSON configuration files. The Cell<C, S, V> is an abstract implementation of the cell behaviour following the Cell-DEVS formalism. The local computation function is a virtual function that must be overwritten to represent the desired behaviour of the cell. If the new state is not equal to the current state, the Cadmium library adds the new state to the output queue that forwards the states to neighbouring cells. The Cadmium library uses a port-based approach when communicating state change messages between cells. Cells send scheduled state change messages that are present in the output queue to neighbouring cells via an output port and receive state change messages from external cells through an input port.

6.3.2 Influencer Model Design

The cells in the influencer model have three possible states, susceptible, influenced, and non-influenced. There are several assumptions that have been made in this model, which are listed below.

1. All cells are vulnerable to being influenced since social media is publicly available and accessible.
2. Individuals in the age group 13–40 are more vulnerable to being influenced.
3. Once an individual is influenced, there is a chance they will unfollow the influencer. Only those who are influenced can become non-followers.
4. Non-influenced individuals cannot refollow an influencer and hence cannot become influenced again. This assumption is made to demonstrate that individuals made the decision to unfollow the influencer purposefully and are unlikely to follow an influencer they have lost interest in based on the fact that there is a multitude of other influencers available.
5. The total population is static; hence, the population in each cell is always the same.
6. Individuals cannot move from one cell to another, hence cell population will never change.

7. The negative opinion can never have a probability of zero. The opinion parameter represents the probability of a positive opinion; hence, the negative opinion is the additive remainder of one. Hence, there can never be a zero probability of a negative opinion since we assume that the probability of a positive opinion will never be one.
8. The frequency of upload by the influencer affects the follower count; more regular uploads keep the audience engaged, while lower frequency values cause the individuals to lose interest overtime. This assumption does not consider the quality of the content released by an influencer.
9. There is no delay between changes in state for a given cell. The probabilistic values for each state are directly impacted on every time step and every interaction.

Following these assumptions, the model is designed to replicate the environment of social platforms and the degree of connections between individuals. The current model's configuration, however, does not simulate social groups, factions or social divisions, or types of social connections that can occur on social platforms, this would require a thorough social network analysis [22] with validated data from current social networks. The model implements a hybrid M&S approach with a combination of network-based information diffusion and agent-based modelling employed during the pre-simulation, and the Cell-DEVS formalism applied for the simulation of the model itself.

1. Pre-simulation: The conceptual model is designed to simulate how individuals (agents) connected within a social network propagate opinion and information. Thus, to illustrate such a configuration of individuals or groups of individuals connected within a network, an agent-based model integrated in cellular automata is employed with agents connected with other agents to create a network structure.
2. Simulation: To achieve results, the conceptual model is converted to use the Cell-DEVS formalism and translated into C++ to conform with the Cadmium tool. The results are analysed, and various scenarios are created to further characterise the behaviour of the model.

In accordance with the Cell-DEVS formalism cells are used to represent individuals, the neighbouring cells, and the social connections between individuals. In the model, each cell represents a static population of 100, which is subdivided into four age groups representing the different generations within the population. The four age groups used are children, 0–16 years, adults, 17–35 years, seniors, 36–50 years, and elders, 51 + years. This configuration is used to represent how in social platforms, an individual's 'friends' (the term used to represent an individual's connections on social platforms) can be vast, whether they be close or distant connections, and these connections have further multitude of connections creating an extensive network of connections. Each cell in the influencer model has four dynamic states, susceptible, influenced, non-influenced, and opinion. The model is probabilistic in nature, with the value presented for each state determined by the values of the corresponding neighbouring cells. Each state is a set of values that represent the probability of that agent being in that current state. Each cell has several static attributes assigned to it that control the personality traits as well as how susceptible the cell is to neighbouring influence, these are, extrovert factor, tech adoption, and free time. The static traits differ per age group consequently varying the probability of how likely the associated age groups are to being affected, thus the number of influenced in a cell differ between age groups and between cells. The model configuration uses a single cell to represent the influencer in the simulation, with that cell having a static population of once since there is only one influencer within the cell. The model has the simulation begin with one influencer and the primary circle of neighbours around the source being 'influenced' for two of the major age groups, which are 17–35 and 36–50.

Each cell has a neighbourhood configuration to simulate the relations and interactions between cells. The model uses a Moore neighbourhood with $r = 4$, with each layer going outward having a reduced 'connection' factor, that is, $r = 1$ having the most favorable influence on the cell in question while $r = 4$ having the least influence. For this model, an extended Moore neighbourhood is implemented to demonstrate that social media increases the range a user can impact others. The number of cells for an extended Moore neighbourhood is calculated using $(2r + 1)^2$, with r being the range of the neighbourhood. The probability of connection from the centre cell to any of its neighbouring cells is based on the Moore radius. Furthermore, to implement assumption 1, the influencer cell is included as a neighbour in every cell's neighbourhood configuration.

The opinion of an individual can vary based on many different factors, for example negative publicity, global information sharing, and influence from friends and family. An individual's opinion on a subject, in this case the influencer in question, would affect how likely the individual will choose to follow the influencer, and following that, how likely they are to unfollow the influencer. A positive opinion increases the chances of the individual becoming a follower; however, opinions can decrease over time and interest in the topic can fade, which would lead to a decreasing opinion, eventually resulting in a negative enough opinion to cause the individual to finally unfollow the influencer in which they have lost interest. There are also the rare cases that became more prominent and are widely referred to as 'cancel culture'. Cancel culture occurs when a popular figure receives a negative response to their online presence, which leads to an avalanche of negative commentary as the opinions spread between individuals, eventually leading to a popular negative opinion to diminish the online support for the influencer.

The main goal of the model is to determine the rate of influencer and non-influencer increase, as well as how other factors affect this rate change. Subsequently, the model implements state transition equations that calculate the evolution of the cells dynamic traits and updates these traits during the simulation. During each time step in the simulation, the state transition equations are used to calculate the evolution of the dynamic traits in every cell, however, if the cell is found to be an influencer, only the opinion dynamic trait is updated, since an influencer cannot follow or unfollow itself, but due to that fact that an influencer is part of every cells neighbourhood, its opinion can affect the corresponding cells opinion evolution, hence the influencers opinion must be updated in accordance with the opinion equation with respect to its own Moore neighbuorhood. This exception is implemented by adding an 'influencer' parameter to every cell, the parameter is Boolean and if set to true, the cell is defined to be an influencer.

The overall flowchart that illustrates the progress of the model is depicted in Fig. 6.1. At each time step, the *localComputation* function is called; the function characterises the cell behaviour for the influencer. Accordingly, if the cell is defined as an influencer, the computation function determines the change in opinion; otherwise, if the cell is a default generated cell, the three dynamic states variables are updated based on the value variables of the cell itself and its influential neighbouring cells.

Finally, in the interest of creating a model that can compare two influencers, the model was designed to work with one or two influencers defined in the environment. This was achieved by including a macro definition within the model. If the macro definition was enabled, the model followed a sequence of calculations with respect to both influencers. The model ensured that individuals were simulated to be able to follow the first influencer, the second influencer, or both influencers at the same time. This was achieved by adding a secondary set of dynamic traits with respect to the second influencer to every cell, however, it is important to note that this addition does not impact the static population of the cell, but rather the summation of each set of dynamic traits is still equal to the static population. Thus, a ratio of the population can be followers of the first influencer while still being susceptible to the second.

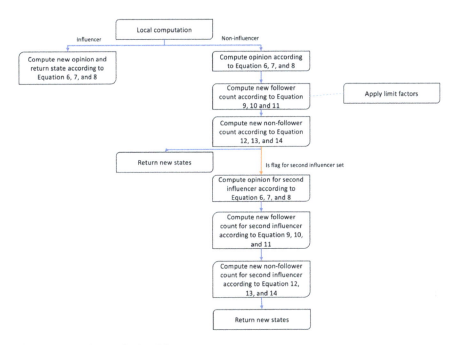

Fig. 6.1 Flow diagram for *localComputation*

Table 6.1 summarises the parameters used in the state transition equations with a brief description of their function, and Table 6.2 summarises the variables that are interdependent with state transition calculations.

The opinion Eq. (6.1) is combined with the sentiment iteration Eq. (6.2) discussed earlier. The 'new follower' calculations are based on those presented in Eqs. (6.3), (6.4), and (6.5), and are represented by the following equations. Equation (6.6) defines the value of opinion for the next time step. This number cannot be greater than 1, since that would equate to having 0 negative opinions, based on the presumption that the negative opinion ratio is calculated using $1 - o^i$.

$$\text{newo}^{i+1} = \min\left(0.999,\ \text{newo}^{i'}\right) \quad (6.6)$$

Table 6.1 Summary of parameters used in the state change equations for the influencer model

Parameter name	Symbol	Description
Follower factor	β	This value is used to vary the probability of becoming a follower
Extrovert factor	χ	This value is based on age and the individual and affects how much influence an individual has on others and vice versa
Random	γ	Random value to mimic unplanned events
Frequency	μ	This is the rate of online uploads by the influencer
Susceptibility factor	ν	Represents how susceptible an individual is to outside influence (not necessarily the influencer)
Non-follower factor	τ	The opposite of the follower factor, it controls the probability of an individual unfollowing
Population	ω	Population of the cell
Fading factor	σ	Used to reduce interest in a topic, affect the opinion on the subject
Limit factors	υ	Used to schedule changes in the environment to imitate global events
Neighbour vicinity	δ	This is the probability of connection between cells and is based on the Moore radius
Tech adoption	ε	The comfort level of individuals using social media
Free time	ϕ	The amount, on average, an individual browses social media

Table 6.2 Summary of variables used in the state change equations for the influencer model

Variable name	Symbol	Description
Global impact	α	Variable used to mimic global trends and is based on the free time and tech adoption of individuals
Opinion	o	Opinion of the individual

The first term of Eq. (6.7) calculates the influence the individual has on themselves in the evolution of their opinion, adjusted by a fading factor (decay in opinion $1 - \sigma^{\frac{o^i}{\gamma}}$), the extrovert factor χ, and the random variable γ, to represent random changes in individuals' thought process. The second term determines the influence of neighbouring opinions and their additive difference $o^i - o^i_j$, while applying the extrovert factor, decay of opinion, a randomised variable, and finally the probability of connection between the cell in question and the current neighbour δ. Finally, the total is multiplied by the susceptibility factor υ, and divided by the population of that cell ω, to determine the ratio of individuals with this new opinion value. That is, the opinion propagation between neighbours depends on the difference between the current cell's opinion and its neighbours; the difference is then added to the current opinion. Additionally, the fading factor considers that with the overwhelming availability of information, the attention span on any given topic deteriorates with time (or is replaced by a more engaging topic).

$$\text{new } o^{i'} = \left(o^i \cdot \left(1 - \sigma^{\frac{o^i}{\gamma}} \right) \cdot \chi \cdot \gamma \right.$$
$$\left. + \sum_{\text{neighbours}=j} \left(o^i - o^i_j \cdot \left(1 - \sigma^{\frac{o^i_j}{\gamma}} \right) \cdot \chi \cdot \gamma \cdot \delta \right) \right) \times \upsilon / \omega \quad (6.7)$$

Equation (6.8) represents the final change in the cell's opinion for the next time step.

$$o^{i+1} = \text{new } o^{i'} \quad (6.8)$$

Equation (6.9) is the newly predicted follower count and takes a similar approach to the new opinion calculation. The current number of susceptible individuals S^i is multiplied by the influence an individual has on themselves, and the influence of the neighbouring cells on the current cell, to define a change in state from susceptible to influenced I. Both terms use the follower rate β, a randomised value that represents unobserved evolving individual behaviour, limit factors υ (if any are assigned for the phase), and the global impact defined as $\alpha = \varepsilon \times \phi$, which describes the individual's propensity towards new technology and the likely amount of time they spend on it. The total is further multiplied by the frequency μ of content release provided by the influencer and the current opinion o^i of the agent.

$$\text{new} f^{i'} = S^i \times \left(I^i \cdot \omega \cdot \beta \cdot \alpha \cdot \upsilon \cdot \gamma \right.$$
$$\left. + \sum_{\text{neighbours}=j} I^i_j \cdot \omega \cdot \beta \cdot \alpha \cdot \upsilon \cdot \gamma \cdot \delta \right) \times \mu \cdot o^i \cdot \upsilon / \omega \quad (6.9)$$

Equation (6.10) takes the result of Eq. (6.9) and ensures that the new ratio of followers is not greater than the current ratio of susceptible individuals. Equation (6.11) illustrates the final predicted number of new followers for the next time step, which subtracts the updated number of non-followers to ensure that the additive ratios of susceptible, followers, and non-followers equate to the static population value of the cell.

$$\text{new} f^{i+1} = \min\left(S^i, \text{new} f^{i'}\right) \tag{6.10}$$

$$I^{i+1} = I^i + \left(\text{new} f^{i+1} - \text{newnon}^{i+1}\right) \tag{6.11}$$

In summary, the new number of influenced individuals is based on the neighbouring cells, the cells current opinion and external factors. An important consideration that is accounted for is that for any cell, the individuals that subscribe to be followers can never be greater than the current susceptible individuals. This ensures that all individuals in the cell have any of the three states, and the summation of all subclasses of individuals per state will equate to the total population of the cell, which is defined to be static.

The calculation of the final two states of the cell, the susceptible and the non-follower numbers are defined below. Similar to White et al. [4], the non-follower calculation adopts a variable that controls the rate of unfollowing an influencer τ— Eq. (6.12). Additionally, the equation includes the negative opinion of the cell by taking the additive inverse of the current cell opinion $(1 - o^i)$. This reflects how a degenerating opinion can increase the likelihood of unfollowing an influencer.

$$\text{newnon}^{i'} = I^i \cdot (1 - o^i) \cdot \tau/\omega \tag{6.12}$$

Similar to the accountability translated in the calculation for the new followers, the number of new non-followers can never be greater than the number of individuals that are currently followers—Eq. (6.13).

$$\text{newnon}^{i+1} = \min\left(I^i, \text{newnon}^{i'}\right) \tag{6.13}$$

Equation (6.14) defines the change in state for new non-followers and simply adds the new non-follower count to the existing number of non-followers.

$$N^{i+1} = N^i + \text{newnon}^{i+1} \tag{6.14}$$

The susceptibility number is the remainder of the ratio of individuals that are neither influenced nor non-influenced—Eq. (6.15), which can be summarised as the total population of the cell $(S^i + I^i + N^i)$ minus the new values of followers and non-followers $(N^{i+1} + I^{i+1})$.

$$S^{i+1} = (S^i + I^i + N^i) - (N^{i+1} + I^{i+1}) \tag{6.15}$$

6.4 Influencer Model Implementation

Having defined the state transition equations and the overall flow of the influencer model, this section describes the implementation of the model in the Cadmium tool. Cadmium is a header only C++ library with the *localComputation* virtual function describing the desired behaviour of the cell which is called every time step. Figure 6.2 shows a code snippet of the actions taken during each cell's *localComputation* call.

The *localComputation* is a Cadmium method that is activated on each cell following the Cell-DEVS formalism. The model incorporates phase changes to simulate how in the real-world external factors may directly affect the follower rate

```
sfn localComputation(sfn state, const std::unordered_map<std::vector<int>,
NeighbourData<sfn, double>>& neighbourhood) const override {
    state.phase = limits->next_phase(clock, state);
    if (state.flag_inf) {
        std::vector<float> new_op_inf = new_opinion(state, neighbour-
hood);
        for (int i = 0; i < n_age_segments(state.flag_inf); i++){
            state.opinion.at(i) = new_op_inf.at(i);
}
        return state;
    }
    for (int i = 0; i < n_age_segments(state.flag_inf); i++) {
        float ratio = state.susceptible.at(i) +
            state.influenced.at(i) + state.noninfluenced.at(i);
        age_ratio.push_back(ratio);
    }
std::vector<float> new_op = new_opinion(state, neighbourhood);
std::vector<float> new_f = new_followers(state, neighbourhood);
std::vector<float> new_n = new_nonfollowers(state);
#ifdef SECOND_FOLLOWER
std::vector<float> new_op2 = new_opinion(state, neighbourhood);
std::vector<float> new_f2 = new_followers(state, neighbourhood);
std::vector<float> new_n2 = new_nonfollowers(state);
#endif
    for (int i = 0; i < n_age_segments(state.flag_inf); i++) {
    state.noninfluenced.at(i) = state.noninfluenced.at(i) + new_n.at(i);
    state.influenced.at(i) = state.influenced.at(i)+new_f.at(i)-
new_n.at(i);
    state.susceptible.at(i) = age_ratio(i) - (state.noninfluenced.at(i) +
        state.influenced.at(i));
    state.opinion.at(i) = new_op.at(i);
    }
 return state;
 }
```

Fig. 6.2 *localComputation* code snippet

Cell_states:	Cell_states_second_influencer:
Susceptible[]	Susceptible_first_influencer[]
Influenced[]	Influenced_first_influencer[]
Non-influenced[]	Non-influenced_first_influencer[]
Opinion[]	Opinion_first_influencer
	Susceptible_second_influencer[]
	Influenced_second_influencer[]
	Non-influenced_second_influencer[]
	Opinion_second_influencer[]

Fig. 6.3 Cell states with one and two influencers

(economy, environment, etc.). Such phase changes are determined by *state.phase* = *limits* → *next_phase (clock, state)*, a limit factor can be applied to affect the calculation of follower number based on global events. Additionally, the limit factors may affect the follower rate for a particular influencer. Then, we check if the cell is an influencer (*if (state.flag_inf)*); in that case, only the opinion will be computed using the *new_opinion* function. The new state of opinion for the influencer is then updated for each age segment (each cell has four distinct age groups; thus, the states of the cell must be updated per age segment) using the function *new_op_inf.at* and the state returned for the next time step. Instead, if the cell is not an influencer, the follower and non-follower counts are calculated. The *ratio* is then calculated as the sum ratio of all ratios for each age segment. This can be defined as the static population of the cell, and it represents $(S^i + I^i + N^i)$ in Eq. (6.15). The *new_opinion, new_follower and new_nonfollower* function calls implement the equations described above: Eqs. (6.7), (6.9), and (6.12) which define the state transitions using the current inputs to the cell. Additionally, if SECOND_INFLUENCER is defined (which means the model includes two influencers), the state changes are re-calculated with respect to the second influencer. To illustrate the difference between the two sets of dynamic states, refer to Fig. 6.3.

We then cycle and update the four distinct age groups. The new states for each cell are then applied for the next time step by equating *state.influenced, state.noninfluenced, state.susceptible, and state.opinion* that embodies Eqs. (6.8), (6.11), (6.14), and (6.15). The opinion calculation implements a pointer variable that is set based on whether the calculation is being done for the first or second influencer. In each case, the pointer points to the correct opinion set required at the time. The same strategy is used for any other state transition functions when required. It is important to note that the 'neighbourhood' variable shown in Fig. 6.2 does not only represent the extended Moore neighbourhood but also includes the influencer/influencers, this embodies assumption 1. An additional constraint implemented in the model is the difference of opinion between the cell and its neighbours, that is, the difference can never be negative, since the opinion of the cell can never be a negative value in this simulation.

6.4.1 The Influencer Model

In this section we show the execution of the influencer model including the previously defined transition functions, a single influencer, and an extended Moore neighbourhood with a radius of four. The neighbourhood around the influencer itself has a ratio that is predominantly influenced, assuming that close friends and families are more willing to follow the influencer. The model is used to find the initial values of the static variables and reflects a natural flow of how the rate of followers and, similarly non-followers would increase over time.

Figure 6.4 shows a spatial visualisation of the Cell-DEVS model execution within a 500-day simulation. The cells are shaded from light to dark with respect to the probability value of becoming a follower. The plot on the left depicts the simulation at the beginning with just the influencer and its closest neighbours being influenced (followers), the middle plot depicts the transition of the simulation at the midpoint of the simulation as the influence of the influencer and its followers affect susceptible cells. The final plot on the right is at the end of the simulation with almost all cells having a probability of being a follower.

Figures 6.5 and 6.6 plot the overall results of the original influencer model and the progression of the follower count within the same time scope as illustrated in Fig. 6.4. Figure 6.5 plots the total number of followers versus non-followers, while Fig. 6.6 plots the relationship between the three dynamic states.

Analysing Figs. 6.5 and 6.6, there is a slight increase in the number of followers in the first days from 0% to 0.1%. During the initialisation of the simulation, several cell ratios are already configured to be influenced (followers) within the Moore neighbourhood of radius $r = 1$ around the influencer cell. This is achieved by setting the 'follower' state of the neighbouring cells to a positive value in the model configuration file. This is purposefully done to model that individuals closely connected to the influencer are far more likely to become a follower. Next, we see a gradual increase in the number of followers, and the non-follower count increases at a slower rate (but parallel to the rate of follower increase). This is an expected feature, as the non-follower count cannot be larger than the current number of followers, thus

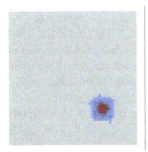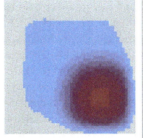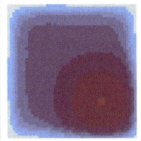

Fig. 6.4 Spread of influence using the DEVS viewer

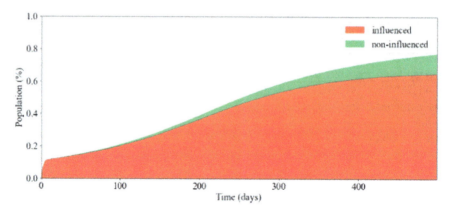

Fig. 6.5 Follower rate increase

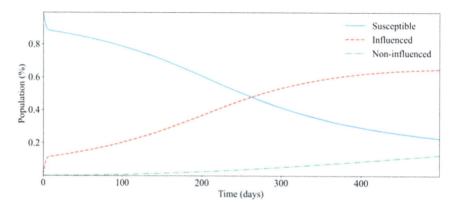

Fig. 6.6 Relationship between susceptibility and follower rate for the default influencer model

a rate of increase in non-followers can only occur when there is a viable number of followers. Furthermore, the susceptibility count is inversely proportional to the number of followers, showing that the current susceptible ratio is determined by subtracting the follower ratio and non-follower ratio from the cell's population.

Although face value validation of the model is not feasible (as data about influencers is not available), we conducted a formal analysis of the design of the model by comparing the behaviour with White et al. [4]. The transition functions for the influencer model and those in Eqs. (6.3), (6.4), and (6.5) are related; the influencer model follower, non-follower, and susceptible can be seen as similar to the infected, recovered, and susceptible for infectious disease (where the state 'deceased' is excluded). The main characteristic in White et al. [4], illustrated in Figs. 6.7 and 6.8, is that as the number of infected increases so does the number of recovered, with the susceptibility being inversely proportional to that of infected + recovered. This is similar to the behaviour represented in Figs. 6.5 and 6.6.

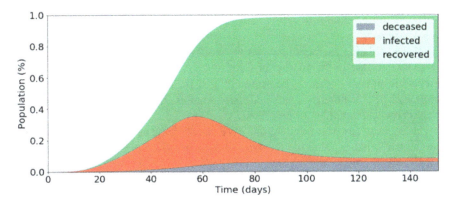

Fig. 6.7 Results for White et al. [4] model

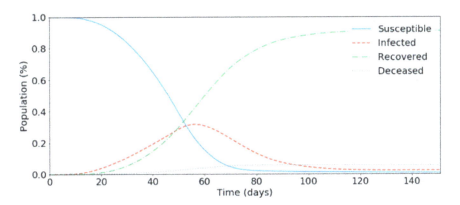

Fig. 6.8 Relationship between susceptibility and infection rate for the White et al. [4] model

Firstly, there is a gradual increase in the number of followers as there is an increase in those infected. Secondly, as the number of followers increases, there is also a gradual increase in the number of non-followers as can be seen in Fig. 6.5 with the increase becoming more prominent after 200 days have elapsed; the same can be said about the number of recovered individuals with respect to the number of infected. As observed in the White et al. [4] model, there is a gradual decline in the number of infected as recovered individuals become immune. However, this behaviour is not present in the influencer model. This is because the influencer model observes a relatively slower growth in the number of non-followers—White et al. [4] model has a 0.9% recovered rate at 70 days, whereas the influencer model does not reach this value even after 500 days, since the influencer model has a lower non-follower rate when compared with the recovered rate in the model presented by White et al. while also taking into account negative opinions, an additional factor that is not included in White et al. model, which results in the number of susceptible cells still being higher than the number of non-followers, hence follower numbers can continue to rise until such a case occurs.

When simulating the model for a longer time (Figs. 6.9 and 6.10), we can observe a gradual decrease in the number of followers at the intersection of the number of susceptible and non-followers depicted in Fig. 6.10. This behaviour is similar to the White et al. [4]: the number of infected also decreases once the number of susceptible intersects with the number of recovered.

The final similarity between the models is the decrease in followers/infected once the number of susceptible has intersected with non-followers/recovered. Furthermore, since the influencer model and the White et al. [4] model have a shared assumption—the influencer model assumes that once an individual has become a non-follower, they can never become a follower again; White et al. [4] assumes that once an individual has recovered, they can no longer be infected—the number of followers/infected cannot increase again once the number of susceptible is lower

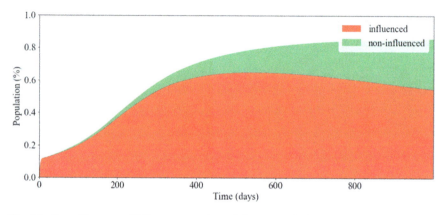

Fig. 6.9 Rate of increase of follower rate and non-follower rate during a longer simulation time

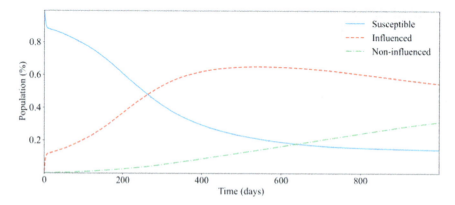

Fig. 6.10 Relationship between susceptibility and follower rate during a longer simulation period

than the number of non-followers/recovered. It is also safe to assume that if the influencer model 'non-follower rate' variable and the calculation for the number of new non-followers identically matched White et al. [4] recovery factor and number of recovered calculations, the rate at which the number of non-followers increased would be extremely similar to the rate at which the number of individuals recovered.

6.4.2 No Opinion Model

To verify that the opinion of an individual plays a key role in determining whether the individual will become a follower or not, a modified 'influence' model was tested in which the new follower evaluation was adjusted to not be dependent on the individual's current opinion. Equations (6.16) and (6.17) illustrate the difference in how new follower and new non-follower count is evaluated without consideration of the opinion.

$$\text{new} f^{i'} = S^i \times \left(I^i \cdot \omega \cdot \beta \cdot \alpha \cdot \upsilon \cdot \gamma + \sum_{\text{neighbours}=j} I^i_j \cdot \omega \cdot \beta \cdot \alpha \cdot \upsilon \cdot \gamma \cdot \delta \right) \times \mu \cdot \upsilon / \omega \quad (6.16)$$

$$\text{newnon}^{i'} = I^i \cdot \tau / \omega \quad (6.17)$$

The resulting rate of increase in follower count can be seen in Fig. 6.11. It is evident that disregarding the opinion of an individual would automatically create a scenario in which any individual would become a follower without the slightest inclination to think otherwise. Hence individuals will assuredly and without thought follow an individual once they are made aware of them. This resolves the importance of having an opinion-based model to interpret the probability of an individual becoming a follower.

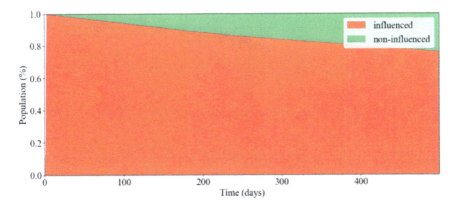

Fig. 6.11 Result of removing the opinion-based calculations from the model

6.5 Varying the 'Follower Factor' Variable

Using the original influencer model the value of the 'follower factor' variable is varied to determine the behaviour of the model in response to this change. A total of five different sets of the follower rate were tested, in each case, each set was randomly generated and simulated. Table 6.3 defines the different sets of follower rates used.

Based on the results obtained using the random variables, the number of new followers is directly proportional to the 'follower rate', this does not come as a surprise, however, the behaviour resulting from the relationship between 'follower rate' and susceptibility can also be observed from this experiment. Building on this point, we must note that for this model, the susceptible ratio of the population per age group is defined as [0.1, 0.61, 0.22, 0.07], which implies that the most susceptible age group is between the ages of 17–35 years. If the follower rate augments the correct age group—in this case the adult age group—there will be a healthy growth in the number of new followers' overtime. However, in the case the follower rate does not augment the age group that is the most susceptible, the rate of increase in the number of followers is less in comparison. We can observe this effect by comparing the results of the random sets 3 and 4 depicted in Figs. 6.12 and 6.13, respectively.

Table 6.3 Random values used for determining the effects of follower rate

	Values
Random 1	[0.269979, 0.745675, 0.227634, 0.26146]
Random 2	[0.587839, 0.777752, 0.724959, 0.336366]
Random 3	[0.009422, 0.777854, 0.252165, 0.614234]
Random 4	[0.713371, 0.468194, 0.774718, 0.647773]
Random 5	[0.23911, 0.214832, 0.274497, 0.512304]

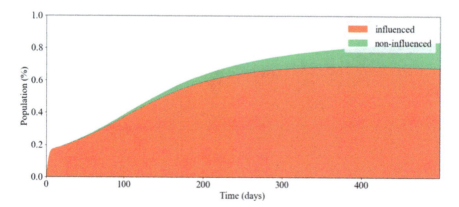

Fig. 6.12 Results of using random set 3 for follower rate

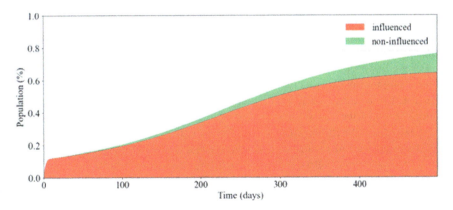

Fig. 6.13 Results of using random set 4 for follower rate

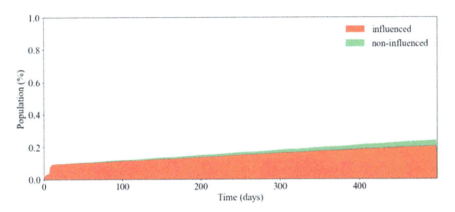

Fig. 6.14 Results for using random set 5 for follower rate

In general, a low 'follower rate' in all age groups results in a weak increase in follower count, and this can be observed in the results of random set 5 visualised in Fig. 6.14.

6.6 Introverts Versus Extroverts

The goal of these tests is to determine whether a particular personality type is more vulnerable to becoming a follower or not. The experiment is carried out using two different methods, the first being two small groups of cells being defined as either introverted or either extroverted, while the second involved having all cells—except from the influencer cell and its Moore neighbourhood—being either introverted or extroverted. Both methods involve changing the 'extrovert factor' of the cell by randomising the values based on a particular range to simulate an introvert or an extrovert personality. For example, in the case of an introvert the randomisation range is set to 0–0.2 while for an extrovert the range is 0.6–0.9.

For the first method, a group of cells was designated as introverted, while in the same configuration file, a secondary group is designated as extroverted. This method was tested three times, each with a random generation of values for the 'extrovert factor'. Based on the results gathered from this first method, there is no definite proof that personality affects the increase in follower count. There are three reasons that could explain why this is the case, the first is that the two groups created were too small to observe a definitive behaviour and thus, the overall behaviour of the other cells overshadowed any relevant behaviour. The second reason could be that the cumulative behaviour of the cells simply results in an increase in the rate of number of followers, overshadowing any observable patterns that would result from different personalities. The final reason could be that personality differences have no effect on the rate of increase in follower count, however, to investigate whether there is a foundation to this reasoning the second method was implemented.

The second method randomises the 'extrovert factor' for all cells, with once scenario having all cells set as introverts, while the second scenario has all the cells set as extroverts. Both scenarios randomly generate the 'extrovert factor', with three different sets, and both scenarios are simulated separately. In both scenarios, no discernible difference was observed in the rate of increase in follower count. Thus, the third reason introduced still holds, and based on these results different personalities do not seem to affect the rate of increase in follower count.

6.6.1 The Pandemic Scenario

Having defined the behaviour of the influencer model, the model is further experimented with by applying it to realistic use cases. This section presents the application of the influencer model towards a scenario with a society that has been affected by pandemic regulations.

The original influencer model is applied to a scenario that has been extended to include limit factors to simulate three general scenarios that appeared to occur during the pandemic. The three phases for this specific model can be described by the assumptions as follows.

1. Before the pandemic—individuals have an active social life, reducing the free time they must have to browse social media. This results in an average interest on influencers, with an average rate of follower increase.
2. During the pandemic—individuals have a reduced social life, with lockdowns in effect and limited physical activities. This results in an increase in free time as individuals try to find more relevant and entertaining forms of digital diversions. With the increased free time, individuals may also engage in personal growth activities, by finding hobbies and influencers that share their interests. Based on

these assumptions, the model is designed to increase the follower rate during this phase to simulate the higher probability of becoming a follower.
3. After the pandemic—as lockdowns and restrictions are lifted, individuals can be assumed to try to make up for the time they lost outdoors. Thus, individuals are more likely to engage in physical and social activities. This assumes that individuals would be mentally exhausted by the lockdown rules and regulations and would be anxious to get back to a sense of normalcy. Based on these statements the follower rate would be lower when compared with during the pandemic phase.

The limit factors and phase change factors are implemented using an incrementing simulation clock. During each cell iteration the simulation clock time is used to calculate which phase the simulation is currently in. Based on the calculated phase, the respective limit factor is applied to the state calculation, which directly impacts the rate at which the follower count increases.

The importance of this model is to demonstrate how external environmental changes can affect how an individual responds to the influencer presence.

The result of the pandemic scenario demonstrates how the follower count can drastically change due to changes in the global environment, as can be observed by the sudden spikes—at time 100 days—and drops—at time 400 days—in the follower count increase presented in Fig. 6.15. Figure 6.16 shows the dynamic behaviour of the states over the shared rate of time.

The pandemic model is controlled by the limit rates and the scheduled phase changes, in this case the phase changes occur at timestamp 100 and 400. If a phase change is scheduled the corresponding limit rates are applied for that phase until the next phase change. For the first phase of this scenario, (time stamp 0–100), the limit factor is 0.5, for the second phase, (time stamp 100–400), the limit factor is 2, and finally, for the last phase the limit factor is 0.5. The effects of the limit factors can be observed in the simulation, in timestamp 100, there is a sharp spike in the rate

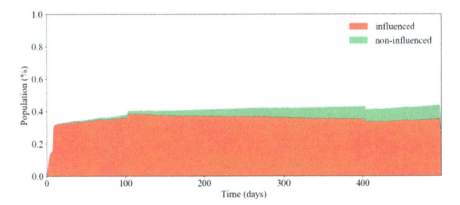

Fig. 6.15 Rate of increase for follower and non-follower rate for the pandemic scenario using the influencer model

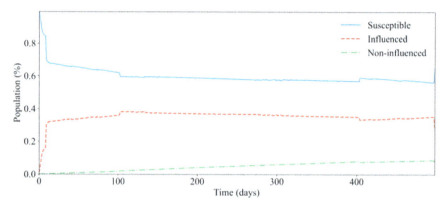

Fig. 6.16 Line plot comparing susceptibility and follower rate for the pandemic scenario using the influencer model

of increase in number of followers as the applied limit factor transitions to 2, and a sudden drop is observed at time stamp 400 as the limit factor transitions to 0.5 illustrating the expected behaviour during each corresponding phase change.

The limit factors are set to simulate the three phases defined for the pandemic scenario. Based on the results, the expected behaviour can be observed for each phase. However, these results only strengthen the assumptions made about social behaviour during a pandemic. This predefined behaviour may not occur and there may be other factors at play. It could be said that after the pandemic (third phase) there was never a drop in interest towards social media, and the effect of remote working further strengthened the attitude individuals had towards pursuing their personal growth objectives and hobbies, thus leading to a continued increasing follower count. It could also be said that during the pandemic individuals have created the habit of following influencers and browsing social media, and this trend continued even after the pandemic. There are several factors that can affect the follower count rate, hence this scenario does not fully represent the reality of the situation but demonstrates how the model can be used for such scenarios if more realistic data is present to corroborate the simulation.

The DEVS simulation for the pandemic model is shown in Fig. 6.17. The affected cells are comparably not as many as the original influencer cell, this is because the pandemic scenario implements limit factors, that is, a reducing limit factor is applied during the first and last phases of the simulation.

6.6.2 The Academy Awards Scenario

This scenario further extends the pandemic scenario applying the limit factors while simulating individual behaviour towards two separate influencers simultaneously.

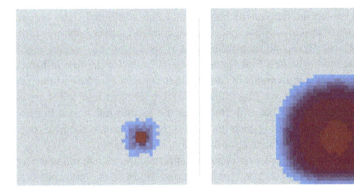

Fig. 6.17 Spread of influence across the cell grid using DEVS viewer

This scenario exemplifies how certain global changes can affect one influencer public status while having no effect on another. As in the case of the Academy Awards, by nominating a certain actor or actress, that influencer in question will be globally publicised, however, other artist may not receive the same amount of attention due to the lack of publicity. This is the reason using two influencers in the simulation is important to compare the differences certain factors that can have on the follower count, and how trends affect the probability of following an influencer.

The scenario will have three phases as described below:

1. Before the Oscars—it is assumed that both influencers in this case have a similar popularity rating, with low public recognition. The rate of increase in follower number will be muted but present.
2. During the Oscars—this statement can be vague without specifying that range of time this takes place. This assumption made here is that the period is between announcements of the nominations to the end of the award ceremony. During this phase, the influencer publicised by the annual awards will experience a spike in follower rate, while the second influencer will continue to observe the same rate of increase.
3. After the Oscars—once the Oscars are over, the promotion and movement to raise recognition for the first influencer loses traction. However, the popularity of the first influencer is still higher than the second. This assumes that, once a popular public personality has gained some amount of global awareness, their presence is widely spread through social media and social interactions.

Overall, this scenario demonstrates how external factors can affect a single influencer instead of all current influencers.

The implementation of this scenario was slightly more complicated when compared with the previous scenarios. Although the state transition equations do not change, the limit factors are applied based on the phase changes as described in the pandemic scenario. However, the possible state for each cell is doubled. Each cell in this case has eight dynamic states, the first four states, influenced, non-influenced, opinion, and susceptible remain unchanged, and four additional states are added to depict the states of the cells with respect to the second influencer, hence the four states added are influenced_second, non-influenced_second, opinion_second, and finally susceptible_second. The first four states are specifically for the first influencer and represent the ratio of individuals in the cell that are influenced/non-influenced by, susceptible to, opinion on the first influencer, while the last four states are equivalent to the first four states but with respect to the second influencer. Furthermore, the calculations applied during each iteration is dependent on if the new state being calculated is for the first influencer or the second influencer.

For the implementation to work on both the first and the second influencer without requiring additional duplicate code for the second follower, the base code is reused, and pointers are applied. The data the pointers point to change based on if the calculation is for the first or the second influencer. During each iteration, all eight new states of the cell are calculated starting with the first influencer and followed by the second influencer. After the states of the first influencer is calculated, a flag is set to ensure the pointers used to point to the state data points to the states for the second influencer. The state transition equations then calculate the new states using the new data that is being pointed to.

This scenario's main goal is to demonstrate that it is possible to simulate two influencers in the same scenario while also demonstrating that it is possible that the attitude of individuals towards different influencers are distinct.

The plot for each influencer is separated and it can be observed that the behaviour of individuals towards the two different influencers are disparate. For the first influencer there is a sharp spike at the beginning of the second phase—100 days—and a progressively increasing follower count after the third phase—400 days, whereas for the second influencer there are no predominant spikes and a mostly stable follower count growth. As previously done with the pandemic scenario there are three phases, and the values of the corresponding limit factors are 0.5 for the first, 2 for the second, and 0.8 for the third to simulate an average following before the Oscars, an increasing popularity during the Oscars after being widely recognised for their accomplishments, and a continued vehemence for the influencer after the Oscars. The third phase assumes that individuals that adore the influencer will continue to propagate awareness of the influencer and endorse their future projects. Additionally, during the second phase the probability of connections between the general population and the influencer is doubled as individuals purposely look for the influencer in question on social media, and news circulations—relevant news associations would want to advertise the Oscar candidates and explore their lifestyle to participate and emphasise the significance of the Oscar's—increase the probability of individuals becoming more aware of the influencer.

Figures 6.18, 6.19, and 6.20 illustrate the expected behaviour during such a global event, however, as mentioned with the pandemic simulation, this behaviour was predetermined and does not reflect the reality of the situation. There are certainly other possible outcomes that can occur, for example the influencer being nominated for the Oscar could suddenly be impacted by negative publicity during the second phase of the scenario, and instead of having an enlarging limit factor, could see a sharp drop—or paradoxically a sharp increase as individuals become absorbed into the scandal—in popularity instead with a reducing follower count after the Oscar's as popularity drops and the scandal fades away.

In conclusion this scenario does accomplish its goal of illustrating how two influencers can have different follower count patterns and are individually impacted by certain scenarios.

The DEVS simulation of the scenario can be seen in Figs. 6.21 and 6.22 for both the influencers. As with the pandemic scenario the cells are shaded from light to dark with respect to the probability value of becoming a follower.

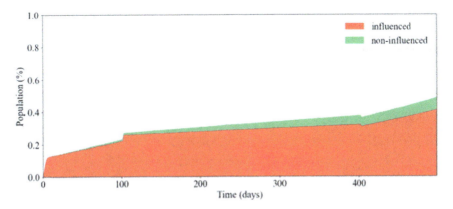

Fig. 6.18 Increase in follower and non-follower rate of the first influencer in the Oscar's scenario using the influencer model

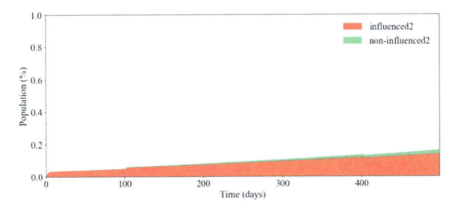

Fig. 6.19 Rate of increase for follower and non-follower for the second influencer in the Oscar scenario using the influencer model

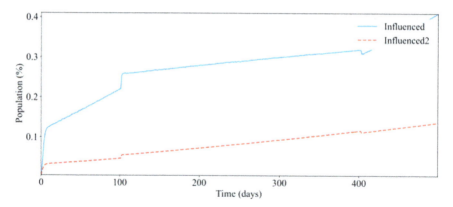

Fig. 6.20 Line plot comparing the rate of increase of the follower rate for the first and second follower in the Oscar scenario

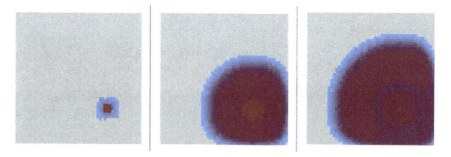

Fig. 6.21 Spread of influence with respect to the first influencer across the cell grid using DEVS viewer

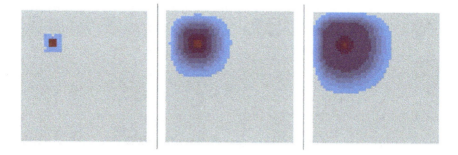

Fig. 6.22 Spread of influence with respect to the second influencer across the cell grid using DEVS viewer

6.7 Conclusion and Future Work

We presented a model that defines a network-based approach with an opinion evolution calculation. This is an important aspect of the model that predicts an individual's opinion towards the influencer, the effects of the opinion towards the individuals' actions, and the impact of the present individual's opinion on its close susceptible neighbouring persons. The evolution of opinion ascertains that the rate of increase in follower count occurs in parallel to the rate of increase in the individual's positive opinion, that is, as the probability of a positive opinion increases, the probability of the individual following the influencer increases. The model also takes into consideration the convenience/accessibility of social media platforms and the constant availability of influencers. Thus, influencers are treated as constant companions to the individuals in the model. Additionally, we provided a relevant example on how the model can be applied to real-world scenarios.

We employ the Cell-DEVS formalism, which has proved to be useful for simulating the spread of infectious diseases, to create a working model that is capable of simulating the rate of increase in follower count pertaining to a certain influencer under various scenarios and possible timed events.

In the future, we hope to extend the influencer model to use a threshold-based approach in determining the impact of neighbuoring agents on the opinion evolution of an individual. Furthermore, a quantitative approach could be taken to explore the real-time validation of the model, by collecting and mining for data that represents the rate of increase in follower count with respect to an influencer under certain conditions and available events. Another interesting approach would be to apply sentiment analysis in combination with machine learning to gather data from various social media platforms that can be used in the real-time validation of the model. By applying sentiment analysis machine learning techniques, we can further refine the behaviour of the opinion simulation in the influencer model. This could be useful—although not an abstract simulation—to further study the impact of influencers on social media.

Finally, modifications can be made to the assumptions taken for this model, for example, removing assumption 4 by modifying the current model to have an additional transition function to determine the count of non-followers that transition to becoming susceptible again with a timed wait period. This would represent individual behaviour that occurs on social platforms, where individuals constantly contradict their previous decisions and are thus likely to refollow influencers given various reasons, such as a change in popular attitude towards the influencer or correcting a previous mistake.

References

1. Bond RM, Fariss CJ, Jones JJ, Kramer ADI, Marlow C, Settle JE et al (2012) A 61-million-person experiment in social influence and political mobilization. Nature 489:295–298. https://doi.org/10.1038/nature11421
2. Behl A, Wainer G, Ruiz-Martin C (2018) Cell_DEVS: an approach to model the influence of social interactions in human behaviour. In: Proceedings of the 50th Computer Simulation Conference. Society for Modeling and Simulation International (SCS). https://doi.org/10.22360/summersim.2018.scsc.030
3. Wang C, Xiao Z, Liu Y, Xu Y, Zhou A, Zhang K (2013) SentiView: sentiment analysis and visualization for internet popular topics. IEEE Trans Human-Mach Syst 43:620–630. https://doi.org/10.1109/THMS.2013.2285047
4. White SH, del Rey AM, Sánchez GR (2007) Modeling epidemics using cellular automata. Appl Math Comput 186:193–202. https://doi.org/10.1016/j.amc.2006.06.126
5. Mustafee N, Godsiff P, Sahnoun M, Baudry D, Smart PA, Louis A (2015) Investigating execution strategies for hybrid models developed using multiple M&S methodologies. Simul Ser 47:78–85
6. Powell J, Mustafee N (2014) Soft OR approaches in problem formulation stage of a hybrid M&S study. Proc Winter Simul Conf 2014:1664–1675. https://doi.org/10.1109/WSC.2014.7020017
7. Wainer GA (2014) Cellular modeling with Cell-DEVS: a discrete-event cellular automata formalism. In: Lecture Notes in Computer Science, pp 6–15. https://doi.org/10.1007/978-3-319-11520-7_2
8. Belloli L, Vicino D, Ruiz-Martin C, Wainer G (2019) Building Devs models with the cadmium tool. In: 2019 Winter Simulation Conference (WSC). IEEE. https://doi.org/10.1109/WSC40007.2019.9004917
9. Bouanan Y, Zacharewicz G, Vallespir B (2016) DEVS modelling and simulation of human social interaction and influence. Eng Appl Artif Intell 50:83–92. https://doi.org/10.1016/j.engappai.2016.01.002
10. Keeling MJ, Eames KTD (2005) Networks and epidemic models. J R Soc Interf 2(4):295–307. https://doi.org/10.1098/rsif.2005.0051
11. Rogers EM (1995) Diffusion of innovations, 4th edn. Routledge, London
12. Katz E, Lazarsfeld PF (2017) Personal influence. Routledge, London. https://doi.org/10.4324/9781315126234
13. Goldenberg J, Libai B, Muller E (2001) Talk of the network: a complex systems look at the underlying process of word-of-mouth. Market Lett 12:211–223. https://doi.org/10.1023/a:1011122126881
14. Radcliffe J, Bailey NTJ (1977) The mathematical theory of infectious diseases and its applications. Appl Stat 26(1):85. https://doi.org/10.2307/2346882
15. Granovetter M (1978) Threshold models of collective behaviour. Am J Sociol 83(6):1420–1443. https://doi.org/10.1086/226707
16. AlFalahi K, Atif Y, Abraham A (2013) Models of influence in online social networks. Int J Intell Syst 29(2):161–183. https://doi.org/10.1002/int.21631
17. Goffman W, Newill VA (1964) Generalization of epidemic theory: an application to the transmission of ideas. Nature 204(4955):225–228. https://doi.org/10.1038/204225a0
18. Wang Y, Li G (2018) The spreading of information in online social networks through cellular automata. Complexity 2018:1–9. https://doi.org/10.1155/2018/1890643
19. Peng S, Yang A, Cao L, Yu S, Xie D (2017) Social influence modeling using information theory in mobile social networks. Inf Sci 379:146–159. https://doi.org/10.1016/j.ins.2016.08.023

20. Khalil H, Wainer G (2020) Cell-DEVS for social phenomena modeling. IEEE Trans Comput Soc Syst 7:725–740. https://doi.org/10.1109/TCSS.2020.2982885
21. Ruiz-Martin C, Wainer G, Bouanan Y, Zacharewicz G, Paredes AL (2016) A hybrid approach to study communication in emergency plans. In: 2016 Winter Simulation Conference (WSC). IEEE. https://doi.org/10.1109/wsc.2016.7822191
22. Cárdenas R, Wainer G (2022) Asymmetric Cell-DEVS models with the Cadmium simulator. Simul Model Pract Theory. 121:102649. https://doi.org/10.1016/j.simpat.2022.102649

Chapter 7
Application of Machine Learning Within Hybrid Systems Modelling

Niclas Feldkamp

Abstract Due to the easy and affordable access to more and more computing power, as well as the continuous improvement and distribution of easy-to-understand and easy-to-use open-source libraries and packages, the popularity of machine learning algorithms has increased significantly in the last years. This has led to a debate in the simulation research community as to whether the discipline of modelling and simulation will eventually be made obsolete by machine learning. However, this apprehension could not be further from the actual reality. In fact, both disciplines can complement each other very well to form a whole that is greater than the sum of its parts, provided that they are combined in a hybrid system in a reasonable way. In this paper, an overview of the existing possibilities is given for each of the three main categories of machine learning, i.e., supervised, unsupervised, and reinforcement learnings. This includes guidelines, application potentials, and use case examples on how to combine simulation and machine learning in a hybrid system.

Keywords Hybrid simulation · Machine learning · Data farming · Simulation analysis

7.1 Introduction

The method of modelling and simulation is one of the most important tools for analysing complex systems. Not exclusively, but especially in the case of systems with stochastic influences [1]. In fact, Lucas et al. even call modelling and simulation a method of first resort when it comes to system analysis [2]. On the other hand, the modelling of complex systems is also very time-consuming and holds many pitfalls [3]. A skilled modeller needs not only a technical understanding of simulation techniques, and programming skills, but also an understanding of processes and procedures and on top of that, social skills to communicate results and findings appropriately.

N. Feldkamp (✉)
TU Ilmenau, Max-Planck-Ring 12, 98693 Ilmenau, Germany
e-mail: niclas.feldkamp@tu-ilmenau.de

© The Author(s), under exclusive license to Springer Nature Switzerland AG 2024
M. Fakhimi and N. Mustafee (eds.), *Hybrid Modeling and Simulation*, Simulation Foundations, Methods and Applications, https://doi.org/10.1007/978-3-031-59999-6_7

The emergence of machine learning has sparked a discussion within the simulation community as to whether or not modelling and simulation will be made obsolete in the long term by machine learning algorithms [4, 5]. The assumption for this is that one no longer needs to understand the underlying process, but only requires a reasonable data basis and can then train a prediction model. As soon as a predictive model is available, it could be used for optimisation.

This assumption is not incorrect in theory, but it fails to recognise that a simulation model offers much more than the purely predictive functions. A good system analyst should master both sides, i.e., modelling and simulation, as well as machine learning, as both have specific advantages and disadvantages. In detail, machine learning is based on training on existing data. Modelling and simulation, on the other hand, are based on creating a model that can then generate data [6]. This admittedly somewhat abstract viewpoint already shows that machine learning and simulation are not at all contradictory but can complement each other very well if properly combined. The key is to combine the advantages of both approaches so that the resulting construct achieves a higher value than the pure sum of its parts [7, 8]. This is called hybrid systems modelling [9]. In this paper, a closer look will therefore be taken at the combination of simulation and machine learning in the context of hybrid systems modelling.

However, the opposite case can also occur, that is that a simulation model is required to be integrated in a machine learning project. So, both directions are possible. There are numerous approaches where simulation makes use of machine learning (to form some type of hybrid system), but machine learning and machine learning-related projects can sometimes in turn be dependent on simulation models to reach their goal. Hence, both disciplines can be said to be mutually reliant on each other, and this leads to two hypotheses: the first hypothesis is that *simulation needs machine learning*, which is the main core of this chapter. This is the case when machine learning approaches are used in one or more stages of a simulation study in accordance with hybrid systems modelling. The second hypothesis, that *machine learning needs simulation*, is also addressed in this chapter.

The structure of this chapter is as follows. After the introduction, Sect. 7.2 provides an overview of the most important basic terms on the topic of machine learning. Using the common three-way categorisation of machine learning algorithms, i.e., supervised, unsupervised, and reinforcement learnings, basic functionalities are explained, and the most important algorithms are introduced. This is followed by some definitions of simulation and hybrid simulation. Sections 7.3 and 7.4 then show possible combinations of simulation and machine learning. For this purpose, a conceptualisation of the topic will be given, and opportunities, potentials, and application possibilities are shown. Section 7.3, representing the main part of this paper, focuses on hybrid systems modelling. In Sect. 7.4, some examples are presented that show how simulation and machine learning are combined and used outside of hybrid systems simulation studies. Section 7.5 finishes the chapter with some concluding remarks.

7.2 Related Work, Definitions

This section introduces the reader to some important concepts and definitions from the related work, which are essential for further understanding of the contribution. First, the term machine learning is examined in Sect. 2.1. For this purpose, the three central concepts of supervised learning, unsupervised learning, and reinforcement learning are discussed in subchapters. Finally, there are some summarising remarks on the term machine learning. Section 2.2 then defines the most important terms about simulation and hybrid simulation.

7.2.1 Machine Learning

Machine learning is fundamentally concerned with training models using data. The model itself or its decision-making processes are not explicitly created on the basis of a set of rules known in advance, like in modelling and simulation. Rather, the model is trained on data, i.e., on existing empirical values. Through this training process, the model adapts to an implicit set of underlying rules, which can then be used for decision-making [10]. With some of the algorithms available for this purpose, this internal set of rules is sometimes no longer even comprehensible to the user. With such machine learning models, we also speak of black-box algorithms [11]. A recent trend in machine learning therefore is the development of explainable artificial intelligence (XAI). Algorithms used in this context can be applied retrospectively to machine learning models in order to explain their decisions and decision-making paths. This can help to make predictions made by black-box machine learning algorithms more transparent and comparable [12].

In general, the term machine learning is very broad and encompasses a variety of different methods and algorithms. Since the underlying landscape is very heterogeneous, it is therefore not at all easy to clearly define and delimit the term machine learning. The methods differ fundamentally in their purpose and area of application, in the way the underlying models are trained, in their requirements for the data basis and, finally, in their performance, required computing capacity, and scalability.

As a common, generally accepted classification, methods, and algorithms in the context of machine learning are divided into three categories: *supervised learning*, *unsupervised learning*, and *reinforcement learning* [13]. Although these three categories share the basic principle of data-driven learning, they are very different in their mode of operation and application. In the following subsections, these three categories will be explained in more detail.

7.2.1.1 Supervised Learning

Supervised learning is the method that is usually the first thing that comes to mind when talking about machine learning. Supervised learning models are prediction models, i.e., they can predict the most probable output Y given X. In addition, we can also distinguish between regression and classification. In regression, numerical values are predicted, whereas classification deals with the prediction of categorical values [14]. Classification also includes very complex prediction tasks, such as image recognition [15]. In classification in the context of image recognition, a probability is predicted with which a recognised object in that image belongs to a certain class, e.g. a dog, a cat, or a certain letter of the alphabet. Supervised learning requires so-called labelled training data. This means that the data for training the model must contain both X and the respective correct Y [14]. Thus, the model to be trained can then approximate the relationship between X and Y. An image classifier that should differentiate cats and dogs therefore needs training data with images that are pre-labelled with one of those classes [13].

When we think of machine learning and supervised learning, we quickly tend to think of very complex algorithms, such as artificial neural networks. However, even a simple linear regression model, for example, technically falls under the definition of supervised learning and thus is also machine learning [16]. However, this view is certainly controversial, and the boundaries between machine learning and classical statistics are fluid. Traditional regression models have always been used in the context of design of experiments and statistical experimental planning in combination with simulation models. It can be implied here that hybrid systems modelling in the form of simulation and machine learning has therefore always existed, at least in this form, but was simply not considered and referred to from this perspective.

As already mentioned, a large number of algorithms are available, and it is obviously advisable to use the simplest but most suitable algorithm for the given project. The complexity of the algorithms increases with the complexity of the prediction task, and this is usually measured by the size of the input and output vectors as well as the complexity of the relationship between X and Y, i.e., whether this is linear, quadratic, or complex. As an example, an image with a size of 100×100 pixels and three colour channels represents an input vector of size 30,000, i.e., 30,000 X-variables. Though modern image recognition and classification methods, such as convolutional neural networks, usually do not pass the raw pixels to the classifier, but separate essential from non-essential features beforehand, thus reducing the input vector for the actual classification [17]. Other popular algorithms for classification are [18]:

- Support vector machines.
- Logistic regression.
- Decision trees.

For regression, popular algorithms are [19]:

- Support vector regression.

- Regression trees.
- Neural networks for regression.

Among the most powerful prediction algorithms (besides artificial neural networks) are so-called ensemble methods. The most commonly known representatives of these are random forests. A random forest thus consists of a combination of many randomised and uncorrelated classification or regression trees. This has the advantage that a single, small subtree can be generated and trained relatively quickly, and on the other hand, large amounts of data with many factors can be processed efficiently in this way. The result of the prediction is then usually an aggregation of the results of the individual parts of the random forest, for example, by calculating an average value (bagging), or by calculating the prediction sequentially (so-called boosting). In boosting, the next predictor is trained on those test data that performed worst in the predecessor. Thus, the aim is for each predictor in the sequence to compensate for the respective prediction errors of its predecessor. Through this mutual error compensation, boosted trees can also pick up and predict very complex patterns in data [20, 21].

7.2.1.2 Unsupervised Learning

While supervised learning is the training of machine learning models with labelled data, unsupervised learning is the opposite, which is training without labelled data [13]. Therefore, there is no inherent relation between X and Y to learn, because there is no Y. Instead, the model learns from the data itself by looking at its structure and its distinct features. The most commonly known and used algorithms that fall in the category of unsupervised learning are therefore clustering algorithms. Cluster analysis is about grouping a set of objects so that the objects in the same group (cluster) are more similar to each other than objects in other groups. Objects in different clusters should therefore be as dissimilar as possible [13, 14]. Possible applications are very versatile, from customer segmentation to genome sequencing. There are also even more innovative ways to use clustering algorithms, for example, for anomaly detection in IT security: everything that is difficult or impossible to match to a certain cluster is likely an outlier and could be a possible anomaly [14].

There are many different algorithms for clustering, and each has different requirements and areas of application. The most used clustering algorithms certainly are k-means and DB-Scan and derivates of those. There are also specialised algorithms, for example, for very high dimensional data [14, 22].

Artificial neural networks can also be used in an unsupervised way. The idea behind this is to train the model on its own input, hence from input X to the same output X. By deliberately narrowing the layers from outside to the inside the net (in contrast to the size of input and output vector), the resulting bottleneck in the middle forces the network to reduce the data to its core features which are needed to describe and identify the data, hence stripping away unnecessary information and reducing the data to its core features [23]. This method is called autoencoding [23]. As already mentioned, this can be used for clustering, but also for generating new data that is based on the features learned from the training data [23]. For example, generative algorithms like variational autoencoders (VAEs) or generative adversarial networks are the basis behind artificial intelligence image generators apps [24].

Finally, supervised and unsupervised machine learning complement each other very well. When we segment our unlabelled, raw data using a clustering algorithm, those cluster assignments can then act as labels or classes, respectively. A supervised algorithm can then use this newly labelled data to train a prediction model by learning the inherent relation between input data and its cluster allocation [25].

7.2.1.3 Reinforcement Learning

Reinforcement learning is quite different from the two previous machine learning categories. Nevertheless, the basic principle of data-driven learning of rules based on examples also applies here.

In a reinforcement learning setup, an agent can see the state of its environment, and it can perform defined actions in order to interact with it [26]. Based on its actions, it receives a reward, and the agent's fundamental intention is to maximise this reward [26]. This can be seen as some kind of guided trial-and-error optimisation approach. If the agent trains for a sufficient amount of time and therefore has seen enough state/action/reward combinations, it can approximate the relation between reward and actions and therefore can derive the best action for each given state. Over all of the possible states, it can therefore find an optimal policy [23, 26].

In the practical implementation, there are many hurdles and challenges to be considered and there are also many different algorithms that handle the specific execution of that general idea very differently.

Reinforcement learning often is the basis for many approaches in the context of artificial intelligence. Artificial intelligence is concerned with algorithms that attempts to solve tasks that would usually require human intelligence. In this respect, AI naturally makes use of the techniques, methods, and algorithms of machine learning and reinforcement learning in particular [27]. If an agent has learned to find an optimal policy for any given situation, it is obviously able to solve given problems. For example, a robot can learn to walk straight by giving it a reward for each fixed distance or timespan that it is able to walk without falling. This way, after successful training, it will know how to use all of its actuators that it controls in order to walk straight without falling, regardless of the given state of the environment [28]. The same principle applies for example to autonomous driving [29, 30].

7.2.2 Simulation, Hybrid Simulation, and Hybrid Systems Modelling

Simulation and modelling are a well-established approach for the analysis of systems, which can be used to understand, plan, improve, and optimise systems of various types [1]. There are several techniques and worldviews under the umbrella term simulation, including discrete-event simulation (DES), agent-based modelling, system dynamics, or Monte Carlo simulation [9].

With increasing complexity of both the analysis task at hand and of the underlying system, the demands on the simulation model also increase. Hybrid simulation offers a good prospect to combine different simulation techniques in one model in order to increase the functionality and accuracy of the model with regard to the system under scrutiny [9]. For the same purpose, *hybrid systems modelling* also offers the possibility to combine different techniques and methods from other disciplines within one simulation model [9, 31].

According to the majority of simulation modelling literature, the stages of a simulation project can be broken down to the following phases (see [1, 9, 32]):

- *Understanding the real-world problem.*
- *Transferring the real-world problem into a conceptual model.*
- *Implementing a computer executable simulation model.*
- *Conducting experimentation and analysis with the goal of solving the real-world problem.*

Hybrid systems modelling is therefore oriented towards the phases of a simulation study. As soon as hybrid systems modelling is used in at least one phase of a simulation project, this is referred to as a *hybrid modelling and simulation study* [9]. So as soon as machine learning is used in at least one phase of a simulation study, this term applies as well.

Based on the concepts and definitions introduced in this section, the next sections will look in detail at how the combination of machine learning and simulation can be beneficial in a hybrid system. As already described in the introduction, two hypotheses will be examined here. The first hypothesis is that *simulation needs machine learning*, which is the main core of this paper. This is the case when machine learning approaches are used in one or more stages of a simulation study in accordance with hybrid systems modelling. The second hypothesis, that *machine learning needs simulation*, is also addressed in this chapter in later sections.

7.3 Simulation Needs Machine Learning

This section explores the hypothesis that simulation needs machine learning, which is the main core of this chapter. This is the case when machine learning approaches are used in one or more stages of a simulation study in accordance with hybrid systems modelling. First, a conceptualisation is presented in Chap. 3.1, which describes how machine learning is generally applicable in the life cycle of simulation-related activities. The next three subsections then go into more detail on how machine learning and simulation can be combined in each of the following simulation stages: *Input Modelling*, *Model Implementation*, and *Output Analysis*.

7.3.1 Conceptualisation

The four main phases of a simulation study or simulation project have already been introduced in the previous section. According to Wilsdorf et al. [32], each phase contains certain activities, which in turn are iteratively linked across the different phases and can be understood as transitions between the phases. This is also referred to as simulation life cycles: problem definition involves formulating and refining the simulation problem under consideration, which is the transition to conceptual modelling. Between the conceptual modelling and the concrete, executable computer model is the activity of implementation. Between the implementation of the executable computer model and the actual problem solution is the phase of experimenting and analyzing of results. Ideally, this leads to solutions that can be implemented in practice in accordance to the real-world problem, so that the life cycle is closed [32].

As has already been pointed out by Elbattah [8], machine learning algorithms can support conceptual modelling, simulation execution at runtime, and simulation output analysis. In the context of hybrid systems modelling, this paper is focusing on activities that are linked to the executable computer model. These are mainly activities in the area of model implementation and the analysis of results.

The first activity in which machine learning can be used as a support is input modelling. Here, the main task is to provide simulation input data and process-relevant data and to prepare them in such a way that they can be used in the simulation model. The simulation model (in the form of the implemented, executable computer model) itself offers two possibilities for the supporting use of machine learning approaches: enhancing the model using machine learning algorithms and replacing model components with machine learning algorithms.

Finally, machine learning approaches can also be used for output analysis, which is probably the most obvious application. From the evaluation of experiments to simulation-based optimisation, there are many possible applications.

Figure 7.1 shows a short summary of possible applications for machine learning along selected phases of a simulation study.

In the following subsections, an overview of the existing possibilities is given for each of these activities on how machine learning approaches can be used. This includes application potentials, use case examples, and general guidelines. The first activity to be looked at in more detail here is the simulation input modelling.

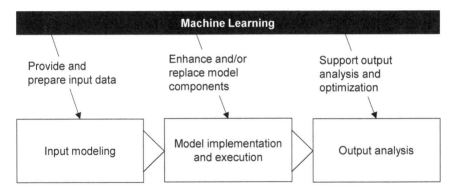

Fig. 7.1 Possible applications for machine learning along selected phases of a simulation study

7.3.2 Machine Learning in Support of Input Modelling

Within the development of a simulation model, input modelling is one of the most challenging tasks. Aspects of input modelling include the determination of probability functions from which samples are drawn, which are usually fitted from historical data [33], but also the initial identification of input processes themselves. This is nowadays usually a very data-driven task, because Big Data generated through data recording systems, sensors, and networks is ubiquitous [33]. The challenge lies in processing this data correctly to use it in a simulation model. As one can easily imagine, this is a task that machine learning algorithms can support very well, since processing data and especially large datasets is at the core of machine learning. In the following, some interesting examples of how machine learning can be used for input modelling are given.

A possible application is the grouping of entities that enter and pass through the system. Clustering algorithms can be used very well to identify categories of entities in historical data from upstream real-world systems and structure them in such a way that they can be reasonably mapped and implemented in the simulation model. These can be, for example, different customer groups or different job types in a manufacturing system. For example, Elbattah and Molloy [34] used k-means clustering to identify pathways of patients in a public healthcare system. More specifically, the paper is about modelling a hip-fracture care scheme for elderly patients. By using clustering, not only were they able to identify reasonable patient cohorts, but they could also derive knowledge about those cohorts by identifying important key factors for cluster allocation. This information could then be used to actually build two simulation models: a system dynamics model for population-level analysis and a more detailed discrete-event model at patient level for simulating and ultimately coordinating patient journeys and points of care [34].

Clustering can also help to improve the estimation of input factors. In the context of large-scale logistics networks, Liu et al. [35] used clustering algorithms to find data patterns among time-dynamic factors in order to convert them into static factors, which make the modelling and implementation of those factors in the actual simulation model much easier [35].

If we go back one step further in the context of input modelling, machine learning can also be used to extract process knowledge in order to obtain a basic impression of the process flow in the system. Akhavian and Behzadan [36] demonstrated this in the context of simulation of construction and construction equipment. They collect data from front-end loaders via various sensors like GPS, accelerometer, and gyroscopes. Those are then assigned to and labelled with certain activities, like *idle*, *scooping*, or *dumping*. They then trained different supervised classifiers, which showed that different actions produce distinct patterns of data in sensors. A trained classifier can then be used to predict those activities, which in turn makes it easier to model duration times for the simulation input modelling. Interestingly, in this case study, k-nearest neighbour classification performed the best, while neural networks showed the worst performance [36].

Of course, machine learning can also be used to help model the needed data distributions for simulation input modelling directly. For example, Li and Ji [37] presented a case study for the simulation of road paving operations to demonstrate how deep learning can be used to predict the parameters of a probabilistic function that can then be implemented in the simulation model. To be more precise, a Bayesian deep neural network was used. This network can be fed with data from multiple data sources including static data like design specs and also dynamic data like time sheets. After training in a supervised fashion, the network was then able to predict the mean and variance parameters for an underlying probabilistic function that could be used in the simulation model to estimate truck hauling durations for different circumstances like weather conditions [37]. Cen et al. [38, 39] investigated how generative neural networks can directly predict and draw samples for simulation input like interarrival times. For this purpose, they developed a framework called Neural Input Modelling (NIM) [38, 39]. To be more precise, they used recurrent neural networks for this purpose. Those networks can process sequence data, like time-based data, as an input, instead of plain vectors. The proposed framework can therefore take existing time-based data, capture even complex stochastic processes from it, and output samples fitting the unknown and underlying distribution [38, 39]. This framework is available as an open-source tool [38].

The next activity within a simulation study that is explored in terms of the applicability of machine learning is simulation model implementation.

7.3.3 Machine Learning in Support of Model Implementation

7.3.3.1 General Considerations

The topic of implementation is considered separately from the topic of input modelling in this chapter, even though the boundaries here are certainly fluid and can be difficult to distinguish from each other. In the previous section, input modelling

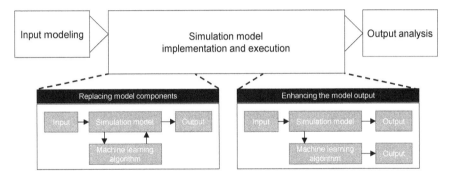

Fig. 7.2 Conceptual overview of simulation life cycle activities that offer machine learning applicability

was understood as the initial extraction of knowledge for the purpose of a better understanding of the process, as well as the more concrete modelling of simulation input in terms of process times, which are usually represented by distribution functions [33].

In this section, the even deeper embedding of machine learning into the simulation model is considered. Here, the machine learning algorithms are integrated into the model in such a way that the actual process flow in the simulation model, which is explicitly created by the modeller, is influenced by machine learning algorithms. Usually, sections of the modelled process flow are replaced by machine learning. The process flows through the simulation model, from input to output, leaves the simulation model, enters a machine learning algorithm, and later enters the simulation model again, which then calculates the final output. Of course, this can be done multiple times, i.e., at different points in the simulation model, or it can be done right at the beginning. In this chapter, this procedure is defined as *replacing of model components*. On the other hand, machine learning can also be used in such a way that it acts in addition to the modelled process flow and thus extends the simulation model by additional output dimensions. The simulation model itself would still be functional without the machine learning algorithm. In the following, this approach is understood as e*nhancing the simulation model*. This is conceptually shown in Fig. 7.2.

In the following two subsections, some possibilities and examples for both approaches are presented.

7.3.3.2 Replacing Model Components

There is a wide range of examples where machine learning algorithms are used to replace internal model components and parts of internal simulation model logic across all types of contexts and use cases. Usually, there are two reasons to do this: data is available, but the underlying process is unknown and can therefore not be modelled explicitly, or it is easier or more time-efficient to use a machine learning algorithm for certain calculations. For example, in the context of data-driven, automated generation of simulation models, the use of machine learning algorithms is indispensable to automatically detect and reproduce the system behaviour purely from data, without external process knowledge of humans, who may have domain knowledge and can explicitly define the knowledge-based rules like decision rules. This is required because the automatic model generation is obviously purely data-driven and, as the name suggests, should run automatically. Only in rare cases, however, such decision rules are explicitly stored in upstream systems, such as enterprise resource planning systems. Therefore, it makes sense to approximate these decision rules using supervised machine learning [40–42]. Bergmann et al. [41, 43] used supervised learning to approximate priority dispatching rules for manufacturing jobs. These dispatching rules, also called sequence rules, are used to determine which job, from a set of jobs waiting in a queue, is to be processed next on a machine. To achieve this, the decision problem was broken down into pairwise comparisons between all waiting jobs. Although this may require many one-on-one comparisons, it has the advantage that it can be performed very flexibly for basically any number of jobs and job types. The actual decision problem in the sense of classification is thus transformed into a simple binary decision of the form *'prefer one job over the other is either true or false'*. For the training data, all available historical data from upstream systems can be used, from individual job properties, such as type of job, delivery time, and so on, to system-wide properties, such as the state of the machines or the time of day. The decision made in each case can be derived from time stamps as to when which job was waiting in the queue or was on the machine. Several common classification algorithms were tested for this purpose, with Cart's decision tree algorithm performing best overall regarding accuracy [41, 43].

Another popular field of simulation modelling, where the internal logic of the simulation model can be replaced by machine learning algorithms, is agent-based modelling [44]. The idea here is to model agents in a network where they can communicate and exchange information, as is usually the case in agent-based modelling, but individual decisions made by agents are completely or partially conducted by machine learning algorithms [45]. For example, Negahban [46] used this approach to enhance consumer diffusion models that are used to model consumer marketplace systems. Here, an agent (the consumer) is influenced by his surroundings and interacts with it via word-of-mouth and decides on whether or not to adopt a product. The individual decision of each undecided consumer is predicted by an artificial neural network that was trained on training products [46]. Dehghanpour et al. [47] used dynamic Bayesian networks, a form of supervised learning that is based on the inference of probability distributions, to implement agents' decision on market price bidding in electrical energy markets [47].

The simulation model and machine learning algorithms can also be arranged in a way where the output of one serves as input to the other in order to get the final output. For example, Mui et al. [48] used this approach to build a hybrid simulation model that can predict energy consumption for cooling residential buildings. They trained an artificial neural network that was able to predict the hourly envelope heat gain of buildings according to a broad range of input parameters like outdoor temperature, time of the year, air temperature, and the various features of the building. The predicted measure of heat gain is then fed into a simulation model for energy consumption that can then predict consumption of cooling energy as the final output of the hybrid model [48]. In a work by Qiao and Yunusa-Kaltungo [49], which also is concerned with the simulation of energy consumption of buildings, the arrangement of simulation model and machine learning algorithm is the other way round. An agent-based model was implemented that represents occupants of a building to simulate their behaviour. This simulation model then feeds into a machine learning model, which then can predict the energy consumption for that building [49].

7.3.3.3 Enhancing the Simulation Model

The simulation of energy consumption also serves as a good exemplary use case where machine learning can enhance an existing simulation model with additional output. This is particularly beneficial in the context of simulation of production and logistics systems, where energy consumption and therefore carbon footprint evaluations get increasingly important. With such approaches, existing models of production systems can be extended to provide output for energy consumption. For this purpose, Wörrlein and Strassburger [50, 51] present an approach using sequence-to-sequence models, which are artificial neural networks that can translate one sequence into another type of sequence, based on recurrent neural networks and autoencoder architectures. The idea is to give the model any kind of production relevant information, like for example the numerical control code for a CNC-drilling machine. Through training the model with historical recorded energy consumption data, the model can then predict the time series for energy consumption for that machine and specific job. This can then be integrated into every machine in the simulation model in order to predict the combined energy consumption for the whole production system as a time series [50, 51]. Zhou et al. [52] combined the simulation model of a production system and the simulation model of an accompanying periphery system that represents heating, cooling, and lighting. Both simulation models can stand alone but were used to feed into a machine learning model that can predict energy demand for heating and cooling [52].

The next stage in the simulation life cycle that is explored in terms of machine learning applicability is output analysis. This stage unquestionably offers the greatest potential for combination with machine learning and the largest variety of application possibilities. This can range from simple regression models all the way to large data farming projects and complex optimisation algorithms.

7.3.4 Machine Learning in Support of Output Analysis

This section covers the use of machine learning for output analysis. The machine learning algorithm uses data generated by the simulation model after simulation runs or simulation experiments. This process can either be designed as a one-way stream for the analysis of experiments that have been carried out once, or as an iterative loop, for example, in the context of simulation-based optimisation, where the simulation model and the machine learning algorithm exchange data multiple times. This is certainly one of the most obvious combinations of simulation and machine learning and has therefore been widely used for a long time. In the following, some selected examples will show how machine learning can be used in the context of simulation output analysis.

7.3.4.1 From Regression Analysis to Data Farming

As already mentioned, the modelling of linear models, e.g. in the context of simulation analysis, has always been an integral part of simulation-based system analysis. If a regression model is regarded as a machine learning algorithm, this could technically be called hybrid systems modelling. Of course, there are also other machine learning algorithms that can be used for this purpose. Decision trees or regression trees are often used in this way since they are easy to apply and can be evaluated rather intuitively. A decision tree can be trained on the experiment data generated by a simulation model by defining the factors of the experiment plan as nodes of the tree and the output of the experiments at its leaves. This then allows if–then rules to be derived at the branches of the tree. For example, Painter et al. [53] use decision trees to derive rules about the relationship between maintenance activities and the maintenance costs incurred using data generated by simulation models in the context of life cycle simulation of aircraft engines [53]. Tang et al. [54] use decision trees to extract rules concerning the damage effect of tanks in battle simulations. Giabbanelli [55] applies classification trees to simulated communication network traffic in order to improve routing in such networks.

One idea behind using supervised learning on data generated by simulation models is to not necessarily use the trained model for prediction, but for learning rules, i.e., for knowledge discovery. The inherent relationship between the factors of the simulation and its output can be better understood and analysed with the help of trained machine learning models, because they always represent some form of approximation of this relation. One popular approach that has taken this basic idea and developed it further is called data farming [56]. Data farming is concerned with the exploration of unknown and surprise in a simulation model [57], which means that data farming is not necessarily concerned to find the one best solution, like in an optimisation setting, but rather with looking for a better understanding of the system in general [58]. This idea was originally developed in the context of combat simulations, where stakeholders were not satisfied with the value and insight of a few simulated scenarios, and therefore preferred to strive for a more complete overview and coverage of the model's response surface, representing a broad range of possible factor value combinations [59]. To accomplish this, obviously a lot of data, specifically a lot of simulation experiments are needed, but smart experiment design methods that have been developed in the context of data farming can help to grow the data in a very efficient and controlled way, which is where the farming-metaphor comes from [60–62]. Still, for complex models and large response surfaces, a large amount of simulation data is needed to be generated [63]. Therefore, an automated, algorithm-assisted analysis is always necessary and an integral part of any data farming study. Machine learning algorithms are therefore ideally suited for this purpose. Feldkamp et al. [64] took a deep dive into the analysis side of data farming and came up with a framework on how to use machine learning for the analysis of large-scale simulation output data for the purpose of knowledge generation. The basic idea here is that if several output dimensions simultaneously and in dependence on each other are under consideration, they need to be structured first. For this purpose, one can use clustering algorithms, i.e., unsupervised machine learning. If there are recognised structures represented by the clusters, these can be fed into classification algorithms in the second step. Classification algorithms (supervised learning) can approximate the connection between factors and the given clusters. In the third step, knowledge and conclusions from these approximated relationships can be derived [64].

Data farming studies have been carried out in a wide range of possible applications, for example, industrial processes [65], military applications [66], logistics [67], or manufacturing simulations [68].

However, a major challenge in using machine learning algorithms to generate knowledge is that many of them, especially the most powerful and accurate algorithms, i.e., those, that can approximate even the most complex relationships, are usually black boxes, as already mentioned in Sect. 2.1. Therefore, the most recent research in the field of data farming output analysis is concerned with the application of explainable AI for this purpose. Here, XAI algorithms are applied on top of black-box machine learning algorithms in order to make their internal decision rules visible and to derive knowledge about the system from them, which yields great potential in terms of automated, AI-assisted simulation output analysis [69–72].

Finally, application possibilities for the verification and validation of simulation models are relatively rare. However, unsupervised machine learning, especially clustering, can also be used for validation of the model, analogous to output analysis. Because a clustering algorithm can structure and group the simulation output, outliers and therefore unexpected and possibly erroneous or faulty model behaviour can be identified relatively quickly.

7.3.4.2 Metamodelling and Optimisation

Metamodelling and optimisation is another very popular use case for applying supervised machine learning algorithms to simulation output data in order to approximate the relation between input and output [73]. However, the approximation capability of the machine learning model is usually not used to derive knowledge from it, but rather to predict non-simulated, i.e., unknown, data. The reason for this is that a machine learning model usually can produce new results much faster than a potentially very complex simulation model, at least after the training is complete. If, for example, a large number of simulation runs are required for an optimisation process, this can be replaced by implementing a machine learning model [73–75]. There is plenty of work using all different kinds of supervised machine learning algorithms for the purpose of metamodelling, including regression [76], random forests and boosted trees [77, 78], or artificial neuronal networks [78, 79].

Machine learning can also be used for the optimisation itself. Reinforcement learning in particular has been gaining popularity for some time. A reinforcement learning agent trains using a directed try-and-error approach. The agent receives a reward for the action it chooses, depending on the state of the environment. After sufficient training time, it learns to maximise its reward accordingly, i.e., it learns to perform the best sequence of actions. This can be seen as the optimal policy for a given state regarding its reward maximisation. By associating the reward with the optimisation goal, reinforcement learning can be utilised for simulation-based optimisation. There are a lot of examples and use cases for this concept. Rabe and Dross [80] presented an approach for decision support in logistics networks, where the reinforcement agent can recommend optimal policies regarding whether to stock keep inventory on site or to replenish goods on demand [80]. Feldkamp et al. used reinforcement learning to dynamically plan the best sequence of stations that autonomous guided transport vehicles in a modular production system should approach to complete a job as quickly as possible [81]. Zhang et al. used reinforcement learning for optimising job shop scheduling [82], as well as for optimising batch sizes [83]. Not only reinforcement learning, but also other machine learning methods can be used for simulation-based optimisation. Feldkamp et al. [84] used generative adversarial networks for robustness optimisation. In robustness optimisation, two experimental plans with decision and system factor configurations are crossed to calculate a robustness measure. For this purpose, two generative machine learning algorithms were used to iteratively generate experiment plans in a turn-by-turn approach to ultimately find the best robust and stable system configuration [84].

In summary, this section gave a detailed look on how machine learning can be combined with simulation in a hybrid simulation system. This was considered from the perspective of different phases of a simulation study, i.e., input modelling, model implementation, and output analysis. In addition, there are other possibilities to combine machine learning and simulation outside the defined framework of phases in simulation studies. This will be considered in more detail in the next section under the hypothesis that *machine learning needs simulation*.

7.4 Machine Learning Needs Simulation

Outside of the well-defined phases of a simulation study, there is also the reverse case: simulation models are used in machine learning and artificial intelligence projects. Here, the simulation is not the leading system, but rather the machine learning algorithm uses the simulation as a supporting tool within a machine learning project.

The reasons for using simulation are probably well known to every simulation expert:

- If there is not enough data available.
- If experimenting on the original model is too slow, expensive, and impractical or simply not possible.
- A reliable and safe sandbox is needed in which things can be tested and tried out safely and risk-free.

In this section, a few examples of this are given.

First, reinforcement learning can obviously be used in this way. As already explained in the previous section, a reinforcement learning agent trains through a directed try-and-error procedure. The agent receives a reward for the action it chooses, depending on the state of the environment. After sufficient training time, it learns to maximise its reward accordingly, i.e., it learns to perform the best sequence of actions, which represents the best strategy, depending on the state of its environment. Therefore, it is advisable to initially provide the agent with a simulation model as a playing/training field. This not only helps to speed up the training, for example through parallelisation, but also to find and correct possible errors and problems, as well as to optimise the hyperparameters and conditions of the training.

In fact, in almost every thinkable (real-world) scenario, the agent must first be trained using a simulation model that provides a controlled and safe training environment, before it can be released into the real world, for example, for autonomous driving [29, 30], autonomously moving robots [85, 86], or human–robot collaboration [87].

Another possible application is the use of simulation models as accurate, data-driven, and real-time updating representations of a real object or system, the so-called digital twin [88]. These can be used to generate data as quickly and efficiently as possible, which can then be used to train all kinds of machine learning algorithms.

Digital twins will play an increasingly important role in the future, both in the planning and manufacture of products and their production lines, production facilities and factories, as well as within the life cycle of the product itself. The digital twin, which ultimately represents a simulation model at its core, produces masses of data that can only be meaningfully processed using machine learning approaches. This is therefore a prime example of the combination of simulation and machine learning. A direct use case of this is the application of digital twins for predictive maintenance. An interesting example of this is the work of Xu et al. [89]. Here, a digital twin of a production system that was still in the planning stage was used to train a machine learning algorithm for the purpose of predictive maintenance. Predictive maintenance is an approach of using machine learning algorithms to predict when a machine is going to fail to plan maintenance more cost-efficiently. The goal is to predict the timing for necessary maintenance, i.e., shortly before a component wears out, for example. This saves costs for regular, interval-based maintenance on the one hand and costs for repair in the event of unforeseen failures and defects on the other hand. In the given example, after the planning and simulation stage and once the real system was up and running, the trained machine learning model could then be rolled out alongside the real system [89].

7.5 Conclusion and Outlook

This chapter took a detailed look at the possible combinations of the two disciplines *modelling and simulation* and *machine learning*. Numerous examples were given to show how both sides can enrich each other and be combined into a whole that is greater than the mere sum of its parts. The concern sometimes expressed in parts of the simulation research community, that machine learning will one day make simulation obsolete, is unfounded, as was shown in this paper. However, it must be noted that machine learning algorithms have become a very powerful and popular tool, and simulation projects, as far as they have reached a certain level of complexity, will probably no longer be able to be implemented without the combination with machine learning in some form or at some stage of the simulation study. However, this by no means makes simulation unnecessary. In this chapter, it was also shown that not only simulation requires machine learning, but that both disciplines are mutually dependent on each other, as illustrated in Fig. 7.3.

Many algorithms and applications in the field of machine learning are also dependent on simulation models. For large-scale, industry-driven simulation projects, it is therefore almost mandatory to use machine learning. On the one hand, to make the best possible use of the available data on the input side during modelling. On the other hand, the evaluation of simulation results can be made more efficiently, as was shown in this chapter based on several examples. It is also worth looking at machine learning for simulation-based optimisation tasks. Machine learning methods, such as reinforcement learning, can nowadays offer a worthwhile alternative or supplement to traditional optimisation methods.

Ultimately, both disciplines will certainly continue to grow together in the future, as well as their dependence on each other, at least for some specific objectives. The concept of the digital twin highlights this path in an exemplary manner. The research and research community in both disciplines should be aware of this. On the other hand, there should also be an understanding of both sides from a practical or user perspective. Data-driven models and data-driven analytics have come a long way and are here to stay in the era of Big Data, whether using simulation, machine learning, or a combination of both. The most important goal is to be able to use the appropriate, required tool for the project at hand, in order to ultimately maximise the benefit for each specific, individual application.

7 Application of Machine Learning Within Hybrid Systems Modelling

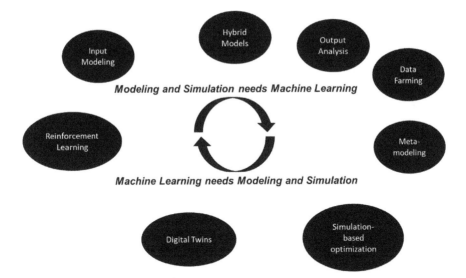

Fig. 7.3 Mutual dependency of modelling and simulation and machine learning through various application possibilities

Of course, there are also some challenges in combining machine learning and simulation. One technical challenge is certainly the creation of a suitable interface between the machine learning code and the simulator. Here, the general settings, requirements, and the possibilities are so diverse that there is no one-size-fits-all solution (yet). This means that, as a rule, a coded, bespoke solution usually needs to be implemented in most practical applications. Some of the modern off-the-shelf simulators already offer machine learning functionalities. However, integration into the simulation model must usually be done by hand and with the help of coding manually. This form of direct integration into simulation software will certainly increase in the future. In addition to this technical barrier, another challenge is centred on the human component, on several levels. First, there may be reservations on the part of the simulation community, simulation experts, and modellers about using machine learning algorithms in a simulation project. This may also stem from the fear that machine learning will make the discipline of building simulation models completely obsolete in the future. This fear has been disproved in this chapter, as it has been shown by means of many examples that, on the contrary, the two disciplines complement each other very well or are even dependent on each other in some application scenarios. Furthermore, some simulation experts may think that they lack the necessary knowledge and skills to use machine learning algorithms. In this respect, it can be stated that there are now many very good, freely available, and also beginner-friendly packages and libraries. Supporting material and tutorials are also numerous and easy to find online. Finally, the integration of machine learning should also be considered and promoted in teaching and training in the discipline of simulation and modelling more effectively.

References

1. Law AM (2003) How to conduct a successful simulation study. In: Chick S, Sanchez PJ, Ferrin D, Morrice DJ (eds) 2003 winter simulation conference; 7 Dec–10 Dec; New Orleans, LA, USA. Institute of Electrical and Electronics Engineers, Piscataway, New Jersey, pp 66–70. https://doi.org/10.1109/WSC.2003.1261409
2. Lucas TW, Kelton WD, Sánchez PJ, Sanchez SM, Anderson BL (2015) Changing the paradigm: simulation, now a method of first resort. Nav Res Logist 62:293–303. https://doi.org/10.1002/nav.21628
3. Law AM (2007) Simulation modeling and analysis, 4th edn. McGraw-Hill, Boston, Mass
4. Schultz MG, Betancourt C, Gong B, Kleinert F, Langguth M, Leufen LH et al (2021) Can deep learning beat numerical weather prediction? Philos Trans A Math Phys Eng Sci 379:20200097. https://doi.org/10.1098/rsta.2020.0097
5. Calzolari G, Liu W (2021) Deep learning to replace, improve, or aid CFD analysis in built environment applications: a review. Build Environ 206:108315. https://doi.org/10.1016/j.buildenv.2021.108315
6. von Rueden L, Mayer S, Sifa R, Bauckhage C, Garcke J (2020) Combining machine learning and simulation to a hybrid modelling approach: current and future directions. In: Berthold MR, Feelders A, Krempl G (eds) Advances in intelligent data analysis XVIII. Springer International Publishing, Cham, pp 548–560. https://doi.org/10.1007/978-3-030-44584-3_43
7. Giabbanelli PJ (2019) Solving challenges at the interface of simulation and big data using machine learning. In: Haas PJ, Mustafee N, Rabe M, Bae K-HG, Szabo C, Lazarova-Molnar S (eds) 2019 winter simulation conference; 8 Dec 2019–11 Dec 2019; National Harbor, MD, USA. IEEE Inc., pp 572–583.
8. Elbattah M (2019) How can machine learning support the practice of modeling and simulation?—A review and directions for future research. In: 2019 IEEE/ACM 23rd International Symposium on Distributed Simulation and Real Time Applications (DS-RT); 7 Oct 2019–9 Oct 2019; Cosenza, Italy. IEEE, pp 1–7. https://doi.org/10.1109/DS-RT47707.2019.8958703
9. Mustafee N, Powell JH (2018) From hybrid simulation to hybrid systems modelling. In: Rabe M, Juan AA, Mustafee N, Skoogh A (eds) 2018 Winter Simulation Conference; 9 Dec 2013 to 12 Dec 2012; Gothenburg, Sweden. Institute of Electrical and Electronics Engineers, Piscataway, New Jersey, pp 1430–1439
10. Abu-Mostafa YS, Magdon-Ismail M, Lin H-T (2012) Learning from data: a short course. AMLbook.com, S.l.
11. Delibasic B, Vukicevic M, Jovanovic M, Suknovic M (2013) White-box or black-box decision tree algorithms: which to use in education? IEEE Trans Educ 56:287–291. https://doi.org/10.1109/TE.2012.2217342
12. Adadi A, Berrada M (2018) Peeking inside the black-box: a survey on explainable artificial intelligence (XAI). IEEE Access 6:52138–52160. https://doi.org/10.1109/ACCESS.2018.2870052
13. Bishop CM (2009) Pattern recognition and machine learning, 8th edn. Springer, New York
14. Han J, Kamber M (2006) Data mining: concepts and techniques, 2nd edn. Elsevier; Morgan Kaufmann, Amsterdam, Boston, San Francisco, CA
15. Krizhevsky A, Sutskever I, Hinton GE (2017) ImageNet classification with deep convolutional neural networks. Commun ACM 60:84–90. https://doi.org/10.1145/3065386
16. Hastie T, Tibshirani R, Friedman JH (2009) The elements of statistical learning: data mining, inference, and prediction, 2nd edn. Springer, New York, NY
17. Saha S (2018) A comprehensive guide to convolutional neural networks—the ELI5 way. https://towardsdatascience.com/a-comprehensive-guide-to-convolutional-neural-networks-the-eli5-way-3bd2b1164a53. Accessed 18 Sep 2023
18. scikit-learn developers (2014) Classifier comparison. https://scikit-learn.org/0.15/auto_examples/plot_classifier_comparison.html. Accessed 2023
19. Han J, Kamber M, Pei J (2012) Data mining concepts and techniques, 3rd edn. Elsevier/Morgan Kaufmann, Amsterdam

20. Opitz D, Maclin R (1999) Popular ensemble methods: an empirical study. JAIR 11:169–198. https://doi.org/10.1613/jair.614
21. Sagi O, Rokach L (2018) Ensemble learning: a survey. WIREs Data Min Knowl. https://doi.org/10.1002/widm.1249
22. Chen F, Deng P, Wan J, Zhang D, Vasilakos AV, Rong X (2015) Data mining for the internet of things: literature review and challenges. Int J Distrib Sens Netw 11:431047. https://doi.org/10.1155/2015/431047
23. Goodfellow I, Bengio Y, Courville A (2016) Deep learning. MIT Press, Cambridge, Massachusetts, London, England
24. Goodfellow I, Pouget-Abadie J, Mirza M, Xu B, Warde-Farley D, Ozair S et al (2020) Generative adversarial networks. Commun ACM 63:139–144. https://doi.org/10.1145/3422622
25. Bandyopadhyay S, Saha S (2013) Unsupervised classification: similarity measures, classical and metaheuristic approaches, and applications. Springer, Berlin, Heidelberg
26. Sutton RS, Barto A (2018) Reinforcement learning: an introduction, 2nd edn. The MIT Press, Cambridge, MA, London, England
27. Kühl N, Goutier M, Hirt R, Satzger G (2019) Machine learning in artificial intelligence: towards a common understanding, pp 5236–5245
28. Schuitema E, Wisse M, Ramakers T, Jonker P (2010) The design of LEO: a 2D bipedal walking robot for online autonomous reinforcement learning. In: 2010 IEEE/RSJ international conference on intelligent robots and systems (IROS 2010); 18 Oct 2010–22 Oct 2010. IEEE, Taipei, pp 3238–3243. https://doi.org/10.1109/IROS.2010.5650765
29. Kiran BR, Sobh I, Talpaert V, Mannion P, Sallab AAA, Yogamani S, Perez P (2022) Deep reinforcement learning for autonomous driving: a survey. IEEE Trans Intell Transp Syst 23:4909–4926. https://doi.org/10.1109/TITS.2021.3054625
30. Zhang P, Xiong L, Yu Z, Fang P, Yan S, Yao J, Zhou Y (2019) Reinforcement learning-based end-to-end parking for automatic parking system. Sensors. https://doi.org/10.3390/s19183996
31. Brailsford SC, Eldabi T, Kunc M, Mustafee N, Osorio AF (2019) Hybrid simulation modelling in operational research: a state-of-the-art review. Eur J Oper Res 278:721–737. https://doi.org/10.1016/j.ejor.2018.10.025
32. Wilsdorf P, Heller J, Budde K, Zimmermann J, Warnke T, Haubelt C et al (2022) A model-driven approach for conducting simulation experiments. Appl Sci 12:7977. https://doi.org/10.3390/app12167977
33. Cheng R (2017) History of input modeling. In: Chan V, D'Ambrogio A, Zacharewicz G, Mustafee N (eds) 2017 winter simulation conference; 3 Dec–6 Dec; Las Vegas. Institute of Electrical and Electronics Engineers; 2017. p. 181–201.
34. Elbattah M, Molloy O, Zeigler BP (2018) Designing care pathways using simulation modeling and machine learning. In: Rabe M, Juan AA, Mustafee N, Skoogh A (eds) 2018 winter simulation conference; 9 Dec–12 Dec; Gothenburg, Sweden. Institute of Electrical and Electronics Engineers, Piscataway, New Jersey, pp 1452–1463
35. Liu Y, Yan L, Liu S, Jiang T, Zhang F, Wang Y, Wu S (2020) Enhancing input parameter estimation by machine learning for the simulation of large-scale logistics networks. In: Bae K-H, Feng B, Kim S, Lazarova-Molnar Z, Zheng Z, Roeder T, Thiesing R (eds) 2020 winter simulation conference; 14 Dec 2020–18 Dec 2020; Orlando, FL, USA. Institute of Electrical and Electronics Engineers, Piscataway, New Jersey, pp 608–619
36. Akhavian R, Behzadan AH (2014) Construction activity recognition for simulation input modeling using machine learning classifiers. In: Tolk A, Diallo SD, Ryzhov IO, Yilmaz L, Buckley S, Miller JA (eds) 2014 winter simulation conference; 7 Dec–10 Dec; Savannah GA. Institute of Electrical and Electronics Engineers, Piscataway, New Jersey, pp 3296–3307
37. Li Y, Ji W (2019) Enhanced input modeling for construction simulation using Bayesian deep neural net-works. In: Haas PJ, Mustafee N, Rabe M, Bae K-HG, Szabo C, Lazarova-Molnar S (eds) 2019 winter simulation conference; 8 Dec–11 Dec; National Harbor, MD. Institute of Electrical and Electronics Engineers, Piscataway, New Jersey, pp 2978–2985

38. Cen W, Herbert EA, Haas PJ (2020) NIM: modeling and generation of simulation inputs via generative neural networks. In: Bae K-H, Feng B, Kim S, Lazarova-Molnar Z, Zheng Z, Roeder T, Thiesing R (eds) 2020 winter simulation conference; 14 Dec 2020–18 Dec 2020; Orlando, FL, USA. Institute of Electrical and Electronics Engineers, Piscataway, New Jersey, pp 584–595.
39. Cen W, Haas PJ (2023) NIM: generative neural networks for automated modeling and generation of simulation inputs. ACM Trans Model Comput Simul 33:1–26. https://doi.org/10.1145/3592790
40. Zhang L, Hu Y, Tang Q, Li J, Li Z (2021) Data-driven dispatching rules mining and real-time decision-making methodology in intelligent manufacturing shop floor with uncertainty. Sensors. https://doi.org/10.3390/s21144836
41. Bergmann S, Feldkamp N, Strassburger S (2015) Approximation of dispatching rules for manufacturing simulation using data mining methods. In: Yilmaz L, Chan WKV, Moon I, Roeder TMK, Macal C, Rossetti MD (eds) 2015 winter simulation conference; 07 Dec–09 Dec; Huntington Beach. Institute of Electrical and Electronics Engineers, Piscataway, New Jersey, pp 2329–2340
42. Lugaresi G, Matta A (2021) Discovery and digital model generation for manufacturing systems with assembly operations. In: 2021 IEEE 17th international conference on automation science and engineering (CASE); 23 Aug 2021–27 Aug 2021; IEEE, Lyon, France, pp 752–757. https://doi.org/10.1109/CASE49439.2021.9551479
43. Bergmann S, Feldkamp N, Strassburger S (2017) Emulation of control strategies through machine learning in manufacturing simulations. JOS 11:38–50. https://doi.org/10.1057/s41273-016-0006-0
44. Platas-López A, Guerra-Hernández A, Quiroz-Castellanos M, Cruz-Ramírez N (2023) Agent-based models assisted by supervised learning: a proposal for model specification. Electronics 12:495. https://doi.org/10.3390/electronics12030495
45. Jäger G (2019) Replacing rules by neural networks a framework for agent-based modelling. BDCC. 3:51. https://doi.org/10.3390/bdcc3040051
46. Negahban A (2017) Neural networks and agent-based diffusion models. In: Chan V, D'Ambrogio A, Zacharewicz G, Mustafee N (eds) 2017 winter simulation conference; 3 Dec–6 Dec; Las Vegas. Institute of Electrical and Electronics Engineers, Piscataway, New Jersey, pp 1407–1418
47. Dehghanpour K, Nehrir MH, Sheppard JW, Kelly NC (2016) Agent-based modeling in electrical energy markets using dynamic bayesian networks. IEEE Trans Power Syst 31:4744–4754. https://doi.org/10.1109/TPWRS.2016.2524678
48. Mui KW, Wong LT, Satheesan MK, Balachandran A (2021) A hybrid simulation model to predict the cooling energy consumption for residential housing in Hong Kong. Energies 14:4850. https://doi.org/10.3390/en14164850
49. Qiao Q, Yunusa-Kaltungo A (2023) A hybrid agent-based machine learning method for human-centred energy consumption prediction. Energy Build 283:112797. https://doi.org/10.1016/j.enbuild.2023.112797
50. Woerrlein B, Strassburger S (2020) A method for predicting high-resolution time series using sequence-to-sequence models. In: Bae K-H, Feng B, Kim S, Lazarova-Molnar Z, Zheng Z, Roeder T, Thiesing R (eds) 2020 winter simulation conference; 14 Dec 2020–18 Dec 2020; Orlando, FL, USA. Institute of Electrical and Electronics Engineers, Piscataway, New Jersey, pp 1075–1086
51. Woerrlein B, Strassburger S (2020) On the usage of deep learning for modelling energy consumption in simulation models. SNE 30:165–174. https://doi.org/10.11128/sne.30.tn.10536
52. Zhou B, Frye M, Sander C, Schmitt RH (2019). A hybrid simulation tool to improve the energy efficiency in production environment. In: 2019 IEEE international conference on systems, man and cybernetics (SMC); 06 Oct 2019–09 Oct 2019. IEEE, Bari, Italy, pp 2103–2108. https://doi.org/10.1109/SMC.2019.8914514

53. Painter MK, Erraguntla M, Hogg GL, Beachkofski B (2006) Using simulation, data mining, and knowledge discovery techniques for optimized aircraft engine fleet management. In: Perrone LF, Wieland FP, Liu J, Lawson BG, Nicol DM, Fujimoto RM (eds) 2006 winter simulation conference; 3 Dec 2006–6 Dec 2006; Monterey, CA. Institute of Electrical and Electronics Engineers, Piscataway, New Jersey
54. Tang Z, Xue Q, Zhao M, Wei Y (2009) Decision tree algorithm for tank damage analysis in combat simulation tests. In: Cui J (ed) ICEMI'2009; 16–19.8; Beijing, China (3-830-3-835). https://doi.org/10.1109/ICEMI.2009.5274185
55. Giabbanelli PJ (2010) Impact of complex network properties on routing in backbone networks. In: 2010 IEEE Globecom workshops; 5–10.12; Miami, FL, USA. IEEE, Piscataway, N.J., pp 389–393. https://doi.org/10.1109/GLOCOMW.2010.5700347
56. Horne GE (2001) Beyond point estimates: operational synthesis and data farming. Maneuver Warfare Sci 2001:1–8
57. Horne GE, Meyer TE (2005) Data farming: discovering surprise. In: Kuhl ME, Steiger NM, Armstrong FB, Joines JA (eds) 2005 winter simulation conference; 4 Dec 2005–7 Dec 2005; Orlando, FL. USA. Institute of Electrical and Electronics Engineers, Piscataway, New Jersey, pp 1082–1087
58. Horne G, Schwierz K-P (2016) Summary of data farming. Axioms 5:8. https://doi.org/10.3390/axioms5010008
59. Schubert J, Johansson R, Hörling P. Skewed distribution analysis in simulation-based operation planning. In: Carson N, Williams A (eds) Ninth operations research and analysis conference; 22 Oct 2015–23 Oct 2015; Ottobrunn, Germany
60. Sanchez SM (2020) Data farming: methods for the present, opportunities for the future. ACM Trans Model Comput Simul 30:1–30. https://doi.org/10.1145/3425398
61. Sanchez SM, Wan H (2015) Work smarter, not harder: a tutorial on designing and conducting simulation experiments. In: Yilmaz L, Chan WKV, Moon I, Roeder TMK, Macal C, Rossetti MD (eds) 2015 winter simulation conference; 07 Dec–09 Dec; Huntington Beach. Institute of Electrical and Electronics Engineers, Piscataway, New Jersey, pp 1795–1809
62. Sanchez SM, Wan H (2009) Better than a petaflop: the power of efficient experimental design. In: Rossetti MD, Hill RR, Johansson B, Dunkin A, Ingalls RG (eds) 2009 winter simulation conference; 13 Dec 2009–16 Dec 2009; Austin, TX. Institute of Electrical and Electronics Engineers, Piscataway, New Jersey, pp 60–74. https://doi.org/10.1109/WSC.2009.5429316
63. Feldkamp N, Bergmann S, Strassburger S (2015) Knowledge discovery in manufacturing simulations. In: Taylor SJE, Mustafee N, Son Y-J (eds) 3rd ACM SIGSIM conference on principles of advanced discrete simulation; 10 June 2015–12 June 2015; London, UK. Association for Computing Machinery, New York, New York, pp 3–12. https://doi.org/10.1145/2769458.2769468
64. Feldkamp N, Bergmann S, Strassburger S (2020) Knowledge discovery in simulation data. ACM Trans Model Comput Simul 30:1–25. https://doi.org/10.1145/3391299
65. Genath J, Bergmann S, Feldkamp N, Spieckermann S, Stauber S (2022) Development of an integrated solution for data farming and knowledge discovery in simulation data. SNE 32:121–126. https://doi.org/10.11128/sne.32.tn.10611
66. Horne GE, Schwierz K-P (2008) Data farming around the world overview. In: Mason SJ, Hill RR, Mönch L, Rose O, Jefferson T, Fowler JW (eds) 2008 winter simulation conference; 07 Dec 2008–10 Dec 2008; Miami, FL. Institute of Electrical and Electronics Engineers, Piscataway, New Jersey, pp 1442–1447. https://doi.org/10.1109/WSC.2008.4736222
67. Hunker J, Scheidler AA, Rabe M, van der Valk H (2021) A new data farming procedure model for a farming for mining method in logistics networks. In: Kim S, Feng B, Smith K, Masoud S, Zheng Z, Szabo C, Loper M (eds) 2021 winter simulation conference; 13 Dec 2021–17 Dec 2021; Phoenix, AZ, USA. Institute of Electrical and Electronics Engineers, Piscataway, New Jersey
68. Lechler T, Sjarov M, Franke J (2021) Data farming in production systems—a review on potentials, challenges and exemplary applications. Proc CIRP 96:230–235. https://doi.org/10.1016/j.procir.2021.01.156

69. Feldkamp N, Strassburger S (2023) From explainable AI to explainable simulation: using machine learning and XAI to understand system robustness. In: Loper M, Jin D, Carothers CD (eds) SIGSIM-PADS '23: SIGSIM conference on principles of advanced discrete simulation; 21 June 2023–23 June 2023; Orlando FL USA. ACM, New York, NY, USA, pp 96–106. https://doi.org/10.1145/3573900.3591114
70. Serre L, Amyot-Bourgeois M (2022) An application of automated machine learning within a data farming process. In: Feng B, Pedrielli G, Peng Y, Shashaani S, Song E, Corlu C et al (eds) 2022 winter simulation conference (WSC); 11 Dec 2022–14 Dec 2022; Singapore. Institute of Electrical and Electronics Engineers, Piscataway, New Jersey, pp 2013–2024. https://doi.org/10.1109/WSC57314.2022.10015513
71. Serre L, Amyot-Bourgeois M, Astles B (2021) Use of shapley additive explanations in interpreting agent-based simulations of military operational scenarios. In: 2021 annual modeling and simulation conference (ANNSIM); 19 July 2021–22 July 2021.: IEEE, Fairfax, VA, USA, pp 1–12. https://doi.org/10.23919/ANNSIM52504.2021.9552151
72. Feldkamp N (2021) Data farming output analysis using explainable AI. In: Kim S, Feng B, Smith K, Masoud S, Zheng Z, Szabo C, Loper M (eds) 2021 winter simulation conference; 13 Dec 2021–17 Dec 2021; Phoenix, AZ, USA. Institute of Electrical and Electronics Engineers, Piscataway, New Jersey
73. Barton RR (2015) Tutorial: simulation metamodeling. In: Yilmaz L, Chan WKV, Moon I, Roeder TMK, Macal C, Rossetti MD (eds) 2015 winter simulation conference; 07 Dec–09 Dec; Huntington Beach. Institute of Electrical and Electronics Engineers, Piscataway, New Jersey, pp 1765–1779.
74. Parnianifard A, Saadi M, Pengnoo M, Ali Imran M, Al Otaibi S, Sasithong P et al (2021) Hybrid metamodeling/metaheuristic assisted multi-transmitters placement planning. Comput Mater Continua 68:569–587. https://doi.org/10.32604/cmc.2021.015730
75. Barton RR (2009) Simulation optimization using metamodels. In: Rossetti MD, Hill RR, Johansson B, Dunkin A., Ingalls RG (eds) 2009 winter simulation conference; 13 Dec 2009–16 Dec 2009; Austin, TX. Institute of Electrical and Electronics Engineers, Piscataway, New Jersey, pp 230–238
76. Kleijnen JPC (2015) Regression and kriging metamodels with their experimental designs in simulation: review. SSRN J. https://doi.org/10.2139/ssrn.2627131
77. Morin M, Paradis F, Rolland A, Wery J, Gaudreault J, Laviolette F (2015) Machine learning-based metamodels for sawing simulation. In: Yilmaz L, Chan WKV, Moon I, Roeder TMK, Macal C, Rossetti MD (eds) 2015 winter simulation conference; 07 Dec–09 Dec; Huntington Beach. Institute of Electrical and Electronics Engineers, Piscataway, New Jersey, pp 2160–2171
78. De la Fuente R, Smith III R (2017) Metamodeling a system dynamics model: a contemporary comparison of methods. In: Chan V, D'Ambrogio A, Zacharewicz G, Mustafee N (eds) 2017 winter simulation conference; 3 Dec–6 Dec; Las Vegas. Institute of Electrical and Electronics Engineers, Piscataway, New Jersey, pp 1926–1937
79. Fonseca DJ, Navaresse DO, Moynihan GP (2003) Simulation metamodeling through artificial neural networks. Eng Appl Artif Intell 16:177–183. https://doi.org/10.1016/S0952-1976(03)00043-5
80. Rabe M, Dross F (2015) A reinforcement learning approach for a decision support system for logistics networks. In: Yilmaz L, Chan WKV, Moon I, Roeder TMK, Macal C, Rossetti MD (eds) 2015 winter simulation conference; 07 Dec–09 Dec; Huntington Beach. Institute of Electrical and Electronics Engineers, Piscataway, New Jersey, pp 2020–2032
81. Feldkamp N, Bergmann S, Strassburger S (2020) Simulation-based deep reinforcement learning for modular production systems. In: Bae K-H, Feng B, Kim S, Lazarova-Molnar Z, Zheng Z, Roeder T, Thiesing R (eds) 2020 winter simulation conference; 14 Dec 2020–18 Dec 2020; Orlando, FL, USA. Institute of Electrical and Electronics Engineers, Piscataway, New Jersey, pp 1596–1607
82. Zhang T, Xie S, Rose O (2017) Real-time job shop scheduling based on simulation and Markov decision processes. In: Chan V, D'Ambrogio A, Zacharewicz G, Mustafee N (eds) 2017

winter simulation conference; 3 Dec–6 Dec; Las Vegas. Institute of Electrical and Electronics Engineers, Piscataway, New Jersey, pp 3899–3907
83. Zhang T, Xie S, Rose O (2018) Real-time batching in job shops based on simulation and reinforcement learning. In: Rabe M, Juan AA, Mustafee N, Skoogh A (eds). 2018 winter simulation conference; 9 Dec–12 Dec; Gothenburg, Sweden. Institute of Electrical and Electronics Engineers, Piscataway, New Jersey, pp 3331–3339
84. Feldkamp N, Bergmann S, Conrad F, Strassburger S (2022) A method using generative adversarial networks for robustness optimization. ACM Trans Model Comput Simul 32:1–22. https://doi.org/10.1145/3503511
85. Inoue T, Magistris G de, Munawar A, Yokoya T, Tachibana R (2017) Deep reinforcement learning for high precision assembly tasks. In: 2017 IEEE/RSJ international conference on intelligent robots and systems (IROS); 24 Sep 2017–28 Sep 2017. IEEE, Vancouver, BC, pp 819–825. https://doi.org/10.1109/IROS.2017.8202244
86. Schwung D, Csaplar F, Schwung A, Ding SX (2017) An application of reinforcement learning algorithms to industrial multi-robot stations for cooperative handling operation. In: 2017 IEEE 15th international conference on industrial informatics (INDIN); 24 July–26 July; Emden, Germany. IEEE, Piscataway, NJ, pp 194–199. https://doi.org/10.1109/INDIN.2017.8104770
87. Liu Q, Liu Z, Xiong B, Xu W, Liu Y (2021) Deep reinforcement learning-based safe interaction for industrial human-robot collaboration using intrinsic reward function. Adv Eng Inform 49:101360. https://doi.org/10.1016/j.aei.2021.101360
88. Kuehner KJ, Scheer R, Strassburger S (2021) Digital Twin: finding common ground—a meta-review. Proc 5CIRP6 104:1227–1232. https://doi.org/10.1016/j.procir.2021.11.206
89. Xu Y, Sun Y, Liu X, Zheng Y (2019) A digital-twin-assisted fault diagnosis using deep transfer learning. IEEE Access 7:19990–19999. https://doi.org/10.1109/ACCESS.2018.2890566

Chapter 8
Simulation and Machine Learning Based Real-Time Delay Prediction for Complex Queuing Systems

Najiya Fatma, Pranav Shankar Girish, and Varun Ramamohan

Abstract Real-time delay prediction involves providing entities arriving at a queue with an estimate of their wait time at the time of their arrival at the queue. This wait-time estimate is generated as a function of the state of the system at the time of the entity's arrival to the queue. In this chapter, we present a hybrid simulation and machine learning-based approach toward real-time delay prediction for complex queuing systems. The approach involves using a discrete-event simulation of the queuing system under consideration to generate system state data that is in turn used to train machine learning methods that generate real-time delay predictions. We provide a case study illustrating this approach that involves generating real-time delay predictions for end-stage renal disease patients registering on kidney transplantation waitlists.

Keywords Hybrid modeling · Discrete-event simulation · Machine learning · Kidney transplantation · Real-time delay prediction

8.1 Introduction

Overcrowding is one of the most important issues faced by the healthcare facilities in India and elsewhere [1]. For example, long queues have become a major concern at state-of-the-art tertiary care centers, with 10,000 patients visiting each day [2]. Limited medical resources relative to the demand and the highly unpredictable nature of the demand for care are a few of the many factors yielding long waits at healthcare

N. Fatma · P. S. Girish · V. Ramamohan (✉)
III-351, Department of Mechanical Engineering, Indian Institute of Technology Delhi, New Delhi 110016, India
e-mail: varunr@mech.iitd.ac.in

N. Fatma
e-mail: mez188287@mech.iitd.ac.in

P. S. Girish
e-mail: me1190823@mech.iitd.ac.in

facilities [3]. Empirical evidence suggests that delays not only lead to poor health outcomes for patients, but also cause unnecessary anxiety and inconvenience to patients and overburden healthcare providers [4, 5]. Delays are also expensive; for instance, among patients with the most acute conditions, a one-hour increase in wait time leads to an approximately 30% increase in medical expenses [6]. One of the most frequent discussions in healthcare operations research is to minimize service delays. Informing arriving patients about their expected delay—*at the time of their arrival to the queue*, as opposed to providing patients with the average wait-time estimate—is a relatively inexpensive technique of reducing wait-time uncertainty for the patient and for service-seeking entities in queueing systems in general. This process is called real-time delay prediction.

Studies on customer psychology in waiting situations reported that real-time prediction of waiting time not only improved patient satisfaction and service quality, but also helped in effective planning of medical resources for healthcare administrators [7]. Effective and accurate prediction tools to forecast demand, predict congestion level, queue-lengths, real-time delays, and lengths of stay have been developed to manage patient flow and improve patient service levels [8–11]. In this work, we present a method for generating real-time predictions of wait times for complex queueing systems for which analytical approaches are likely to be intractable. These typically include queueing systems whose queue disciplines are not among those commonly encountered, such as first-come first-served, last in first out, and so on. In such systems, data-driven approaches are often used. For example, queue log data recording the state of the system at or near the point in time when each arrival occurs is used to train statistical and/or machine learning (ML) models for predicting the delay. However, in many queueing systems, such queue log data is either not available or difficult to record. Our approach is suitable for such systems, and its development was motivated by our experience in predicting real-time delays for patients seeking admission to a neurosurgery ward in a large tertiary care hospital where such data was not available [12].

Our approach is hybrid: We use a combination of discrete-event simulation (DES) and ML to generate these predictions. This approach first involves developing a validated DES of the queueing system under consideration and then generating data capturing the state of the system at the point in time at which a service-seeking entity arrives in the system along with its delay prior to starting service. Such system state data can involve, for example, the number of entities already in the queue and the elapsed service time of the entity currently receiving service. This data is then used to train an ML model for generating real-time estimates of delay for each new arriving service-seeking entity.

In the realm of hybrid simulation (HS) and hybrid modeling (HM) literature, the above approach aligns with the HM paradigm. HS primarily centers on the integrated use of methods originating within the modeling and simulation domain, involving the simultaneous application of various simulation techniques to represent the system more effectively under examination. In contrast, HM is concerned with the integration of simulation methodologies (including DES, system dynamics, agent-based simulation, or HS) with other modeling and optimization techniques derived from the broader fields of operations research and management sciences. In other words, HM serves as an extension of HS, enhancing its capabilities [13–16]. The healthcare industry is experiencing a growing trend of employing hybrid models, driven by their ability to effectively depict complex service systems.

To demonstrate our HM approach, we consider a case study in the Indian context involving real-time delay prediction for end-stage renal disease (ESRD) patients registering on a kidney transplantation waitlist. We first predict whether patients registering on the waitlist will receive a transplant or not. If they are predicted to receive a transplant, we then predict the wait time before the organ is allocated. This approach can be used by both patients as well as medical care providers (in case doctors deem it appropriate to not disclose this information to patients) in their decision-making regarding seeking care at the queueing system under consideration. For instance, if a patient is predicted to not receive an organ, they may immediately start exploring options for receiving a kidney from a living donor. For those predicted to receive an organ, they can discuss plans for continuing dialysis for the foreseeable future. For such patients, having an estimate of the actual wait time (the regression problem) will be useful. Finally, the wait times for many patients are in the order of months, and both types of predictions are likely to be particularly useful for such patients.

Overall, the HM real-time delay prediction approach that we propose is suitable in the context of complex queuing systems where the queue log data often lacks the granularity required to generate accurate real-time delay predictions. The patient waitlisting and kidney allocation system that we model is such a queueing system. Further, the proposed hybrid approach is advantageous in comparison to real-time simulation for generating real-time delay predictions, especially in terms of speed of generation of the delay prediction. The proposed approach requires a single function evaluation, whereas the real-time simulation may need to be executed multiple times to generate the average real-time delay prediction. Note that in both cases—the proposed hybrid approach as well as the real-time simulation case, the simulation driving the delay prediction would have to be reprogrammed if the queueing system configuration changes.

The remainder of the chapter is organized as follows. In Sect. 8.2, we provide a more detailed introduction to the concept of real-time delay prediction and discuss the related literature. In Sect. 8.3, we briefly describe the kidney allocation process and discuss the development of the simulation, including input parameter estimation. We later illustrate the use of the validated simulation model of the waitlist registration and kidney allocation process to generate real-time delay predictions for ESRD patients. We make concluding remarks in Sect. 8.4 wherein we summarize this work, describe its advantages and limitations, and discuss avenues of future research.

8.2 Background and Literature Review

8.2.1 Real-Time Delay Prediction: Overview and Proposed Approach

Real-time delay prediction involves providing each entity with an estimate of their expected wait time—at the point in time at which they arrive at the queueing system—until the start of the service [17]. These estimates may be generated by either: (a) analyzing the delay history data for previous arrivals to the queue; (b) utilizing the system state information to develop closed-form expressions for mathematically tractable queuing systems (where possible); (c) using queue log data where such data is available to train statistical/ML models; or (d) as proposed here, using validated simulation models of the queueing system in question to generate system state data for training statistical/ML delay prediction models. To illustrate the concept of real-time delay prediction, we take the example of the simple $M/G/1$ queuing system with a generally distributed service time. The real-time delay is estimated by the expression below [18].

$$P(T \le t|x) = \frac{P(x \le X \le t+x)}{P(X \ge x)} \Rightarrow G(t|x) = \frac{G(t+x) - G(x)}{1 - G(x)}. \quad (8.1)$$

In Eq. (8.1), T is the random variable representing the remaining service time of the entity currently in service given the elapsed service time x; that is, it is the delay assuming no other entities are in the queue, and t is a realization of T. X is the random variable representing the service time itself and $G(x)$ represents its cumulative distribution function (cdf). Now the expected remaining service time given an elapsed service time of x can be found as the expected value of T, given by:

$$E[T] = \int_{S_T} t g(t|x) dt. \quad (8.2)$$

In Eq. (8.2), S_T represents the support of T. Once $E[T]$, referred to henceforth as w for economy of notation, is estimated, then the real-time predicted delay for the arriving entity, denoted by d, can be found as follows:

$$d = w + L_q E[s]. \quad (8.3)$$

In Eq. (8.3), L_q represents the length of the queue at the time the delay prediction is generated, and $E[s]$ is the expected value of the service time random variable X. As an example, for uniformly distributed service times with parameters $U(a, b)$, w can be estimated as follows:

$$w = \int_0^{b-x} \left(\frac{1}{b-x}\right) t \, dt = \frac{b-x}{2}. \quad (8.4)$$

Depending upon the functional form of $G(x)$ in Eq. (8.1), the computation of w may be straightforward or tedious. For example, computing w for the triangular distribution requires working with its piecewise cdf, and for the Gaussian distribution, one has to work with integrals of the Gaussian error function, which require numerical evaluation. Fatma and Ramamohan [8] propose an approximate real-time delay predictor that is agnostic to the specific service time distribution as long as it is symmetric and unimodal, while still using the distributional information of the service time. This predictor is given below in Eq. (8.5).

$$w = \begin{cases} G^{-1}(0.5) - x, & 0 \leq x < G^{-1}(0.5) \\ G^{-1}(0.75) - x, & G^{-1}(0.5) \leq x < G^{-1}(0.75) \\ \frac{G^{-1}(\text{ext}) - x}{2}, & G^{-1}(0.75) \leq x \leq G^{-1}(\text{ext}) \end{cases} \quad (8.5)$$

In Eq. (8.5), G^{-1} represents the quantile function of the service time and $G^{-1}(\text{ext})$ represents an extreme right quantile. The logic underlying the development of the above predictor and the exact expressions of w for few symmetric and unimodal service time distributions can be understood from [8]. We also refer readers to Table 8.5 in [8] to learn about the wide variety of other queuing systems for which analytical expressions for real-time delays have been developed.

Developing analytical expressions for real-time delay prediction can become challenging for complex queueing systems where queue disciplines are complex (for example, if it is not FCFS or LIFO), or significant non-stationarity is present in multiple queueing aspects—for example, if balking and/or reneging behavior in addition to arrival/service processes are non-stationary. For such systems, queue log data, if available, may be used to train statistical/ML methods for predicting delays. However, in cases where queue log or system state data is not available or not captured in a manner suitable for training statistical/ML methods, the hybrid approach that we propose in this chapter may be used. This requires the development of a simulation of the queueing system in question, validating the simulation, and then using the validated simulation to generate system state data for each arriving entity at the time of its arrival. This system state data, along with the actual delay information of the service-seeking entities for which the data has been generated, is used to train statistical/ML methods for the purpose of real-time delay prediction. An example of such a queueing system can be found in Baldwa et al. [12], where the multi-class queueing system represented by the admission, surgery, and recovery stay processes at the neurosurgery ward in a large public tertiary care hospital uses an algorithm based on patient severity to determine admission to the neurosurgery ward. Real-time delay prediction for this queueing system was accomplished by first simulating the admission and the patient stay processes at the neurosurgery ward. Then, the validated simulation was used to generate data to train ML algorithms to predict whether the patient will be admitted or not as a function of the state of the simulation at the time the patient arrives seeking admission to the ward.

The key steps involved in generating real-time delay predictions using our proposed hybrid simulation and machine learning technique are given below.

Develop a DES of the queueing system under consideration.

Use the DES in steady state to record for each of, say, M entities (e.g., the kth entity) the following information:

The state of the simulation at the time the kth entity arrives in the system (denoted by S_k).

If there exists a prespecified threshold wait time T_k prior to which entity k must receive service before it exits the queue (i.e., a form of reneging, which is common for most healthcare service systems where the patient's condition may deteriorate), record whether the said entity receives service within T_k as a binary variable V_k ($V_k = 1$ if service is received prior to T_k and 0 otherwise).

For entities receiving service prior to T_k (where applicable), record the wait time w_k before the start of their service.

For queueing systems where the reneging threshold T_k is not applicable, record w_k for every service-seeking entity k.

Construct training sets (S, V) and/or (S_w, w) (note that $S_w \subseteq S$) as applicable using the data recorded for all M entities.

Train and validate ML methods f and f_w on (S, V) and (S_w, w), respectively.

For each new service-seeking entity $k\text{(new)}$, record $S_{k\text{(new)}}$ at the time of its arrival to the queueing system and predict $\hat{V}_{k\text{(new)}}$ as $f(S_{k\text{(new)}})$. If $\hat{V}_{k\text{(new)}} = 1$, then predict $w_{k\text{(new)}}$ as $f_w(S_{k\text{(new)}})$.

In the above procedure, it is assumed that if an entity does not receive treatment prior to T_k, they exit the queueing system (i.e., renege or leave the queue). In the context of the kidney transplantation system modeled in this study, this implies that the patient dies, or their condition deteriorates to the extent that they become ineligible for a transplant.

Recording the system state data S for each service-seeking entity precisely at the time of its arrival to the queueing system may be possible if a sufficiently comprehensive information technology infrastructure is available to capture the required data. For example, in the neurosurgery ward case mentioned above, which had a large number of servers (beds in the ward), key system state variables involved the duration of occupancy of each bed at the time a new patient arrived at the ward seeking admission. This information is likely already tracked by the billing system (using the time of admission for each patient currently occupying a bed) and hence can be leveraged in the deployment (if not development) of the above approach.

8.2.2 Literature Review

We now provide an overview of the literature around real-time delay prediction, beginning with a very brief discussion of the use of DES in modeling healthcare delivery. Subsequently, we briefly discuss HM literature.

DES is one of the most commonly used methods for modeling healthcare delivery operations across the world, and we refer readers to [19, 20] for a comprehensive discussion of the relevant literature. Specifically, since we use the DES of a kidney transplantation system to illustrate our approach, we refer to [21–23] for examples of the application of simulation and operations' research methods to analyze and optimize different aspects of organ transplantation systems and allocation policies' systems in multiple countries.

Researchers investigated delays using queueing theory, game theory, and data-driven approaches at call centers, airports, construction sites, retail industries, healthcare facilities, and others [24–29]. Primarily, two types of approaches have been adopted: (a) analytical approaches grounded in queueing theory and (b) data-driven statistical learning approaches that are trained on queue log data [17]. With regard to queueing-theoretic approaches, most studies focused on developing system state and/or delay history-based predictors for queuing systems ranging from $M/G/1$ to $M(t)/GI/s(t) + GI$ systems. System state-based predictors estimated real-time delays using queue length, elapsed service time, number of servers, or the quantiles of the service time distribution. One of the earliest studies on the application of system state-based delay estimation was conducted by Whitt in 1999 [18], where customers were communicated information on expected delays in single and multi-server queues. Additionally, information about various other system parameters such as the arrival rate, the abandonment rate, and the number of servers were considered by Whitt in [30]. Nakibly studied ways to predict waiting times based on information about the system state upon arrival mainly for queueing models with priority [31]. Fatma and Ramamohan [8, 32] developed novel approximate system state-based delay predictors for simple queueing systems with symmetric and unimodal service time distributions and used the predictions for diverting patients in a healthcare facility network. Delay history information such as the delay of the last entity to receive service, wait time elapsed at the head of the line, etc., were discussed in [33–35]. Ibrahim and Whitt [36–38] highlighted the better performance of queue length-based system state delay prediction methods over the delay history-based estimators in simple and complex queueing systems.

The limitations of queueing-theoretic analyses such as assumptions that are often used to make the analysis mathematically tractable led to recent interest in data-driven methods such as ML algorithms and data mining techniques for complex queuing systems [26, 29, 39, 40]. ML-based predictors, which consist of classification and regression methods trained on queue log data, were discussed in Senderovich et al. and Thiongane et al., respectively [41, 42]. Ang et al. [43] and Arora et al. [29] combined process mining and queueing-theoretic results for predicting waiting times in the emergency departments (EDs) of the healthcare facilities. Baldwa et al. [12] proposed a hybrid simulation and machine learning method for real-time delay prediction where adequate queue log data for training a predictor is not maintained. Further, robust ML-based prediction models were developed for predicting real-time lengths of stay of patients, which is a metric of quality, efficiency, and hospital performance [44, 45]. The effectiveness of predictors was quantified using mean absolute deviation, mean absolute percentage error score, and other metrics via computer simulation models.

With respect to the HM literature, Harper and Mustafee [46] demonstrated the applicability of an HM approach involving DES and time-series forecasting in a real-life setting to support short-term decision-making in an urgent care network. Similarly, other studies, such as those by Ordu et al. [47] and He et al. [48], developed hybrid frameworks that incorporated DES, system dynamics, and optimisation techniques like integer programming to address both the operational (short-term) and strategic (long-term) objectives of healthcare facilities. However, for delay prediction, except for the study by Baldwa and colleagues [12], there has been limited exploration of hybrid methods.

The majority of the empirical work on delay prediction involved using historical data regarding the queuing system under consideration and training statistical/ ML predictors using this data. However, in situations where such data is not available, or sufficient information regarding system state data required for training an accurate prediction is not maintained in the queuing system logs, the data may be generated from a DES model of the system. From our review of the literature, only one previous study [12] has incorporated DES with ML for real-time delay prediction. This is summarized in Table 8.1. An example of such a system is the kidney transplantation system that we consider as a case study to illustrate our proposed approach. In this chapter, we build upon the work by Baldwa et al. [12], which to our knowledge is the only study that uses a DES of the queueing system under consideration for generating the system state data for training data-driven predictors. However, unlike their approach, we do not use predetermined reneging thresholds for service-seeking entities (patients); we instead use patient-specific 'personalised' reneging thresholds that are based on patient characteristics. Our approach may be used by healthcare providers to help advise ESRD patients on the best course of action from the standpoint of obtaining a kidney transplant.

Finally, our approach resembles metamodeling methods to some extent. A metamodel f has an explicit form, deterministic output, and once fitted, is computationally inexpensive to evaluate as they serve as proxies for evaluation, thereby replacing the need for conducting computationally expensive and stochastic simulation runs [49]. A relevant study by Fatma et al. [50] explored the use of stochastic metamodels in developing primary healthcare delivery network systems, resulting in reduced execution times while maintaining comparable results. In this work, similar to metamodeling, we use DES to generate a dataset for training classifiers and regressors. However, in the case of metamodeling, the system simulation (e.g., a DES) is executed multiple times with different sets of input parameters, while we generate a dataset only once with a single set of input parameters for training the delay predictors.

8.3 Case Study

We discuss a case study through which we illustrate our HM approach for real-time delay prediction. The case study involves predicting whether a patient registering on the kidney transplant waitlist—at the time of registration on the waitlist—will receive a kidney transplant or not, and if predicted to receive a transplant, then their wait time to allocation of a kidney is also estimated.

Table 8.1 Studies utilizing machine learning algorithms and simulation methods for real-time delay prediction

Study	Problem description	Methodology	Predictor variables
Baldwa et al. [12]	Prediction of whether a patient seeking admission receives admission within a prespecified duration	Simulation, ML (ensemble bagged trees, gradient boosted trees, neural network, decision tree)	Patient type, waitlist-related features and operational system state features such as number of empty beds
Balakrishna et al. [51]	Estimation of average taxi-out times at airport	Stochastic dynamic programming with reinforcement learning	Features describing the airport and runway state
Arora et al. [29]	Estimation of probability distribution of individual patient wait times	Quantile regression using decision trees	Calendar effects, demographics, staff count, ED workload, severity of patient condition
Ang et al. [43]	Wait-time prediction	Data mining, queuing theory (Q-Lasso technique)	Patient visit data, mode of arrival, triage level
Senderovich et al. [39]	Delay prediction for single-class setting (homogeneous customers) and multi-class setting (different class of customers)	Queue mining, regression-based predictors	Time of event occurrence, instance of service process, service transition, customer class
Arik et al. [40]	Prediction of the time to meet with the first provider at hospital	Supervised learning (congestion graphs of two types—heavy traffic approximations of congested systems and Markovian state representation of queues)	Clinical state of patients and congestion-related features
Chocron et al. [52]	Prediction of the wait time for service at the time of arrival	ML models, queueing theory	Arrival-related features, service-related features, queue-related features, short-term history-related features

8.3.1 Kidney Transplantation System: Problem Introduction

Large urban public tertiary care hospitals in India typically face significantly more demand than their available capacity. Kidney transplantation is the most effective long-term treatment option for ESRD patients undergoing maintenance dialysis. The substantial shortage of donated kidneys in India causes an increasingly long waitlist of ESRD patients awaiting a transplant. According to the Indian Ministry of Health, the number of Indian ESRD patients who need kidney transplants ranges between 200,000 and 300,000, with only 6000 donors available [53]. A kidney patient who gets on the state government's waiting list typically waits for at least four years to get a cadaveric donor transplant, thereby aggravating uncertainties among patients regarding whether they will receive a transplant before their health deteriorates to a critical level [54]. A first step toward alleviating the uncertainty is to provide ESRD patients and/or their medical care providers with information regarding whether they will receive a transplant or not—at the time of their registration on the kidney transplantation waitlist—and if the patient is predicted to receive a transplant, the wait time of the candidate before receiving the transplant. This will help patients and/or their medical care providers make an informed decision about whether they should wait or seek treatment elsewhere (e.g., seek living donors). We list the specific objectives associated with the case study below.

1. Development of a DES of the patient arrival and registration, organ arrival and organ allocation processes.
2. Classification of waitlisted patients to predict whether they will receive a transplant within a 'personalised' patient characteristic based duration.
3. Regression to estimate the time to allocation of an organ for patients predicted to receive a transplant.

We use publicly available domestic data based on real-world reports from kidney transplantation organizations in Indian states, namely Rajasthan, Kerala, and Tamil Nadu, for developing and parameterizing the DES of the kidney transplantation system in the South Indian state of Kerala. We now describe the development of this DES, beginning with a description of the cadaveric donor kidney allocation process. We note here that considering kidney transplantation from living donors is beyond the scope of this work.

8.3.2 Patient Registration, Organ Arrival, and Organ Allocation Process Simulation Development

The allocation of cadaveric kidneys to transplant candidates is a complex process determined by a variety of organ and patient characteristics, including time spent on the waitlist. As per kidney allocation guidelines published by the Indian government's National Organ and Tissue Transplantation Organization (NOTTO) [55], ESRD patients eligible for registering on the waitlist must be aged less than 75 years at the time of registration, must have undergone regular maintenance dialysis for at least three months, and must be registered in a single approved transplantation center. Upon registration in a transplantation center, the patient is assigned a kidney allocation priority (KAP) score after registering in the respective state and district waitlists that determine the position of the patient on the transplantation waitlist. The KAP score is computed based on a scoring algorithm provided in NOTTO's kidney allocation guidelines [55]. If a cadaveric kidney is retrieved in a government hospital, then patients registered in the government transplant centers within the state are given higher priority for allocation and patients registered in private hospitals are considered for allocation only if a suitable recipient is not found on the government hospital waitlist. The same recipient selection process is followed if the kidney is retrieved from the deceased donor in a private hospital, but in reverse order. Therefore, whether the patient is registered with a government or a private hospital affects their probability of transplant.

Further, patients registered in transplant centers within the district where the organ was retrieved are given higher priority. In other words, organs are first considered for allocation to patients registered in the same district where the organ is retrieved, and patients registered in other districts in the state are considered for allocation only if a suitable recipient is not found in the district of retrieval.

In the event that the kidney is retrieved from a donor aged less than 18 years, then patients aged less than 18 years are first considered for allocation. Finally, for each 'subwaitlist' (i.e., district and then state waitlist), donor/recipient matching is done on the basis of the blood groups of the donor and the patients. A blood group O (universal donor) kidney is first matched to a recipient with group O, then to the other compatible blood groups—first, it is allocated to group A, then to group B, and finally to a blood group AB (universal recipient) patient. Group A or B kidneys are allocated to patients with the same blood groups; else it is allocated to a group AB patient. An AB group kidney is only allocated to an AB patient. More details regarding the kidney allocation process, including an algorithmic representation of the above process, can be found in Shoaib et al. [56].

The advancement of the DES of the kidney allocation process is dependent on three principal events: patient arrival, removal of patients due to death, and organ arrival. Organ arrival drives removal of patients via organ allocation and transplantation. The mechanisms that determine the removal of patients are as follows: either the patient receives a transplant, or the patient dies, implying that we do not consider balking in our model. We represent districts in each state by their district headquarters, and hence, travel times between districts (for calculating organ transport times) are also calculated between the district headquarters. We now describe the estimation of two primary types of model parameters: (a) those related to patients and (b) those related to organs.

8.3.2.1 Patient-Related Parameters

The patient's position on the waitlist (district as well as state, government as well as private 'subwaitlists') is determined by the KAP score. Thus, a key set of parameters that need to be estimated with respect to patient characteristics is those that constitute the KAP score. These parameters include the following:

1. Time spent on dialysis.

2. Whether the patient has had a previous immunological graft failure, and if so, the number of such failures.
3. Age of the recipient.
4. Patient with all failed arteriovenous (AV) fistula sites.
5. Patient with failed AV graft after all failed AV fistula sites.
6. Panel reactive antibody level.
7. Whether the patient under consideration has previously donated a kidney or not.

The time spent on dialysis, which is a key driver of the KAP score, is estimated from the data available for this parameter from the waitlist data for the state of Rajasthan, because similar data was not available on the waitlist for Kerala. Because this is a clinical parameter, its distribution is assumed to not change substantially across states in India. The exponential distribution was found to fit the time on dialysis data best under a χ-squared goodness-of-fit test, with a p-value of 0.581. Parameter 7 was not considered in our analysis since it was highly unlikely to encounter a patient with this characteristic in the kidney transplantation process, based on discussions with clinicians involved in organ transplantation. Other parameters—i.e., parameters 2 through 6—were estimated from the clinical literature, and the sources, along with the parameter estimates, are given in Table 8.1. The patient interarrival time was estimated from the patient waitlist data available on the state organ transplantation authority website (Kerala Network for Organ Sharing (KNOS), [57]). The interarrival time was found to follow the exponential distribution with a mean of 1.382 days, which was estimated by applying the χ-squared goodness-of-fit test on the KNOS waitlist data, using information regarding the date of patient registration. Patient blood group data was estimated from waitlist data available on the neighboring Tamil Nadu state organ transplantation authority website (TNOS, the Tamil Nadu Network for Organ Sharing) [58]. Indian census data was used to assign the district in which a newly arriving patient was registered, and the transplant center of registration was also assigned based on the transplant hospital set available in the KNOS dataset.

A critical patient parameter in this context is the patient removal time due to death, occurring due to unavailability of a cadaveric kidney. We used survival data of ESRD patients from the clinical literature [56] and the KAP score computation process to estimate this parameter. This computation process is described below.

Algorithm: Computation of patient removal time due to death.

- Input: KAP score data of a waitlisted patient
- Output: Removal time of the patient under consideration
- Generate a large sample of KAP scores by generating multiple random realizations of the KAP score components and combining them according to the KAP score computation algorithm given in the NOTTO kidney allocation guidelines.
- Find the distribution that best fits this sample. This was determined to be a beta distribution with $\alpha = 0.89$; $\beta = 33.99$ (best fit out of alternatives).
- For each new patient registering on the waitlist, using the distribution of the KAP score estimated using steps 1 and 2, do the following:

Compute the patient's KAP score based on their randomly assigned KAP score component values.

Find the percentile of the KAP score for a given patient from its distribution estimated in Steps 1 and 2. Let this percentile be x.

A patient in the x th percentile of KAP scores is likely to be in the $(100 − x)$th percentile of removal times. Thus, the patient's removal time percentile = $100 − x$.

- Estimate the mean removal time μ_{rt} of the patient by determining the $(100 − x)$th quantile of the removal time distribution, which we assume to a beta distribution with limits $a = 3$ months and $b = 67$ months and mean $= 40.31$ months, yielding $\alpha = 4.38$ and $\beta = 3.51$ for this distribution. The removal time distribution was estimated based on the mean and standard deviation of the mean survival time of patients on hemodialysis as reported in Lakshminarayana et al. [59] (40.31 months and 26.69 months, respectively).
- Using this estimate of the mean removal time, define another beta distribution with limits $a = 0.67 \times \mu_{rt}$ and $b = 1.33 \times \mu_{rt}$. α and β for this beta distribution were estimated via the beta-PERT three-point estimation procedure. This was done in order to avoid making the removal time a deterministic function of the KAP score.
- The assigned removal time value for the patient is a random sample from the beta distribution for the removal time estimated in Step 5.

8.3.2.2 Organ-Related Parameters

All organ-related parameters were estimated using data pertaining to kidney transplantation alone. Key organ-related parameters involve the interarrival time of kidneys, the district in which the organ originates, the deceased donor's blood group, and the number of kidneys retrieved from an organ (i.e., 1 or 2). With regard to the organ interarrival time, a parameter critical to the analysis, precise data regarding the dates of arrivals of organs were not publicly available on organ-sharing websites. Hence, we estimated the parameters of the interarrival times of the kidneys from deceased donors using the published annual aggregate organ donation data after assuming it to be exponentially distributed. According to the aggregate organ donation data published on the KNOS website [57], the number of organs donated in the years 2016, 2017, and 2018 were 113, 34, and 14, respectively, and hence, the mean interarrival time in days was estimated as 365 divided by the average of the number of organs arriving in those three years (this average amounts to approximately 33 kidneys being donated every year). Thus, the estimate of the mean interarrival time, assuming an exponential distribution, was 11.17 days per organ. We estimated the other organ-related parameters, such as the donor blood group and age, which are required to determine the kidney allocation, according to the proportions of various blood groups and age ranges in the population of the entire state. We estimated the probabilities of retrieving a kidney in a public or private hospital in a given district based on the proportion of each type of hospital in each district. We list all the patient-related and the organ-related parameters, along with their distributions and estimates and corresponding sources in Table 8.2.

Table 8.2 Patient and organ-related parameters

Parameter	Distribution	Estimate	Source
Patient-related parameters			
Patient interarrival time	Exponential	Mean = 1.382 days	[57]
District origin (14 districts in the state of Kerala)	Discrete	$P(0) = 0.128, P(1) = 0.096, P(2) = 0.023, P(3) = 0.041, P(4) = 0.018, P(5) = 0.100, P(6) = 0.055, P(7) = 0.064, P(8) = 0.091, P(9) = 0.050, P(10) = 0.055, P(11) = 0.164, P(12) = 0.091, P(13) = 0.013$	[57]
Age	Gaussian	$\mu = 49.74; \sigma = 7.423$	[57]
Blood group	Discrete	$O = 0.458, A = 0.238, B = 0.224, AB = 0.079$	[58]
Time on dialysis	Exponential	Mean = 260.3 days	[57]
Removal time	Beta	$\mu = 40.31; \sigma = 26.69$ Random sampling (beta) $\alpha = 0.66 * \mu; \beta = 1.33 * \mu$	[59]
PRA level	Discrete	$P(\text{PRA level} = 0) = 0.65, P(1–20) = 0.05;$ $P(21–79) = 0.136, P(80–100) = 0.158$	[60]
Probability of a previous immunological graft failure within 3 months of transplant With failed all AV fistula sites With failed AV Graft after failed AVF sites	Discrete Discrete Discrete	$P(\text{yes}) = 0.020, P(\text{no}) = 0.980$ $P(\text{yes}) = 0.052, P(\text{no}) = 0.948$ $P(\text{yes}) = 0.031, P(\text{no}) = 0.968$	[61] [62] [62]
Organ-related parameters			
Donor interarrival time	Exponential	Mean = 11.17 days/organ	[63]
Donor blood group	Discrete	$P(AB) = 0.069, P(A) = 0.192, P(B) = 0.254, P(O) = 0.485$	[58]
District in which organ originates	Discrete	$P(0) = 0.067, P(1) = 0.098, P(2) = 0.033, P(3) = 0.075,$ $P(4) = 0.039, P(5) = 0.078, P(6) = 0.059, P(7) = 0.092,$ $P(8) = 0.123, P(9) = 0.084, P(10) = 0.035, P(11) = 0.099,$ $P(12) = 0.093, P(13) = 0.024$	[64]
Number of kidneys retrieved from an organ	Discrete	$P(1) = 0.777; P(2) = 0.223$	[63]

μ : mean; σ : standard deviation; PRA: panel reactive antibody, a screening test to identify the immunological sensitization of a transplant recipient for estimating likelihood of finding a compatible donor.

8.3.3 Simulation Model Outcomes

Using the parameters reported in Table 8.1 and the kidney allocation process described earlier, we programmed the simulation on the *Python* computing platform. We ran the simulation on a workstation with an Intel *i7* 10th generation processor system with 16 gigabytes of memory. We used a warm-up period of 12 years of simulation time before collecting the results over a period of 18 years. We performed 20 replications for collecting and reporting the results. The length of the warm-up period was determined by observing when the average simulation outcomes became stable. A data collection period spanning 18 years was selected as it provided enough time to calculate all the outcomes to the necessary precision. The decision to use 20 replications was influenced by the observation that the variances of the outcomes changed minimally when the number of replications was increased beyond 15. These choices were also influenced by the availability of computational resources. The key outputs collected from the simulation include (for all patients, by blood group and by the type of hospital (government or private) in which the patient is registered): (a) the probability of receiving a kidney transplant; (b) the average wait time to allocation for those who received an organ; (c) the average number of deaths; and (d) the average number of allocated organs.

The probability of receiving a transplant, which is possibly the most critical output from a patient and provider standpoint, was estimated as follows. For a set of patients who register in a given year (the registration year), we record the proportion of those patients who receive a transplant over the next year, the second year after the registration year, and so on for a period of five years. We then average these probabilities for patients registering every year after the warm-up period to obtain the within replication estimate of the average probability of receiving a transplant within 1 year, 2 years, and so on up to 5 years. The across replication average values of these probabilities are then calculated by averaging the within replication average probability estimates. These probabilities are depicted in Fig. 8.1a, b.

From Fig. 8.1a, the 5-year probability of receiving a transplant, which is negligibly different from the overall probability of receiving a transplant (before death from ESRD), is approximately 21%. It is also evident that patients with blood group AB and blood group B are the most likely to receive a transplant, with 5-year probabilities exceeding 25% (nearly 40% for blood group AB). Patients with blood group AB have the highest transplant probabilities because they are universal recipients. Patients with blood group O have the lowest probability of transplant likely because of the large volume of patients on the waitlist compared to the lower number of organs of the blood group being donated. This low likelihood reflects the disparity between the organ (including kidneys) donation rate and the number of patients needing

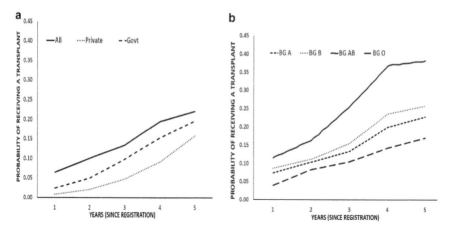

Fig. 8.1 a Year-wise probabilities of receiving a transplant for all patients and by the type of hospital where they are registered. **b** Year-wise probabilities of receiving a transplant by patient blood group

transplants. Further, patients registered in a private hospital are significantly less likely to receive a transplant than those registered in a government hospital. This reflects the interaction between the number of government transplant centers versus the number of private transplant centers and the number of government organ retrieval centers versus the number of private retrieval centers.

We report the other outcomes collected from the simulation model in Table 8.3. Average organ transport time is defined as the average time required to transport an organ from the organ retrieval location to a transplant center where the recipient is registered. The average time to transplant was calculated only for patients who received a transplant during the post warm-up (steady-state simulation) period. We calculated the number of deaths by counting those removed from the waitlist without receiving a transplant.

Based on the results in Table 8.3, we observe significant variations in the allocation based on the patient's blood group and the hospital type. This supports the outcomes in Fig. 8.1b. For example, patients with blood group O receive the highest number of allocations compared to those with other blood groups. However, because the number of patients with the blood group O is the highest (it is the most common blood group in India), as evident from the highest probability of a patient being of blood group O, the probability of receiving a transplant for patients with blood group O is the lowest.

We now discuss validation of the model outcomes prior to describing the process of generating real-time delay predictions from the DES. Validation of the DES of the kidney transplantation system in Kerala is challenging due to the lack of data regarding waitlist and transplantation outcomes for the state and in general for the Indian transplantation system. For example, in our knowledge, no data suitable for validating simulation results for outcomes such as time to allocation or the probability

Table 8.3 Simulation outcomes for Kerala

Outcomes (year)	Average (95% CI)
Average number of deaths per year	203.81 (1.57)
Average organs allocated per year	58.31 (1.14)
Average wait to allocation (hours)	19,175.10 (215.43)
Average time to transportation (minutes)	200.73 (4.21)
Average number of allocations in government hospitals	24.45 (0.63)
Average number of allocations in private hospitals	33.68 (0.93)
Average number of allocations to blood group A patients	14.34 (0.46)
Blood group AB allocations	8.17 (0.34)
Blood group B allocations	14.91 (0.49)
Blood group O allocations	20.72 (0.61)
Average number of unallocated organs	0.35 (0.09)

of receiving an organ is available. Since kidney transplantation outcomes data was not available for validation, we decided to validate the patient and organ arrival outcomes against a portion of the publicly available patient arrival data that we reserved for this purpose. We recall here that publicly available waitlist data was available for the years 2016–2019 and that we used data from 2016 to 2018 for calculating patient arrival rates. A similar approach was followed for organ and donor arrivals as well; however, we must recall that organ arrival data was much more limited in comparison to patient arrival data. In fact, for 2016–2018, we only had access to three values; that is, the number of organs donated and donors arriving in each year. Hence, we validated our patient arrival and organ outcomes—i.e., the number of patients arriving in one year against the actual values for 2019. We provide the results of this validation exercise in Table 8.4.

From the results from the above validation exercise, we can see that the patient arrival process is reasonably well represented by our DES. This is particularly the case when patients from all blood groups are considered together, and even when breaking

Table 8.4 Simulation validation outcomes

Parameters	Actual [57]	Simulation (UL, LL)
Number of patients registered in 2019	264	261.61 (265.90, 257.32)
Patient in blood group A	82	61.56 (63.71, 59.40)
Patient in blood group AB	13	21.22 (22.37, 20.08)
Patient in blood group B	51	59.43 (61.29, 57.56)
Patient in blood group O	118	119.41 (122.60, 116.22)
Organ arrived in 2019	32	57.61 (60.45, 54.77)
Donor arrived in 2019 [63]	19	32.70 (34.32, 31.09)

UL is upper limit, *LL* is lower limit

Table 8.5 Input features for the machine learning models

Feature type	Features
Continuous	Clinical: age, KAP score, PRA level, time on dialysis,
	Waitlist related: position on waitlist, patients above this patient, A patients above, B patients above, O patients above, AB patients above, total patients on the waitlist, total A patients, total B patients, total O patients, total AB patients
Categorical	Operational: district name, hospital name, hospital type
	Clinical: blood group, PRA type, AVG, AVF, PIGF

AVG arteriovenous grafts, *AVF* arteriovenous fistula, *PIGF* placental growth factor

down arrivals by blood group, we see that for blood groups with larger numbers of patients, such as O and B, the simulation outcomes and validation outcomes are reasonably close. Note that while we report the confidence intervals for the simulation outcomes, formal statistical inference based on these CIs and the actual simulation outcomes may not be advisable because only a single value of the actual number of organs is available for each year for each blood group. In other words, the actual value is also a realization of a random variable (the number of organs arriving in a year), and given the small sample size (3), sufficient data is not available to conduct formal inference—for example, a two-sample nonparametric test for equality of means.

We also perform a comparison between the probabilities of allocation from the simulation versus an approximate value of this outcome calculated from the organ and patient arrival data. The average probability of receiving a transplant calculated across years 1–5 from the simulation data is approximately 14% (ranging from 7 to 21% from 1 to 5 years and taking their average), and the value of this outcome from the validation data is approximately 12.2% (obtained by dividing the yearly organ arrivals with the patient arrivals). We computed the average probability of transplant in this manner because while the information regarding the average number of patients registering is known, it was unclear when the patients who registered in these three years (2016, 2014, and 2013) will undergo the transplant. Therefore, we took the average of the yearly probabilities to make a comparison between the simulated data and the actual data. This indicates that our DES of the kidney transplantation system appears to approximate the actual system to an acceptable level.

Overall, it is clear that the probability of receiving a transplant, even at 5 years on the waitlist, is low. However, it is also clear that there is significant variation based on patient and operational characteristics such as blood group and the type of transplant hospital where the patient is registered. This motivated us to develop a classification model that will predict whether a patient on the waitlist will receive a transplant at the time of registration on the waitlist—i.e., real-time delay prediction of transplant registration outcomes. We discuss this now.

8.3.4 Real-Time Delay Prediction for Waitlisted ESRD Patients: Classification

As described in previous sections, we now use the DES of the ESRD patient registration, waitlisting, and organ allocation process to generate the dataset required to train ML-based methods to predict whether a patient will receive an organ at the time of their registration on the waitlist. At the point in time at which a patient registers on the waitlist, we record three types of features for this patient: clinical, operational, and waitlist-related features. Clinical features primarily include those that determine the patient's KAP score and their blood group, and operational features include those features such as the district of registration, the hospital type, the hospital name, and so on. Waitlist-related features include the total number of patients above the current patient on the waitlist and the numbers of patients of different blood groups that are above the current patient on the waitlist. The label for the classification exercise was whether the patient received a transplant or not before they were removed from the waitlist due to death. Note that data was recorded only for patients who were allocated an organ or those who were removed due to death. There could be patients still in the model with neither of these outcomes at the end of the model time horizon, but the data for such patients are not recorded. For those who did receive an organ, we recorded the time to allocation from their time of registration. This formed the label for the regression exercise. The feature set consisted of both continuous as well as categorical features, and we list them in Table 8.5.

The training dataset thus consisted of a total of 23 features representing the patient characteristics and the queueing system state at the time of registration of the patient and the classification/regression labels. We split the input data into training and test sets with a 75/25 split ratio. The data was scaled—after it was split to prevent data leakage—using the MinMaxScaler function of the *scikit-learn* ML package in the Python programming platform. The input dataset consisted of 929 patients who received an organ out of a total of 3945 patients, which indicated a dataset imbalance. Hence, we used the Synthetic Minority Oversample Technique (SMOTE) to balance the dataset so that it does not negatively impact the accuracies of the classification models [65]. SMOTE is an oversampling technique that allowed us to generate synthetic samples for the Label 1 class (minority class, those who received an organ) until the number of samples became equal to the majority class. Note that SMOTE was applied only on the training dataset and the validation (test) dataset was left untouched by the dataset balancing technique.

We trained several ML methods for the classification exercise such as support vector machines, decision tree methods such as bagging and random forest classifiers, and artificial neural networks to find the method that performed best on the dataset. Before training the model, we optimized the hyperparameters of each of these classifiers via the GridSearch hyperparameter tuning method. For example, the gradient boosting classifier implementation contained 100 boosting stages with a learning rate of 0.1, while the bagging classifier implementation contained 1000 estimator trees. We also trained the artificial neural network with two hidden layers using the *adam* optimizer with a batch size of 10 [66].

To measure the performance of the classification models, we used the Receiver Operator Characteristics-Area Under Curve (ROC-AUC) and F1 performance measures. The ROC is a probability curve that plots the true-positive rate against the false-positive rate at various threshold values. The AUC score, which is used as a numerical characterization of the ROC curve, is the measure of the ability of the classifier to distinguish between classes. We compute the $F1$ score instead of using the classifier accuracy alone as it calibrates the trade-off between sensitivity and specificity at the best-chosen threshold. The $F1$ score is the harmonic mean of precision and recall. Precision is the number of correct positive predictions relative to total positive predictions, while recall is the ratio of the number of correct positive predictions relative to the total number of actual positives. The $F1$ score provides a measure of both the Type I (false-positive) and Type II (false-negative) errors in the model. We provide the classification results in Table 8.6, where the mean and the standard deviation of the AUC score of the model along with the precision, recall, and $F1$ scores for both cases where a transplant is received (target $=1$) and a transplant is not received (target $=0$) are listed. We also show the ROC curve for the gradient boosted trees classifier in Fig. 8.2.

The mean and the standard deviation of the performance measures were generated by training and validating each model over ten random permutations of the dataset. From these results, we see that the classifiers, especially the decision tree ensembles (bagging and random forest) and gradient boosting techniques, are performing well in classifying the data. We achieve over 80% precision for patients receiving a transplant

Table 8.6 Classification results for status of transplant

Mean	LR	SVM	Bagging classifier	Random forest classifier	ANN	Gradient boosted trees
AUC score	0.91 (0.010)	0.91 (0.010)	0.90 (0.009)	0.90 (0.008)	0.90 (0.018)	0.91 (0.011)
Precision (target $=0$)	0.97 (0.007)	0.97 (0.006)	0.96 (0.007)	0.96 (0.005)	0.97 (0.018)	0.96 (0.007)
Recall (target $=0$)	0.91 (0.013)	0.91 (0.012)	0.93 (0.011)	0.94 (0.007)	0.92 (0.031)	0.94 (0.011)
F1 score (target $=0$)	0.94 (0.007)	0.94 (0.007)	0.95 (0.006)	0.95 (0.005)	0.94 (0.011)	0.95 (0.007)
Precision (target $=1$)	0.76 (0.030)	0.76 (0.0280)	0.80 (0.028)	0.82 (0.021)	0.77 (0.063)	0.81 (0.030)
Recall (target $=1$)	0.91 (0.019)	0.91 (0.015)	0.87 (0.016)	0.87 (0.012)	0.89 (0.057)	0.87 (0.017)
F1 score (target $=1$)	0.83 (0.018)	0.83 (0.018)	0.83 (0.016)	0.84 (0.013)	0.82 (0.021)	0.84 (0.019)

LR logistic regression, *SVM* support vector machine, *ANN* artificial neural network

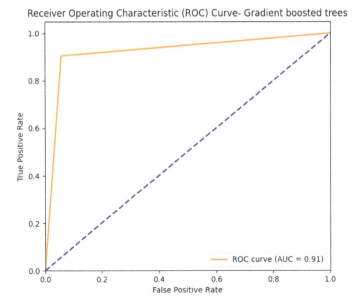

Fig. 8.2 ROC curve for gradient boosted tree classifier

(varied between 76 and 82% for different classifiers) and around 90% recall (varied between 87 and 91% for different classifiers), implying that around 90% of those who receive a transplant are identified correctly.

After generating the above predictions, we performed additional computational experiments around generating similar predictions for: (a) whether patients receive transplants within 2 years or 5 years (referred to subsequently as the 2-year and 5-year analyses) and (b) individual patient blood group-based classification. We also performed feature selection analyses by estimating the variance inflation factor (VIF) of all features and by identifying features with statistically significant association with the classification label via logistic regression models trained on the entire dataset. We estimated the accuracy metrics for successfully receiving transplants in 5 years and 2 years for different blood groups.

The results for the 5-year organ transplant status classification were similar in comparison to the overall organ transplant status results, reflecting a low chance of receiving a transplant in time (before removal) if not received in 5 years. We observed that the precision and recall values for the 2-year analysis crossed 90%, while the AUC score exceeded 95% for patients receiving an organ. Note that the classification label for these analyses was determined by whether a patient received an organ or not within the period of interest. With regard to the classification exercises for datasets with a single blood group, we observed that the AUC score of the blood group AB was the highest, but with the lowest precision, while blood group O had the lowest AUC score with greater precision. This can be attributed to the difference in organ and patient arrival rates for the blood groups, which affects the size of the training dataset for these blood groups.

With regard to the feature selection analysis, we observed that the precision and recall values for the 2-year analysis crossed 90%, while the AUC score exceeded after removing: (a) the age and KAP score features based on the R^2 estimate of the classification models and (b) the age, KAP score, and hospital name, respectively, for the VIF and logistic regression feature selection analyses. The former result can be attributed to the fact that the KAP score is a function of its components, which are also features in the dataset.

We now discuss the real-time prediction of the time to allocation for patients predicted to receive a transplant.

8.3.5 Real-Time Prediction of Time to Allocation

In order to predict the time to allocation for patients predicted to receive a transplant, we used a subset of the simulation-generated dataset developed for classification, obtained by restricting the dataset to only those cases where a transplant was successful. We used the time to allocation as the label for the same feature set. Before training, we preprocessed and balanced the input data and later conducted hyperparameter tuning to optimize the hyperparameters, similar to the exercise conducted earlier for classification. Once again, we applied several ML methods, including standard support vector regressors, decision tree methods such as bagging and random forests used as regressors, and artificial neural networks, to find the best-performing method. Once a model was trained, we estimated the coefficient of determination (R^2 error), root mean squared error (RMSE), and mean absolute percentage error (MAPE) scores on the validation dataset to compare the relative performance of the regressors. We present the accuracy metrics of the regression models in Table 8.7.

LR logistic regression, *SVM* support vector machine, *ANN* artificial neural network

It is evident that the regression models do not perform well, as evidenced by the high MAPE values (MAPE values up to 20–25% are considered acceptable in the regression literature). Further, the negative R^2 value for the ANN indicates significant overfitting. The ensemble bagging decision tree classifier yielded the best results, without the presence of any negative predictions.

Table 8.7 Regression results for time to allocation

Accuracy metrics	SVM	Bagging	Random forest	ANN
R^2	0.87	0.87	0.84	−1.42
RMSE	193.29	193.62	208.61	856.4
MAPE	80.31	89.01	190.39	91.49

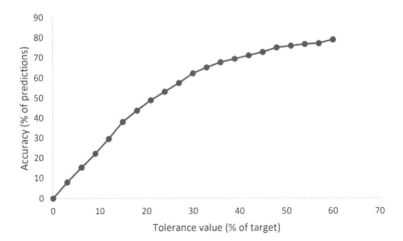

Fig. 8.3 Tolerance range analysis for real-time prediction of time to allocation of an organ

In order to better understand the regression performance, we created tolerance ranges around the prediction label (i.e., the time to allocation for a patient predicted to receive a transplant) to determine the tolerance range at which we observe reasonable performance. For example, a 20% tolerance range implied that we measure whether our predictions are within 20% of the actual time to allocation. This is then treated as a classification problem, wherein a prediction within the tolerance range is treated as 'acceptable' and predictions outside this range are 'unacceptable'. The results of this exercise are depicted in Fig. 8.3.

As expected, larger tolerance values yielded better performance. We see that at a tolerance range of 25%, more than 50% of predictions are acceptable, and at a tolerance range of 40%, more than two-thirds of the predictions are acceptable.

Following this analysis, we performed an outlier detection and elimination exercise on the dataset and retrained the ensemble bagging decision tree classifier on the revised dataset. This retrained model yielded significantly better MAPE scores. Removing these outliers caused the MAPE score to improve to 22%. As part of the outlier detection process, we determined that the worst-performing prediction cases were predominantly due to disproportionately fast transplants—in all such cases, the target time to allocation was orders of magnitude less than the predicted time to allocation. We provide the blood group-wise and total MAPE and MAD scores after outlier removal in Table 8.8.

We describe the above process to highlight the importance of outlier detection in training regression models on the dataset generated by the simulation for real-time delay prediction.

Table 8.8 Blood group-wise and entire dataset regression performance after outlier removal

Blood groups	MAPE	MAD
A	19.830	121.779
AB	29.862	186.055
B	18.049	160.645
O	18.532	162.548
Grand total	22.041	154.409

MAD mean absolute deviation, *MAPE* mean absolute percentage error.

8.4 Discussion and Conclusions

In this study, we present a hybrid modeling approach for real-time delay prediction. This approach is suitable for generating real-time delay predictions for complex queueing systems which are not amenable to analytical approaches and also do not maintain adequate queue log data required for directly training ML/statistical methods for delay prediction. HM involves combining research approaches or methods from other disciplines with one or more stages of the simulation modeling process [16]. Our approach toward real-time delay prediction for complex queueing systems involves developing a DES of the queueing system in question, validating it, and then using this validated DES to generate system state and other relevant data (e.g., characteristics of the service-seeking entity) and train a statistical/ML method on this dataset as a real-time delay predictor. Our approach, given its combination of DES and ML, fits in well within the hybrid modeling paradigm.

A natural question that arises with regard to our approach is this: Given that the ML real-time delay predictor is trained and validated on a synthetic dataset generated by the DES, how does simulation error affect the 'real-world' performance of the real-time delay predictor? Answering this question definitively is beyond the scope of this study; however, our preliminary analytical work on the first question yields the answer that simulation error and ML method error add linearly to form the total error associated with the prediction. Further, we have shown that for a validated simulation wherein 'validated' implies that the expected value of the simulation error is zero, then the expected value of the total error of the real-time delay predictor then depends only on the expected value of the ML method error. Another important question is as follows: While the above result holds for the average total error of prediction (that is, the average total error of the prediction, averaged across the total errors of individual predictions, will tend to zero for validated simulations), how does simulation error affect an individual prediction? These questions and more will need to be answered as this approach matures.

The approach proposed in this study also will need appropriate IT infrastructure for deployment. For example, in the kidney transplantation case, the state of the waitlist will have to be queried each time a new patient registers on the waitlist for generation of the real-time delay prediction. Thus, a suitable software set up for generating the feature set required for input into the ML real-time delay predictor will be required. The process of setting up the IT infrastructure required to record the input data for each service-seeking entity may lead to the generation of adequate queue log data capturing the system state so that an ML model can directly be trained. In that case, the hybrid DES-based approach may eventually be phased out. This approach will be of use until such queue log data is generated; however, prior to its phasing out, comparing the performances of both approaches may be useful as the hybrid approach may outperform the direct ML approach depending upon the extent of the inaccuracy in the recording of the queue log system state data.

Note that the DES of the queueing system in consideration may not need to be developed specifically for this purpose—the DES may a priori be developed for routine operational analysis of the system and can be repurposed for generating the synthetic dataset. On the other hand, even if it is developed de novo for the real-time delay prediction purpose, it can later be repurposed for routine operational and policy evaluation analyses.

References

1. Sharma R, Prakash A, Chauhan R, Dibhar DP (2021) Overcrowding an encumbrance for an emergency health-care system: a perspective of Health-care providers from tertiary care center in Northern India. J Educ Health Promot 10(January):1–6
2. newslaundary, Queue for queues, packed online slots, and the endless wait to be treated at AIIMS Delhi. https://www.newslaundry.com/2023/04/21/queue-for-queues-packed-online-slots-and-the-endless-wait-to-be-treated-at-aiims-delhi. Accessed 9 July 2023
3. Diwas Singh KC, Scholtes S, Terwiesch C (2020) Empirical research in healthcare operations: past research, present understanding, and future opportunities. Manuf Serv Oper Manag 22(1):73–83
4. Awoke N, Dulo B, Wudneh F (2019) Total delay in treatment of tuberculosis and associated factors among new pulmonary TB patients in selected health facilities of Gedeo Zone, Southern Ethiopia, 2017/18. Interdiscip Perspect Infect Dis 2019
5. Dharmawan Y, Fuady A, Korfage IJ, Richardus JH (2022) Delayed detection of leprosy cases: a systematic review of healthcare-related factors. PLoS Negl Trop Dis 16(9):1–14
6. Woodworth L, Holmes JF (2020) Just a minute: the effect of emergency department wait time on the cost of care. Econ Inq 58(2):698–716
7. Sun Y, Teow KL, Heng BH, Ooi CK, Tay SY (2012) Real-time prediction of waiting time in the emergency department, using quantile regression. Ann Emerg Med 60(3):299–308
8. Fatma N, Ramamohan V (2023) Patient diversion using real-time delay predictions across healthcare facility networks, vol 45, no 2. Springer, Berlin
9. Fatma N, Ramamohan V (2022) Outpatient diversion using real-time length-of-stay predictions. In: Proceedings of the 11th international conference on operations research and enterprise systems (ICORES 2022), 2022, pp 56–66. ISBN: 978-989-758-548-7; ISSN: 2184-4372
10. Xu K, Chan CW (2016) Using future information to reduce waiting times in the emergency department via diversion. Manuf Serv Oper Manag 18(3):314–331
11. Deo S, Gurvich I (2011) Centralised vs. decentralised ambulance diversion: a network perspective. Manage Sci 57(7):1300–1319
12. Baldwa V, Sehgal S, Ramamohan V, Tandon V (2020) A combined simulation and machine learning approach for real-time delay prediction for waitlisted neurosurgery candidates Vaibhav. In: Proceedings of the 2020 winter simulation conference, 2020, vol 21, no 1, pp 956–967

13. Mustafee N, Harper A, Fakhimi M (2022) From conceptualisation of hybrid modelling & simulation to empirical studies in hybrid modelling Navonil. In: Proceedings of the 2022 winter simulation conference, pp 1199–1210
14. Mustafee N, Harper A, Onggo BS (2020) Hybrid modelling and simulation (MS): driving innovation in the theory and practice of MS. In: Proceedings of winter simulation conference, vol 2020-Dec, no September 2023, pp 3140–3151
15. Mustafee N, Powell JH (2018) From hybrid simulation to hybrid systems modelling, pp 1430–1439
16. Tolk A, Harper A, Mustafee N (2021) Hybrid models as transdisciplinary research enablers. Eur J Oper Res 291(3):1075–1090
17. Ibrahim R (2018) Sharing delay information in service systems: a literature survey. Queueing Syst 89(1–2):49–79
18. Whitt W (1999) Predicting queueing delays. Manage Sci 45(6):870–888
19. Jun JB, Jacobson SH, Swisher JR (1999) Application of discrete-event simulation in health care clinics: a survey. J Oper Res Soc 50(2):109–123
20. Vázquez-Serrano JI, Peimbert-García RE, Cárdenas-Barrón LE (2021) Discrete-event simulation modeling in healthcare: a comprehensive review. Int J Environ Res Public Health 18(22)
21. Bertsimas D, Farias VF, Trichakis N (2011) Fairness, efficiency and flexibility in organ allocation for kidney transplantation dimitris bertsimas fairness, efficiency and flexibility in organ allocation for kidney transplantation. Business
22. Yahav I, Shmueli G (2014) Outcomes matter: estimating pre-transplant survival rates of kidney-transplant patients using simulator-based propensity scores. Ann Oper Res 216(1):101–128
23. Cechlárová K, Hančová M, Plačková D, Baltesová T (2021) Stochastic modelling and simulation of a kidney transplant waiting list. Cent Eur J Oper Res 29(3):909–931
24. Jouini O, Akşin Z, Dallery Y (2011) Call centers with delay information: models and insights. Manuf Serv Oper Manag 13(4):534–548
25. Chan CW, Farias VF, Escobar GJ (2017) The impact of delays on service times in the intensive care unit. Manage Sci 63(7):2049–2072
26. Gondia A, Siam A, El-Dakhakhni W, Nassar AH (2020) Machine learning algorithms for construction projects delay risk prediction. J Constr Eng Manag 146(1):1–16
27. Salari N, Liu S, Shen ZJM (2022) Real-time delivery time forecasting and promising in online retailing: when will your package arrive? Manuf Serv Oper Manag 24(3):1421–1436. https://doi.org/10.1287/msom.2022.1081
28. Zhao X, Wang Y, Li L, Delahaye D (2022) A queuing network model of a multi-airport system based on point-wise stationary approximation. Aerospace 9(7):1–14
29. Arora V, Taylor JW, Mak H-Y (2023) Probabilistic forecasting of patient waiting times in an emergency department. Manuf Serv Oper Manag Publ 1–20
30. Whitt W (1999) Improving service by informing customers about anticipated delays. Manage Sci 45(2):192–207
31. Nakibly E (2002) Predicting waiting times in telephone service systems
32. Fatma N, Ramamohan V (2021) Patient diversion using real-time delay predictions across healthcare facility networks
33. Armony M, Shimkin N, Whitt W (2009) The impact of delay announcements in many-server queues with abandonment. Oper Res 57(1):66–81
34. Ibrahim R, Whitt W (2011) Real-time delay estimation based on delay history in many-server service systems with time-varying arrivals. Prod Oper Manag 20(5):654–667
35. Dong J, Yom-Tov E, Yom-Tov GB (2019) The impact of delay announcements on hospital network coordination and waiting times. Manage Sci 65(5):1969–1994
36. Ibrahim R, Whitt W (2009) Real-time delay estimation in overloaded multiserver queues with abandonments. Manage Sci 55(10):1729–1742
37. Ibrahim R, Whitt W (2009) Real-time delay estimation based on delay history. Manuf Serv Oper Manag 11(3):397–415

38. Ibrahim R, Whitt W (2011) Wait-time predictors for customer service systems with time-varying demand and capacity. Oper Res 59(5):1106–1118
39. Senderovich A, Weidlich M, Gal A, Mandelbaum A (2015) Queue mining for delay prediction in multi-class service processes. Inf Syst 53:278–295
40. Arik S, Weidlich M, Gal A (2017) Feature learning for accurate time prediction in congested healthcare systems, pp 189–190
41. Senderovich A, Weidlich M, Gal A, Mandelbaum A (2014) Queue mining—predicting delays in service processes. Lect Notes Comput Sci (including Subser Lect Notes Artif Intell Lect Notes Bioinf) 8484:42–57
42. Thiongane M, Chan W, L'Ecuyer P (2020) Delay predictors in multi-skill call centers: an empirical comparison with real data. In: ICORES 2020—proceedings of the 9th international conference on operations research and enterprise systems, no 1999, pp 100–108
43. Ang E, Kwasnick S, Bayati M, Plambeck EL, Aratow M (2016) Manufacturing & service operations management. Manuf Serv Oper Manag 18(1):141–156
44. Baril C, Gascon V, Vadeboncoeur D (2019) Discrete-event simulation and design of experiments to study ambulatory patient waiting time in an emergency department. J Oper Res Soc 70(12):2019–2038
45. Daghistani TA, Elshawi R, Sakr S, Ahmed AM, Al-Thwayee A, Al-Mallah MH (2019) Predictors of in-hospital length of stay among cardiac patients: a machine learning approach. Int J Cardiol 288:140–147
46. Mustafee N, Powell JH, Harper A (2016) RH-RT: a data analytics framework for reducing wait time at emergency departments and centres for urgent care. In: Proceedings—winter simulation conference, 2019, no 2016, pp 100–110
47. Ordu M, Demir E, Tofallis C, Gunal MM (2021) A novel healthcare resource allocation decision support tool: a forecasting-simulation-optimisation approach. J Oper Res Soc 72(3):485–500
48. He Y, Li M, Sala-Diakanda S, Sepulveda J, Bozorgi A, Karwowski W (2013) A hybrid modeling and simulation methodology for formulating overbooking policies. In: Proceedings of the 2013 winter simulation conference, 2013, p 10
49. Barton RR, (2009) Simulation optimisation using metamodels. In: Proceedings of winter simulation conference, pp 230–238
50. Fatma N, Mohd S, Ramamohan V, Mustafee N (2020) Primary healthcare delivery network simulation using stochastic metamodels. In: Proceedings of winter simulation conference, vol 2020-Decem, no June 2021, pp 818–829
51. Balakrishna P, Ganesan R, Sherry L, Levy BS (2008) Estimating taxi-out times with a reinforcement learning algorithm. In: Proceedings of AIAA/IEEE digital avionics systems conference, pp 1–12
52. Chocron E, Cohen I, Feigin P (2022) Delay prediction for managing multiclass service systems: an investigation of queueing theory and machine learning approaches. IEEE Trans Eng Manag 1–11
53. NarayanaHealth (2019) The current scenario of kidney transplants in India. NH Narayana Health: health for all for health. https://www.narayanahealth.org/blog/kidney-transplants-in-india/. Accessed 9 July 2023
54. TimesofIndia, Waiting time for kidney transplant is 4 years in Karnataka; 5,000 on list_Doctor. https://timesofindia.indiatimes.com/city/bengaluru/waiting-time-for-kidney-transplant-is-4-years-in-karnataka-5000-on-list-
55. NOTTO (2018) Allocation criteria for deceased donor kidney transplant (guidelines)
56. Shoaib M, Prabhakar U, Mahlawat S, Ramamohan V (2022) A discrete-event simulation model of the kidney transplantation system in Rajasthan, India. Heal Syst 11(1):30–47
57. KNOS (2018) Kerela Network for Organ Sharing. http://knos.org.in/Aboutus.aspx
58. TNOS (2013) TRANSTAN | Transplant Authority Government of Tamil Nadu, Government of Tamil Nadu | Statistics. https://transtan.tn.gov.in/statistics.php
59. Lakshminarayana GR, Sheetal LG, Mathew A, Rajesh R, Kurian G, Unni VN (2017) Hemodialysis outcomes and practice patterns in end-stage renal disease: experience from a tertiary care hospital in Kerala. Indian J Nephrol 27(1):51–57

60. Cecka JM, Kucheryavaya AY, Reinsmoen NL, Leffell MS (2011) Calculated PRA: initial results show benefits for sensitised patients and a reduction in positive crossmatches. Am J Transplant 11(4):719–724
61. Abraham G, John GT, Sunil S, Fernando EM, Reddy YNV (2010) Evolution of renal transplantation in India over the last four decades. NDT Plus 3(2):203–207
62. Sreeramareddy CT, Qin ZZ, Satyanarayana S, Subbaraman R, Pai M (2014) Delays in diagnosis and treatment of pulmonary tuberculosis in India: a systematic review. Early Hum Dev 18(3):1–24
63. MohanFoundation (2018) Transplant Centres in India. https://www.mohanfoundation.org/transplant-centres/index.asp
64. Census Commissioner of India. Ministry of Home Affairs. Ministry of Home (2011). https://censusindia.gov.in/census.website/
65. Chawla NV, Bowyer KW, Hall LO, Kegelmeyer WP (2002) SMOTE: synthetic minority oversampling technique. J Artif Intell Res 16:321–357
66. Kingma DP, Ba JL (2015) Adam: a method for stochastic optimization. In: 3rd International conference on learning representations, ICLR 2015—conference track proceedings, 2015, pp 1–15

Chapter 9
Combining SD and ABM: Frameworks, Benefits, Challenges, and Future Research Directions

Susan Howick, Itamar Megiddo, Le Khanh Ngan Nguyen, Bernd Wurth, and Rossen Kazakov

Abstract System Dynamics (SD) and Agent-Based Modelling (ABM) are two commonly used simulation methods with different characteristics and benefits. When tackling a complex problem, the use of one of these methods may be insufficient and, instead, a combination of the two methods in a hybrid simulation may be required. To support modellers in the development of SD-ABM hybrid simulations, this chapter provides a comprehensive overview of methodological and practical considerations. Frameworks are presented to facilitate the implementation of hybrid SD-ABM models including the development of a conceptual SD-ABM hybrid model. The chapter then presents key benefits associated with SD-ABM hybrid modelling, which include being able to model an appropriate level of complexity, facilitate communication of the model design, enhance confidence building and reduce compute intensity. Two case studies are used to illustrate these benefits. Although there are many benefits, there are also key challenges associated with the development of a SD-ABM hybrid model, and these are discussed. The chapter concludes with a discussion of opportunities and areas for future research.

Keywords System dynamics · Agent-based modelling · Hybrid simulation frameworks · Hybrid simulation designs

---

S. Howick (✉) · I. Megiddo · L. K. N. Nguyen · B. Wurth · R. Kazakov
Strathclyde Business School, 199 Cathedral Street, Glasgow G4 0QU, UK
e-mail: susan.howick@strath.ac.uk

I. Megiddo
e-mail: itamar.megiddo@strath.ac.uk

L. K. N. Nguyen
e-mail: nguyen-le-khanh-ngan@strath.ac.uk

B. Wurth
e-mail: bernd.wurth@glasgow.ac.uk

R. Kazakov
e-mail: R.Kazakov@hw.ac.uk

9.1 Introduction

System dynamics (SD) and agent-based modelling (ABM) are two simulation approaches that are used to provide understanding about the behaviour of a system. As simulation modelling methods, they both enable the user to test out scenarios in a risk-free environment and evaluate their impact without making changes to the real system.

Whilst both methods have a long history of individually providing support for decision-makers, they can also be used together as a hybrid simulation. A hybrid simulation is a modelling approach that combines two or more simulation methods [1]. As with the broader area of mixed method modelling where two or more modelling methods are used together [2], a hybrid simulation allows the benefits of multiple simulation approaches to be combined to tackle a complex problem. The need for hybrid simulations arises as the complex nature of real-world problems can mean that a single simulation method may not be sufficient to tackle the problem and, instead, multiple simulation methods are required to address different aspects of the problem.

There has been a significant increase in the use of hybrid simulations in recent years, with publications covering both applications and theoretical considerations [1]. Several factors contribute to this trend, such as the growing exposure of modellers to more than one simulation method [3, 4] and the development of software platforms like AnyLogic that facilitate the development of hybrid simulations with minimal coding skills, thereby making them more accessible to a wider audience [5]. Furthermore, in the era of big data and open data, the amount and the disaggregated nature of data that is available to researchers and practitioners open up new opportunities for using hybrid simulations to model complex problems [4]. Whilst hybrid simulations may include the use of SD, ABM or discrete-event simulation (DES), this chapter will focus on hybrid simulations that combine SD and ABM.

Brailsford et al. [1] highlighted healthcare as the most dominant application area for hybrid simulation models and, with respect to SD-ABM hybrid models, examples include the spread of a disease [6], prospective health technology assessment [7] and application to the pharmaceutical industry [8]. In addition to this, examples of other application areas include supply chain management [9], sustainable mobility [10], automotive industry [11], project management [12], and entrepreneurial activities in universities [13]. These examples demonstrate the broad applicability of hybrid SD-ABM simulation models.

Despite numerous examples of the application of SD-ABM hybrid simulations, work focussing on methodological and practical considerations has been more limited, in particular with respect to frameworks that provide guidance to support modellers when combining SD and ABM.

This chapter seeks to provide a comprehensive overview of methodological and practical considerations associated with combining SD and ABM in a hybrid simulation in addition to considering the associated benefits and challenges. The next section will provide a short overview of SD and ABM and a comparison of the two methods. Section 9.3 will then provide guidance on how SD and ABM can be used together in a hybrid simulation by presenting frameworks to support and guide their combination. Section 9.4 will then discuss the benefits associated with using SD and ABM together in practice and will use two case studies to illustrate these benefits. Section 9.5 will then conclude the chapter by considering some key challenges and opportunities that arise when combining SD and ABM before ending with potential areas for future research.

9.2 Introduction to System Dynamics and Agent-Based Modelling

This section will introduce both SD and ABM followed by a comparison of the two methods.

9.2.1 System Dynamics

SD was pioneered by Forrester [14] and originally emerged in efforts to explain fluctuations in production, inventories and hiring in a household appliance plant [15].

SD is based on two key concepts of systems theory related to (i) the principles of feedback loops, created from interactions between components in the system and (ii) that the structure of a system determines a system's behaviour. When using SD to explain the behaviour of a complex dynamic system, it seeks to identify the effects of feedback, including the impact of time delays and bounded rationality that lead to nonlinear and often counterintuitive behaviour [16, 17].

SD is a continuous simulation modelling method that represents the structure of systems as accumulations (stocks), rates (flows), feedback, and time delays [16]. Stocks are defined as accumulations of inflows and outflows over time, and delays represent the time it takes to measure and report information, make decisions or update stocks that cause outputs to lag behind inputs. SD models are normally deterministic, with the focus being on understanding the underlying patterns in a system rather than being concerned with randomness around these patterns. SD has been criticised for this determinism, but Lane [18, p. 18] argues that an approach based on determinism and "the modelling of causal laws, which transcend human subjectivity, is a reasonable position because of the level of aggregation of models".

The SD modelling approach can contain both qualitative and quantitative models. Two commonly used qualitative models are causal loop diagrams (CLDs) and stock flow diagrams (SFDs). These qualitative models may either be used when conceptualising a problem or to help explain the behaviour that arises from a quantitative model [19]. CLDs capture and communicate key feedback loops that are important to the behaviour of a system [16], with a key advantage of being generally easy to comprehend. SFDs differ from CLDs by also identifying the accumulations in a system and the inflows and outflows that affect them. However, both types of models have recognised limitations [19]. In particular, the behaviour that rises from the structures they capture can only be inferred, and it is only by building and running a quantitative SD simulation model that behaviour can be deduced, particularly for a complex system containing many feedback loops [19]. When developing a quantitative SD model, model conceptualisation often involves defining reference modes of behaviour. After this stage, the quantitative SD model is formulated and then simulated to test out different policies before transferring the insights through implementation [17, 20, 21].

Whilst SD was originally created to explain an industrial problem, it has since been applied to an extensive range of areas. Examples include healthcare [22], project management [23], transportation [24], supply chains [25], public policy [26], environment [27], and economics [28].

9.2.2 Agent-Based Modelling

ABM emerged in several fields around the same time, with applications seeking to understand natural and social complex systems and phenomena. The modelling methodology draws from complex adaptive systems (CAS) theory [29, 30], advocating for a bottom-up approach to building and explaining systems.

The system-level outcomes in an ABM emerge from the interaction between individual-level entities called agents and their environment, and these outcomes may have feedback effects, leading to self-organisation and influencing subsequent behaviours and interactions. ABMs are typically stochastic, and simulating the model multiple times with the same parameters will yield different realisations each time. This inherent variability represents first-order uncertainty and, along with the presence of negative and positive feedbacks from agent interactions, can lead to path dependence, where small random deviations can result in significantly different system states and outcomes [31]. Time is usually represented by discrete time-steps.

There is no consensus among researchers on the essential elements for defining an entity as an agent and thus for a model to constitute an ABM. In accordance with Macal and North [32], we define agents as discrete entities with potentially heterogeneous characteristics, behaviours, and decision-making capabilities. Agents are autonomous, self-directing their decisions and behaviours based on rules that inform how they adapt to what they sense about their own state and the environment. Their rules can be associated with goal-oriented behaviour aimed at achieving individual benefit and guided by behavioural decision theory principles such as satisficing, anchoring, and adjustment, rather than necessarily assuming utility-maximising rational entities with complete information [33–35]. Agents' rules may range from simple if–then statements to complex decision-making models. Additionally, agents are social, interacting with other agents, with these interactions potentially mediated through their environment, when it is explicitly modelled. The network of agent interactions may also implicitly represent the environment. Agents may also learn from experiences and change the rules representing their behaviour, although the use of ABMs with learning has been limited to date.

Scientific methods for conceptualising and reporting ABM's structure have been gaining traction in recent years. Prior to this, the parallel development of ABM and contributions across multiple fields has resulted in diverse and ad-hoc approaches to developing and reporting on ABMs, hindering understanding of their structure and assumptions. Typically, the modelling process follows the bottom-up approach, where the actors, their characteristics, behaviours, and interactions relevant to a particular question or problem are identified. These elements are translated into computer code using either ABM-dedicated or non-ABM-dedicated programming languages and software. One of the major advancements in formalising ABMs originated in ecology, the Overview, Design concepts, Details (ODD) protocol [36, 37]. The protocol was developed as a tool to clarify the structure of ABMs, making the modelling process more scientific and replicable. It has been adopted by modellers in various fields, including social sciences [38]. The ODD protocol can also assist in conceptualising the model, organising its structure, and ensuring that the ABM design concepts have been thoroughly considered.

ABMs are applied to explore problems and systems in many different areas. These include studying ecology [39], healthcare [40, 41], economics [42], and operational research among others.

9.2.3 Comparing SD and ABM

Whilst both simulation methods seek to provide understanding about the behaviour of a system, as indicated in the summaries above, they represent systems at different levels of abstraction and have fundamental differences. A few studies have previously compared SD and ABM, including Lättilä et al. [43], Scheidegger et al. [44] and Nguyen et al. [45]. Table 9.1 is adapted and extended from Nguyen et al. [45] and highlights some of the key differences between the two methods. These differences have implications for the data required, time and skills needed to implement a model, outputs produced, and the process of validation or building confidence in the model. The different insights these modelling methods offer impacts their suitability for addressing specific questions or problems. For instance, SD models can be used to examine systems at the strategic level, while ABMs provide insights into operational-level and individual-level adaptive behaviour and interactions. Affecting or modifying individual behaviour can have ripple effects on the overall system behaviour. ABMs can also account for the effects of the environment on behaviour, including the spatial environment, leading to system-level behaviour. On the other hand, SD models are used to analyse system structure and potentially questions about their redesign. The different insights that these methods offer suggest that combining them in hybrid models can be fruitful to be able to take account of both the high-level feedback structures of a complex system and the interactions between agents in the system. However, careful consideration is required with respect to how SD and ABM are integrated. The next section will therefore consider how different frameworks that have been reported in the literature can be used to support modellers when combining SD and ABM.

Table 9.1 A comparison of System dynamics (SD) and agent-based modelling (ABM)

	SD	ABM
Modelling approach	Top-down	Bottom-up
Building blocks	Stocks, flows, feedback loops, and delays	Agents, environment, and interactions between agents
Entities	Represented by homogenously mixed stocks	Heterogeneous agents
System behaviour	Determined by system structure	Emergent from individual decisions/rules and interactions, with potentially evolving system structure
Time representation	Continuous	Discrete
Stochasticity	Normally deterministic	Normally stochastic
Qualitative model	Causal loop diagrams (CLDs) and stock/flow diagrams (SFDs)	
Quantitative model	Equation-based	Logical instructions in computer code
Conceptual model presentation	CLDs and SFDs	Overview, Design Concepts, Details (ODD) protocol; various other suggested tools such as Unified Modelling Language (UML) but not used widely across fields
Data dependency	Objective data at aggregate levels supplemented by judgemental, subjective data, and informational links	Depending on simulation aims and level of complexity or abstraction, these methods can be highly data-dependent because they model entities at the individual level and try to describe variations in their characteristics and other inputs
Outputs	Deterministic time series of population/stock levels and flows and insight into behaviour of the system	Stochastic (typically) time series of population and sub-population outputs such as number of entities in a specific state, frequency of actions, and frequency of events as well as state of the environment; insights into the system emergence behaviour; tracking individual entities
Insights	System structure and behaviour; strategic-level	System emergence; operational- and individual-level behaviours and interactions; effect of environment

9.3 Designs and Frameworks for Combining SD and ABM

This section will discuss theories, designs, and frameworks which can be used to support the combination of SD and ABM.

9.3.1 Theories Supporting SD and ABM Hybridisation

When considering frameworks for combining SD and ABM, Kazakov et al. [8] present a qualitative problem structuring approach that brings together a resource-feedback and agent-based perspective. This approach can also be used to support the hybridisation of quantitative SD and ABM. The approach is intended to provide a comprehensive resource/agent perspective on complex adaptive systems.

To develop their approach, Kazakov et al. consider four additional theories: resource dependence theory [46, 47], resource-based theory [48–50], behavioural decision theory [51, 52], and anticipatory systems theory [53, 54].

The resource-based theory aligns with SD theory and practice [55] by employing concepts such as resources, stocks and accumulation. In particular, the resource-based theory of the firm focusses on an internal perspective, suggesting that differences in firm performance stem from specific internal capabilities, unique organisational assets, or resources [56], which should lead to a sustainable competitive advantage. Complementing this perspective, resource dependence theory offers insights into how external resources influence an organisation's behaviour and how the organisation takes actions to manage external resource interdependencies.

To support the agent perspective, behaviour decision theory is employed to examine how agents follow behavioural patterns and are rationally bounded due to incomplete information and imperfect cognition. Agents are assumed to behave according to heuristic principles to reduce judgement and choice complexity [51, 57]. In addition to behaviour decision theory, anticipatory systems theory proposes that agents build "forward models" of themselves [54], enabling them to simulate different future paths with different corresponding outcomes. In the context of anticipatory systems, agents therefore possess anticipations related to behaviourally dependent payoff and/or future states of a system, which guide their optimal action selection [58].

By using these four theories together, Kazakov et al. [8] develop a resource/agent map which serves as a qualitative modelling approach that can be used to identify and explore system scenarios. This enables resource structures and agent behaviour to be considered together, forming a foundation for the hybridisation of quantitative SD and ABM models.

9.3.2 Methodological Designs for Hybrid SD and ABM Simulation

Consideration of the theories in Sect. 9.3.1 emphasises the complementary nature of SD and ABM, highlighting the benefits of their hybridisation in capturing both aggregate-level behaviour and individual-level interactions and offering a more comprehensive understanding of complex systems. Building on this theoretical foundation, methodological and practical frameworks provide a structured approach for combining SD and ABM. By bridging the gap between theory and application, frameworks facilitate the implementation of hybrid SD-ABM models, enabling researchers and practitioners to effectively leverage the strengths of both approaches and gain deeper insights into the dynamics of complex systems. The next two subsections summarise key frameworks that are detailed in the literature.

An increasing number of studies discussing hybrid SD-ABM approaches have showcased the diverse array of designs (also referred to as, for example, classes) for combining these two methods [3, 43, 59–73]. Several frameworks that offer guidance for combining SD and DES, combining analytical and simulation modelling methods, or combining methods in general, can also be applied to combining SD and ABM [74–78]. These studies provide high-level designs for hybridising SD and ABM, categorising these designs based on the direction of interaction between SD and ABM submodels, frequency of interaction, separability, and dominance of the submodels. In general, six different designs are discussed in the literature: parallel (genuinely independent), sequential (loosely coupled), interaction, dynamic, enrichment, and integration (inseparably coupled). Due to the diverse terminology in the literature for similar designs, we will provide an overview of the general concepts for each design without delving into the specific terminology. For more comprehensive explanations, readers are encouraged to refer to the cited papers.

Parallel design corresponds to Class I as described by Shanthikumar and Sargent [74], comparison mode according to Bennett [75], interfaced class as proposed by Swinerd and McNaught [3, 69], and parallel as outlined in Morgan et al. [78]. In this design, SD and ABM are used to construct independent models. These models serve either to address distinct aspects of the same problem, leveraging the strengths of each method for a specific problem, or they directly represent the same problem for the purposes of cross-validation and triangulation of outputs. The outcomes generated by these models are eventually combined to resolve the same problem or compared to enhance confidence in their respective outputs. Whether this design is considered a true hybrid model is subject to debate.

Sequential design has been documented in several works, such as Class III and IV as outlined by Shanthikumar and Sargent [74], Scenario Explanation or Crisis Response in Martinez-Moyano et al. [66], Sequential class in Swinerd and McNaught [3, 69], Cyclic interaction according to Chahal et al. [77], and Sequential design in Morgan et al. [78]. The term "sequential" inherently conveys the logical sequence of processes within this design. This design also includes distinct SD and ABM submodels, in which one model informs the other. The first simulation is executed to generate the necessary output, which subsequently serves as input for the second simulation. Once the first simulation terminates, the second simulation begins. Data are transferred strictly in a unidirectional manner and only once from the first simulation to the second. The output of the second simulation represents the final output of the hybrid model.

Interaction design considers SD and ABM submodels as equally important, with cyclical interaction between them during runtime. The interactions occur multiple times, enabling bidirectional exchanges. Sequential design is a special case of the Interaction design, where the interaction occurs only once and in one direction. Several publications have outlined the design principles for a hybrid simulation model that aligns with Interaction, including the Hierarchical format discussed by Chahal and Eldabi [76], Parallel interaction examined by Chahal et al. [77], and Interaction design presented by Morgan et al. [78].

Integration design represents an approach aimed at seamlessly combining distinct simulation modelling methods into a unified hybrid model, where explicit differentiation between SD and ABM components becomes challenging. This design, described by Swinerd and McNaught [3] and Brailsford et al. [67], offers a comprehensive perspective on the problem, facilitating the continuous flow of information and feedback and enabling the capture of interactive effects within the system. Various terms have been employed to refer to this design, including Shantikumar and Sargent's Class II, Bennett's Integration, Martinez-Moyano et al.'s Intertwined models, Chahal and Eldabi's Integration mode, Brailsford, Desai, and Viana's "Holy Grail," Swinerd and McNaught's Integrated class, and Morgan et al.'s Integration. These designations essentially denote the concept of integration.

While there is general consensus on the definition of Integration design across multiple studies, Swinerd and McNaught [3] provide detailed insights into the different approaches for developing an integrated hybrid model. They propose three sub-designs within the integrated class, including agents with rich internal structures, stocked agents, and parameters with emergent behaviour. For agents with rich internal structure, SD is utilised to model the decision logic and cognitive structure of agents within an ABM, influencing their behaviours, decision rules, and characteristics (Fig. 9.1). This integration sub-design has been suggested by Parunak et al. [60], Akkermans [61], Schieritz and Größler [62], Borshchev and Filippov [63], Lorenz and Jost [64], and Vincenot et al. [68], and Wallentin and Neuwirth [73]. Stocked agents involve individual agents within an ABM interacting with a single SD model, in which an aggregate measure of an ABM is bounded by a stock level of an SD module (Fig. 9.1). For parameters with emergent behaviour, an SD model parameter is informed using an aggregate observation or measure from an ABM (Fig. 9.1).

Enrichment design has received limited discussion in the literature, with notable mentions in Bennett [75] and Morgan et al. [78]. Chahal and Eldabi's Process Environment format can be argued to be a specific case of the Enrichment design, where SD components hold greater dominance than ABM components [76]. This design aims to integrate different simulation modelling methods into a unified hybrid model, with one method taking precedence and benefiting from elements of the other. As enrichment and integration designs exhibit significant similarities, there

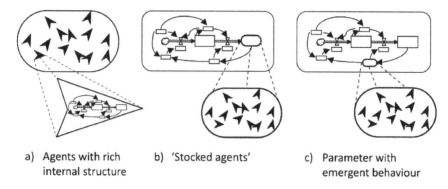

Fig. 9.1 Swinerd and McNaught's three concepts of integrated SD-ABM hybrid design (reproduced from Ref. [3]. Copyright 2012 by Elsevier B.V. Reprinted with permission)

exists a continuum ranging from enrichment to full integration in hybrid simulation modelling designs, dependent on the relative dominance of the adopted simulation approaches.

Dynamic Switching design, exemplified in the studies of Bobashev et al. [65] and Vincenot et al. [68], introduces the capability for dynamic transitions between SD and ABM within a model's structure. Wallentin and Neuwirth [73] further expand on this concept by proposing four different representations for entities within a system. Consequently, several SD-ABM model configurations arise, where distinct entities may be represented by stocks or agents. The model dynamically switches among these various SD-ABM configurations, guided by a threshold that determines the population size at which the influence of heterogeneity among individuals within each entity type becomes either significant or negligible. The authors justified the utilisation of this design to optimise the balance between predictive accuracy and computational performance.

Whilst providing guidance to modellers who wish to combine SD and ABM, the designs discussed above exhibit two primary limitations. Firstly, they lack explicit instructions regarding the steps and considerations that modellers should undertake to make informed decisions about the design of a hybrid model. Secondly, they remain abstract and high-level, limiting their practical applicability. For instance, Enrichment, Interaction, and Integration designs share many similarities and differ only in terms of the separability and dominance of the SD and ABM submodels. The relative nature of these characteristics makes it challenging to select an appropriate design for a hybrid model. In addition to these limitations, the majority of existing hybrid simulation modelling studies focus on addressing domain-specific issues, such as inter-organisational network development in Akkermans [61] or supply chain management in Schieritz and Größler [62]. Consequently, there is a need for more comprehensive and detailed guidance that specifies when, why, and how to effectively combine SD and ABM approaches in a broader context that cuts across domains.

9.3.3 Framework for Combining SD and ABM: Moving Towards More Practical Guidance

In an effort to address the gap in practical guidance, Nguyen [79] presents a framework for combining SD and ABM that outlines the necessary steps for constructing a conceptual hybrid simulation model. The framework focuses on the conceptual phase of the modelling process, which has been identified as "the least developed stage in the modelling cycle, despite its importance" [1, p. 1].

The framework consists of four stages. In the first stage, modellers delve into the problem of interest by defining the modelling objectives, scoping the problem, and specifying its characteristics, following the conventional approach employed in single-method simulation modelling studies. Moving on to stage 2, modellers assess whether single simulation methods or a hybrid simulation is the most suitable approach to model the problem based on the exploration conducted in stage 1. Stage 3 involves designing the modules that constitute the hybrid model, and stage 4 entails establishing linkages between these modules. Throughout the framework description, Nguyen [79] also considers the activities that modellers should undertake to enhance confidence in the conceptual model.

This framework also proposes specific elements that a modeller should describe to provide a comprehensive representation of a conceptual hybrid model, catering to other modellers and stakeholders for understanding and reproducibility. These elements include (i) modules' names and descriptions of what parts of a system they represent, (ii) abstraction level, rationale for the chosen modelling methods, and content associated with each module, and (iii) the linkages among modules, including information flows, interfaces, and updating rules. This comprehensive depiction of the overall hybrid model at the conceptual level communicates the model design to stakeholders and the broader research communities, fostering confidence and facilitating the verification and development of computerised simulation models [77, 80].

As part of the wider framework in Nguyen et al. [79], Nguyen et al. [81] described new practices for designing interfaces between SD and ABM modules in a hybrid simulation model. The interface between modules defines how the information is passed from a generating module to a receiving module during the runtime of the hybrid model. Nguyen et al. [81] classified interface designs between SD and ABM modules and explained how information is generated and handled for each design (Table 9.2). These interface designs include additional feedback mechanisms that were not implemented in previous hybrid models: (i) the SD module generates information that shapes the environment of the agents or influences their decision-making, and (ii) the aggregation of the agents' characteristics or actions serves as a representation of a stock or parameter in the SD module. In addition to interface designs consolidated from existing literature, two novel interface designs are introduced: (i) a stock level that defines the network topology of the agent, and (ii) the agents' state variables that impact the flows within the model. The framework does not represent an exhaustive list of interface designs. This is primarily attributed to the "art" of modelling, whereby different modellers may opt for distinct representations of a given situation. Nevertheless, the framework serves as a guiding structure that can be expanded upon in the future research.

With respect to the flow of information from an ABM to an SD module (predominantly relevant for interfaces 5–7 from Table 9.2), traditionally, information flows have been based on individual actions or states of agents. However, ABMs are a means for modelling multi-level (emergent) dynamics [93], with an example being 'complex events,' that are defined by multiple constituent actions or state changes of agents and the relationship between these events [94]. SD is based on homogeneously mixed stocks, and it is not able to track individual entities or actions through the system. Therefore, the ability to model complex events in ABMs provides an additional feature that modellers can consider when combining ABMs with SD in hybrid simulations. Wurth [13] uses the concept within a hybrid simulation by analysing complex events that result from the simulation. These events are composed of specified interactions between agents linked in time (i.e. specific interactions must occur in a defined order). Complex events can also be aggregated into an SD module as part of a sequential design (e.g. basing a parameter in an SD module on the complex events arising from an ABM module) and in an integrated design. Lastly, the occurrence of a complex event (or a pre-defined number of occurrences) could also be used in an 'event-reconfigurable' (3) or dynamic switching hybrid simulation design.

9.3.4 Summary

This section has described work that seeks to provide modellers with practical guidance on combining SD and ABM. Further work in this area would be beneficial to modellers, offering them more detailed guidance in choosing and implementing appropriate designs for their hybrid SD-ABM models.

Alongside practical support combining SD and ABM, modellers should be aware of the benefits and challenges associated with such modelling. This awareness can help them leverage the two methods when conceptualising a model and navigate the obstacles they may encounter. The next two sections will therefore discuss each of these in turn.

9.4 Benefits of Combining SD and ABM

SD and ABM provide distinct but complementary approaches to the study of complex systems. Combining these two approaches in a hybrid simulation provides several benefits, facilitated by progress across methodological, and practical dimensions of combining both simulation methods. We illustrate some key benefits by drawing predominantly on two case studies, supported by additional literature. These two case studies are briefly described next before then using them to illustrate the benefits.

Table 9.2 Interfaces between SD and ABM modules in a hybrid model (a detailed description, visual presentations, and examples for each design are provided in Nguyen et al. [81])

Interface	Description	Example studies
Information flows from SD module to ABM module		
(1a) Stock levels define agent-specific state variables	Stock levels of the SD module embedded in each agent determines the agent's state variables	Caudill and Lawson [82]
(1b) Generating agents from stocks	Individual agents with distinct characteristics can be generated from stocks that represent larger population using distribution functions	Djanatliev and German [83] and Kolominsky-Rabas et al. [84]
(2) Stock levels define behaviours of individual agents	Stock levels determine the corresponding behaviours that individual agents will execute. SD characterises the environment in which agents live or their internal decision-making process	Jo et al. [85] and Cernohorsky and Voracek [86]
(3) Stock levels bounds aggregate measures of agents	Stock levels serve as upper limits for aggregated measures of agents, ensuring that these measures do not surpass the specified levels. In interface design (2), individual agents' behaviours are directly influenced by the stock levels, whereas in this design, the stock levels indirectly affect behaviours by considering the collective measures of agents	Verburg and Overmars [87] and Robledo et al. [88]
(4) Stock levels define agents' network topologies	Stock levels determine the corresponding spatial relationships and/or interacting network topologies among agents	N/a
Information flows from ABM module to SD module		
(5) Agents' state variables affect flows	Agents' state variables can undergo changes during a simulation as they execute behaviours and interact with other agents/the environment, affecting flows	N/a
(6) Behaviours of agents affect flows	Agents' behaviours can affect flows by increasing/decreasing parameters used in equations for flows	Mazhari et al. [89] and Chen and Desiderio [90]
(7) Aggregated measures of agents affect flows	Collective measures of agents can affect flows. When SD and ABM modules represent different parts of a system, the physical transfer of agents from ABM to SD is symbolised as an inflow into a stock	Swinerd and McNaught [91] and Jo et al. [85]
(8) Network topologies affect flows	Spatial/social relationships and/or network topologies of agents can affect flows	Vincenot and Moriya [92]

The first case study is reported in Nguyen et al. [6]. A hybrid SD-ABM model is developed whose aim is to understand the interdependencies between intra- and inter-facility transmission of Covid-19 in a network of multiple heterogeneous care homes (Fig. 9.2). The complexity and multi-scale characteristics of the problem presented a challenge to the use of SD or ABM on its own. Thus, the two simulation methods were combined in a hybrid model to achieve a comprehensive representation of such complexity. The hybrid model consisted of three modules constructed using either SD or ABM: Network (of Care Homes) (ABM), Temporary Staff (ABM), and Intra-facility (transmission in individual care homes) (stochastic SD). The temporary staff (also known as bank or agency staff) act as a conduit, transmitting the infectious agent across the network of care homes, with each individual care home population being treated as homogenous. Therefore, it was crucial to consider temporary staff in a separate module. The model was developed using information gathered in interviews and discussions with stakeholders, including staff, managers, and policy makers in HSCP Lanarkshire and the Scottish Government.

The second case study is reported in Wurth [13]. A hybrid SD-ABM model is developed with the aim of understanding how resources, organisational level quality, and the entrepreneurial reputation of UK universities affect their engagement with businesses. Universities are a key driver of innovation and crucial for fostering socio-economic development in regions through research, education, and knowledge exchange. University-industry interactions exemplify inter-organisational collaborations, where one partner (the university) collaborates with a large number of other

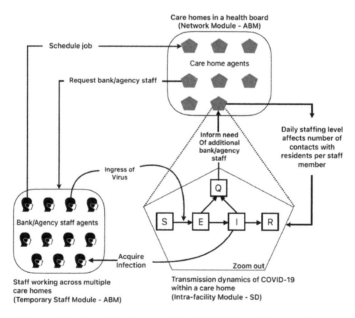

Fig. 9.2 Architectural design of the integrated hybrid SD-AB model comprising three modules (reproduced from Ref. [6]. Copyright 2022 by Elsevier B.V. Reprinted with permission)

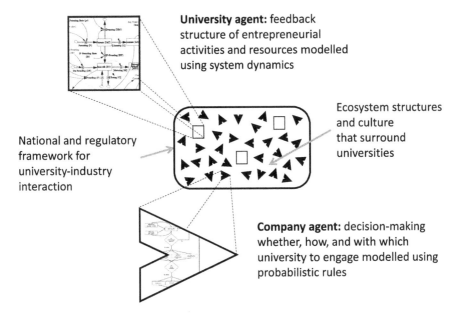

Fig. 9.3 Stylised architectural view of the integrated SD-AB model with three university agents (squares) whose behaviour is modelled using SD and several company agents (black arrow heads) that follow probabilistic rules [13]

organisations (here, companies) through a number of different channels such as licencing technologies, consulting, collaborative and contract research. The overall design of the model is an ABM, in which universities' behaviours are modelled using SD, and company agents follow probabilistic rules (Fig. 9.3). The simulation is based on a triangulation of data from multiple surveys, interviews with key university decision-makers, and the academic literature.

9.4.1 Modelling the Appropriate Level of Complexity

Since the world as a whole and the problems tackled by modellers are highly complex, single modelling approaches cannot always cope with this complexity and richness to sufficiently address specific questions [95]. For example, decision-making problems often span multiple levels of aggregation (e.g., organisational structures, spatial scales) and combining these within a single simulation model is where hybrid simulations are often superior to single-method simulations [3, 4]. While SD focuses on averages and homogenous populations, ABM can complement SD when heterogeneity or interactions between entities is essential for understanding the dynamics of the system. This includes capturing spatially distributed dynamics or network structures among others [96, 97].

For case study 1, an ABM was chosen to model the temporary staff due to the heterogeneous temporal and behavioural patterns of individuals' movement across a network of care homes that drives the spread of Covid-19. Events such as a number of care homes experiencing outbreaks in communities with low infection prevalence may emerge from the collective movement actions of temporary staff agents. In addition to individual staff preferences, the pattern of movement depends on individuals' working record as care homes desire to utilise the same temporary staff. Furthermore, using ABM to model temporary staff offers more flexibility for explicitly incorporating the restriction of movement on these staff within a group of care homes. These behavioural rules and interventions are difficult to model in SD as it would consider these individuals homogeneous. By contrast, interactions between individuals (i.e., staff and residents) within a care home are modelled at a higher level of aggregation using SD as how these interactions drive the intra-facility transmission, and the risk of outbreaks are not the focus of high-level policymakers whose decisions this model aimed to support.

For case study 2, two issues were crucial in addressing the gap in the literature on academic entrepreneurship and university-industry interactions and advancing understanding of their dynamics: including multiple universities with different characteristics in the same model and explicitly modelling firm heterogeneity [13]. Universities do not operate in isolation, but they contribute to the innovative climate, (regional) absorptive capacity of the industry, and, therefore, the environment in which other universities operate. Absorptive capacity and resources more broadly are not distributed evenly among companies. The ABM module captures this heterogeneity and allows a modeller to 'zoom in' on relevant parts of the system where required. On the other hand, SD captures the relevant organisational level dynamics within universities at an aggregate level. This 'zooming out' allows modelling potential interventions and strategies by university decision-makers without increasing the number of assumptions by breaking down their impact to individual academics or colleges, schools, or departments.

The value derived from interactions between universities and companies typically comes from repeated interactions, i.e. partnerships. These partnerships build trust, as well as organisational and technological proximity that facilitates the exchange of knowledge and know-how. Therefore, universities want to stimulate and facilitate repeated interactions with the same company. This cannot be modelled easily using SD because of its focus on aggregation and homogeneously mixed stocks. Beyond the capability to integrate heterogeneity among firms in the hybrid model, the design used in the study also allows identifying 'complex events,' temporal patterns of behaviour of single or multiple agents, as a post-simulation analysis.

Brailsford et al. [67] argue that although using one simulation method is possible to represent problems at a macro- and micro-level at the same time, they described this approach as "a case of hammering in a screw" because it forces modellers to use a simulation method that may not be suitable for all components of the problem. Morgan et al. [78] agree that it may not be possible to develop a model using one method to obtain all the intended objectives without the need for additional assumptions, which risks making the model less representative of reality.

In hybrid simulation research, the primary methodological objective is to incorporate elements that utilise and build on the core concepts of individual paradigms while exploring innovative ways to combine them at various levels of granularity [96]. By leveraging the strengths of each method in a combined manner, the complexity of the problem can sometimes be reduced as fewer assumptions that are driven by the tools rather than the context or problem need to be made [77]. This also helps address a common dilemma that modellers face with using single simulation methods. On the one hand, they may choose to model the whole problem using a single method, but this approach potentially leads to invalid assumptions or oversimplification due to limitations imposed by the tool. On the other hand, they can focus solely on those parts of the problem that are suitable for their chosen method and acknowledge that the remaining aspects lie beyond the model's scope [1]. By adopting a hybrid simulation approach, modellers can navigate the dilemma and strike a balance between comprehensiveness and methodological suitability, without being unduly constrained by the limitations imposed by a single tool.

9.4.2 Facilitate Communication of the Model Design with Stakeholders

Hybrid simulations are often better suited for modelling multiple levels of aggregation, making them well-suited to engage a range of stakeholders who operate across these levels of aggregation [72]. In addition, they provide the ability to direct stakeholders' attention to an appropriate level of detail at different parts of the system, which is particularly useful when eliciting stakeholders' views to inform model design.

For example, in case study 1, the model supported decisions at a network level and ABM was used to consider how temporary workers moved around a network of care homes. While the entire model could have been modelled using ABM, as indicated in Sect. 9.5.1, interactions between individuals within a care home were modelled using SD. This choice was made because the interactions that drive the intra-facility transmission and the risk of outbreaks were not the focus of the work and thus the model. Modelling the interactions within care homes using ABM could have been distracting, potentially pulling decision-makers who are viewing the model into unnecessary levels of detail for understanding the specific problem at hand. Detail, however, was necessary for understanding the pattern of movement of temporary staff across care homes. The uniqueness of how individual staff moved between the care homes affected the spread of disease across the network. Modelling this element as an ABM encouraged discussions between the modellers and stakeholders, including care home managers and staffing agencies, on the heterogeneity in these staff and the details of scheduling decisions that describe which care home they work in on a given day. This allowed a low-level representation of the pattern of movement that was informed by the stakeholders. In essence, the modellers were able to guide the decision-makers through the model, zooming in and out of different parts of the system as appropriate. Moreover, hybrid simulation facilitated collaborative decision-making between the modellers and stakeholders/decision-makers during the process of gathering information and developing the model. It enabled the modellers to implement shared decisions on modelling assumptions and model simplifications that emerged from their interactions.

The benefits of communicating the model design to stakeholders are similar for case study 2. The dynamics within universities are modelled at an aggregate level because the relevant characteristics are at the organisational level, as described in Sect. 9.4.1. Breaking this down to individual academics or departments would have introduced complexity and could distract from the main focus of the model and the problem it addresses. Higher levels of detail regarding the types of companies that engage with their university, such as size and innovativeness (details that are modelled using ABM), are however relevant to decision-makers within universities.

9.4.3 Enhance Confidence Building

Hybrid simulation models provide a comprehensive representation of the system being studied as SD and ABM have different strengths in representing different aspects of the system. This allows stakeholders to have a more complete understanding of the system, fostering increased confidence in the model. This impacts trust between the stakeholder, the modeller, and the model, affecting model acceptability and results implementation [98]. Also, by using both SD and ABM, a model can show how the interactions between agents and the system, as whole impact outcomes, provide stakeholders with clearer insights into the system's behaviours and the underlying mechanisms driving those behaviours. This enhances stakeholders' ability to grasp the dynamics of the system, leading to more effective communication and informed decision-making processes.

In case study 1, the use of stochastic SD for modelling the intra-facility dynamics of transmission within each care home in a multi-agent network helped build confidence in the model among stakeholders. While alternative approaches existed, they presented certain challenges. One potential approach to capture the transmission dynamics would have involved directly incorporating related variables such as the number of infected resident and staff members in isolation as state variables of the care home agents. However, this approach would have necessitated making multiple additional assumptions about the behaviour of these variables over time. Convincing stakeholders and modellers of the validity and credibility of such assumptions and this level of abstraction would have required considerable effort.

Another alternative option would have been to model the intra-facility transmission using ABM exclusively. However, aside from other considerations, gathering sufficient data to model the microstructures and interaction patterns within each care home in the network would have been challenging. This limitation would have resulted in an ABM module that closely resembled the stochastic SD module or,

9 Combining SD and ABM: Frameworks, Benefits, Challenges …

alternatively, necessitated additional assumptions about the structural and operational characteristics of care homes and resident and staff behaviours. Such assumptions would have faced significant scrutiny from individual care home managers.

A decision was, therefore, made to adapt a well-established epidemiological SD model, the Susceptible-Infected-Recovered (SIR) model [99, 100], to effectively capture the transmission dynamics within each care home. By abstracting the detail, SD required less data and assumptions. Also, as our stakeholders involved teams in the Scottish Government and the UK Government that included infectious disease modellers external to our project and decision-makers, they were familiar with and trusted the epidemiological SD model.

There are, again, similar benefits in this regard in case study 2. The problem could have been addressed either with a pure SD or a pure ABM. However, both options would have come with additional challenges. Building confidence in the model is based on both the structure of the model as well as its outputs. In a pure ABM, it is difficult to model the dynamics within universities. This would require modelling individual academics as counterparts for collaborating businesses. However, both self-reported data from universities and primary data from interviews with decision-makers are mostly concerned with the organisational level. Breaking this down further to the individual level would require further assumptions to model these dynamics at a finer level of granularity. In addition, university leaders (as the main stakeholders for this model together with policymakers) informed the design of the causal structure of the SD module through interviews. They are, therefore, able to recognise aggregate metrics such as the number of patents, licences, and the volume of collaborative research that they are used to working with. On the other hand, a pure SD model would model the interactions with firms as stocks and flows that are affected by rates. This does not correspond to the perspective of the main stakeholders, who look at collaborations with distinct entities in their daily work. Translating this into rates would be an abstraction that is potentially hard to understand and would also complicate the model as multiple stocks and flows need to be created for the same type of interaction (e.g., licensing) for different types of firms. Therefore, the design of the hybrid simulation, coupled with the active involvement of stakeholders in the model development process, built confidence in the simulation and, ultimately, the results.

9.4.4 Reduce the Compute Intensity

Hybrid simulations also provide a means to reduce the computational intensity to find solutions or simulate the system behaviour over time [74]. Different simulation methods have varying levels of computational performance. SD typically falls on the lower end, as simulating differential equations is relatively quick with modern computers. In contrast, ABM can easily become very computationally demanding when dealing with a large agent population and complex agent rules and interactions. ABM also often requires a vast amount of data and several simulations to generate a distribution of results that accurately represents the system dynamics rather than extreme realisations. This is time-consuming to collect and execute. Using SD to model parts of a system that do not rely on heterogeneity and low levels of aggregation could significantly reduce the required computational intensity.

For case study 1, a comparison of the runtimes was conducted between two parallel models: the hybrid model described in Nguyen et al. [6] and an ABM implementation, where the stochastic SD Intra-facility module of the hybrid was developed using ABM. Both models consisted of 12 care home agents, with an average of 65 residents and 80 permanent staff members. The runtime for 1000 simulations over 90 time-steps in the hybrid model was approximately 30 seconds, whereas the parallel ABM model took 10 minutes to complete. Both simulations were performed on the same computer with a 16-core processor. Although a runtime of 10 minutes may not be excessively long and remains feasible, the relative duration in the ABM-only simulation is significantly longer compared to the hybrid model. When simulations involve a larger number of agents, state variables and actions or require running for more time-steps, the time to execute simulations would increase substantially.

For case study 2, a comparison of the runtime of the hybrid SD-ABM simulation and a pure ABM was not performed. However, for the three universities, 2034, 1126, and 494 individual academics would need to be modelled, respectively in a pure ABM model. In addition to the increased number of agents, these academic agents would interact with individual companies, which makes modelling the interaction between universities and companies much more complex. Therefore, this would have significantly extended the runtime of the simulation.

Mustafee et al. [72] demonstrate the benefits of hybrid simulation modelling methods with respect to balancing between computational intensity and predictive performance, using Djanatliev's [101] healthcare study as an example. They develop three models for the same problem using SD, ABM, and a combination of both. The SD model proves to be fast but less accurate, while the ABM model is more accurate but takes 1.5 hours to run and cannot handle populations larger than 20,000 agents. To overcome these limitations, they then developed a hybrid simulation model by using ABM to represent specific parts of the original SD model in which greater detail was of interest to the problem they considered. In addition, they aggregated agents with similar properties in the original ABM into one "super-agent" in the hybrid model. This hybrid simulation model generated results comparable to those of the ABM in an acceptable runtime. When weighing result accuracy and model simplicity, it is important to emphasise that the level of accuracy is dependent upon the research problems and objectives. For example, estimating the costs of an infection prevention and control intervention for resource planning and allocation would require a higher level of result accuracy than evaluating the clinical effectiveness of the intervention for directing further research.

The ecology dynamic hybrid SD-AB model in Wallentin and Neuwirth [73] similarly illustrates the use of hybrid simulation that balances the trade-off between the predictive and computational modelling performance. The model dynamically switches between SD-ABM configurations, where, for instance, one entity may be represented by stocks and another entity by 11 agents. The switching point is informed by a threshold determined by the size of the population of interest. This results in heterogeneity and spatial networks among individuals of each entity type having more or less impact on the model's outcomes. A similar approach is adopted in Bobashev et al.'s [65] epidemiological modelling study. The model adopts an ABM approach when the number of infected people is small and individual variation is critical for studying the dynamics. It switches to an SD model after the infected population becomes large enough to apply a population-averaged approach [65].

9.5 Summary

We have outlined several benefits of developing hybrid SD-ABM simulations. These benefits are neither mutually exclusive nor can all of them be maximised simultaneously. For example, while modelling larger parts of the system using ABM can increase the level of detail and complexity, this also leads to higher computational demands. Therefore, hybrid SD-ABM simulations and their benefits should not be judged against individual criteria but a set of desired benefits in the light of the problem and its context, taking into account the purpose of the model and the stakeholders involved.

Whilst there are a number of benefits in using hybrid SD-ABM models, there are also challenges associated with their use. The next, and final, section discusses some of these challenges, but at the same time highlights opportunities that arise from these challenges before concluding with suggested directions for future research.

9.6 Challenges and Opportunities

A modeller may face a number of challenges when building and using a hybrid SD-ABM simulation. Whilst this section does not claim to be fully comprehensive, it aims to highlight some key challenges that may be faced and propose opportunities for further research to tackle them.

9.6.1 *General Challenges of Mixing Methods*

Whilst this section will focus on the challenges associated with combining SD and ABM, it is worth briefly noting some of the practical challenges associated with combining any two modelling approaches, whether they are simulation approaches or other types of modelling approaches. As compared to using a single modelling method, there can be issues associated with explaining a mixed modelling approach and the need for increased time and resources. There is also the need for additional knowledge and expertise due to applying two different modelling approaches [102,

103]. However, these challenges have not prevented researchers and practitioners from adopting a mixed-method approach due to the benefits that can be gained from bringing together the strengths of different modelling approaches to represent and understand a complex problem or system. The remainder of this section will focus on specific challenges associated with combining SD and ABM.

9.6.2 Differences in Characteristics

Referring back to Table 9.1 in Sect. 9.2, the differences in characteristics such as entities, time representation and stochasticity create challenges when the two approaches attempt to communicate with one another. For instance, there may be a need to change from discrete values to a distribution (when moving from a homogeneous to heterogeneous population), from a deterministic to a stochastic approach or from discrete to continuous time and also a need to confirm that different levels of aggregation correspond to one another.

For example, case study 1 in Sect. 9.4 can be used to demonstrate the challenge of integrating the deterministic and stochastic nature of SD and ABM, respectively. It can also be used to illustrate the use of stochastic SD to address this challenge. In this hybrid model, a stochastic SD module is embedded within each care home agent within the network, capturing the intra-facility transmission dynamics. Section 9.4 discussed the rationale for using SD rather than ABM for this module. However, a deterministic SD approach would fail to capture the stochastic nature of intra-facility transmission dynamics and virus extinction [104, 105], factors that significantly influence the disease transmission dynamics, including the risk of outbreaks within care homes and the subsequent spread across the network. To overcome this limitation, a comparative analysis was conducted by incorporating parallel deterministic SD, stochastic SD, and ABM modules, each offering a distinct representation of the same system at different levels of abstraction (Appendix F in Nguyen [79]). The resulting hybrid models, incorporating these diverse intra-facility transmission modules, were evaluated based on average outputs of infection numbers over time and cumulative infection numbers. These outputs were comparable in all three models with stochastic SD and ABM modules providing insights into output uncertainties. However, the model with the deterministic SD module failed to capture the risk of outbreaks and the spread patterns. When using deterministic SD, the hybrid model outputs remained relatively stable despite stochasticity in staff movement. This also affected the dynamics. Thus, when considering further research opportunities, the implications of the use of stochastic vs deterministic versions of SD and ABM on the outputs of hybrid models can be explored.

9.6.3 Specific Designs

Whilst Sect. 9.3 discussed a number of designs for combining SD and ABM, the designs provided guidance at a high level, and challenges can still arise when implementing these designs. One example of this is when aggregating from ABM to SD as part of an integrated design.

Interactions between agents in an ABM can lead to behaviours that cannot be readily captured using state aggregation of the agent population. For example, in case study 2 in Sect. 9.4, the partnership between universities and industry, or recurring interactions between different organisations more generally, requires examining multiple actions for the same agent over time. Handling sequences of interactions poses an additional challenge as it may not be initially known whether a specific interaction will result in a partnership or complex event/(emergent) multi-level behaviour more generally. In case study 2, this issue was avoided as complex events were only extracted post-simulation and not aggregated into the SD model. However, when aiming to include complex events into an integrated hybrid simulation, knowledge of the sequence of events that should be classed as a 'partnership' is needed from the first interaction. This raises a general question about the relationship between (sequences of) events (ABM) and homogeneously mixed stocks and aggregate parameters (SD) when combining both approaches in a hybrid simulation. Further, guidance would be beneficial with a need to developing frameworks and guidance for aggregating beyond individual agents' states to an SD module within a hybrid simulation.

9.6.4 Standardisation of Documentation for Model Transparency and Replicability

Effective communication and documentation tools for SD and ABM are essential for ensuring model transparency, replicability, and validation. Equations, CLDs, and SFDs communicate and document the structure and assumptions of an SD model concisely. Doing this for ABMs is more difficult since they comprise logical instructions implemented in code that are often long, and unlike mathematical equations, the (programming) language used is not consistent. The ODD protocol was designed to address this challenge and overcome the difficulty for non-expert software engineers to interpret UML diagrams [36]. Although not as concise as tools for reporting on SD, the ODD protocol provides a consistent and comprehensive documentation tool to report on ABMs' structure, assumptions, and design rationale. Stakeholders with a basic but sound understanding of SD and ABM should be able to understand the model using their respective communication tools. The STRESS guidelines can also be used for documenting system simulations [106]. These guidelines provide checklists for SD, ABM and DES. Whilst the ODD protocol focusses on documenting a model, the STRESS guidelines go beyond this and reports on all stages of the simulation study. They are less prescriptive than the ODD protocol, providing a checklist of elements to document and consider without suggesting a specific structure. Each of the above tools also provide sufficient information for model replication, a central tenet of the scientific process. Further, a transparent well-communicated and documented model is also necessary for confidence building and validation. Modellers cannot evaluate the validity of an opaque model, and without a shared understanding among stakeholders, it would be difficult for stakeholders to evaluate and agree on whether a model is valid for a particular purpose.

Hybrid models lack tools that are used consistently for transparently documenting and communicating their structure, assumptions, and rationale. This hinders understanding about existing models, learning from them, and moving the field towards a scientific approach to model development. With respect to the STRESS guidelines that include individual checklists for SD and ABM, these checklists could be brought together for a hybrid SD-ABM models, although would not cover elements specifically related to hybrid models. Although the ODD protocol was designed for ABM, parts of it are relevant for simulation methods other than ABMs [107], and a few hybrid SD-ABM model studies have adopted the protocol and used its submodels section to describe SD modules, including the care home model we describe in Sect. 9.4 [6, 92, 108, 109]. Nonetheless, the original ODD protocol is designed for ABMs and, similar to the STRESS guidelines, does not consider elements that are essential for understanding hybrid models. A couple of examples come to mind. First, to replicate a hybrid model, detailed information must be provided about how the models are combined. A couple of applied studies have extended the ODD protocol by adding a description of how the models are combined to the submodels section [92, 108]. Second, in the ODD protocol design concepts section, modellers explain design elements specific to ABMs and provide a rationale. A recent review noted the lack of rationale provided for hybrid models [45]. It may be beneficial to extend the design concepts section to include concepts specific to hybrid modelling, such as the type of hybrid design (sequential, enrichment, integration, interaction, or parallel), which modelling method is dominating for the relevant designs, and the types of interfaces. Similar amendments have been previously made in the ODD + D protocol to more explicitly describe decision-making in human agents [110].

The hybrid modelling research community has an opportunity to come together as hybrid model use is exploding and develop or decide how to extend existing tools to consistently report hybrid models. The community could increase adoption with consensus on a tool. We do not suggest that an ODD protocol or its extension or the STRESS guideline are necessarily the best approach for describing hybrid models, but either of these tools could gain traction. We do believe that a transparent communication tool for hybrid models could improve learning for future hybrid modelling and support their transparency and validation.

9.6.5 Verification and Validation

Understanding a model is necessary but not sufficient for assessing its validity. A recent review highlighted the dearth of methods for validating hybrid models, with most hybrid model studies failing to report on verification and validation, and the few that do only verify and validate the single-method submodels [1]. Even this restricted approach poses a significant challenge because, for example in the case of SD-ABM hybrid models, it requires a solid understanding of both SD and ABM validation methods. Though validation methods for SD and ABM overlap, they are distinct because of the unique characteristics of SD and ABM, and in the case of ABM, research on validation lags the other systems modelling methods [111].

Further, while it is tempting to think that combining single systems modelling methods' validation approaches is sufficient, only utilising these neglects the overarching hybrid model and the links between the submodels. Systems simulation methodology and complex adaptive system theory suggest the systems' parts do not necessarily provide insight into how the whole system will behave. There is no reason to believe that this principle does not hold true for combining submodels using different methods in hybrid models.

Future research should prioritise developing new methods and a framework for holistically verifying and validating hybrid models. These models necessitate verifying and validating the interfaces that facilitate the exchange of information between the different-method modules, each of which handles inputs and aggregates outputs differently. The impact of different interface designs on model outputs remains unexplored, underscoring the need for developing new methods to verify and validate them. A well-defined framework will guide researchers in verifying and validating the single-method submodels, the interfaces, and the hybrid model as a cohesive whole.

9.6.6 Software

Modellers considering developing a hybrid SD-ABM model have a variety of software options, and their choice can significantly impact the modelling process. Broadly, they can opt for (i) software packages specifically designed for systems simulation that support both SD and ABM, (ii) a combination of an SD software package and an ABM one, or (iii) coding in general purpose programming languages. Option (ii) can also combine dedicated software, likely for SD, with option (iii), likely for ABM.

In option (i) available choices are limited and include *Anylogic* and *NetLogo*, with *NetLogo* primarily focussing on ABMs. Option (i) simplifies the modelling process as users are not required to juggle multiple software packages or transfer information between them. Further, in the case of *AnyLogic*, modellers can get by without significant coding knowledge by relying on its graphical user interface. However, this user-friendliness comes at the cost of flexibility, which may be necessary for complex models.

With a similar sacrifice of flexibility, option (ii) can also simplify coding, depending on the specific software package, but particularly for SD packages. However, option (ii) may entail developing code to facilitate information transfer between the software packages. Both options (i) and (ii) also often offer dynamic visualisation, which is valuable for model exploration and communication. Further, while options (i) and (ii) reduce flexibility compared to option (iii), they likely reduce errors in model implementation and steer modellers towards good modelling practice (e.g., randomising the order agents implement an action to avoid first-mover effects).

Option (iii) provides the greatest flexibility but generally demands building models from scratch, requiring substantial effort. However, a number of programming languages, such as Python, now offer libraries for specific systems simulation modelling methods, simplifying their development and combination in hybrid models. Moreover, option (iii) likely enhances modellers understanding of the underlying methods and their models compared to options (i) and (ii), which tend to abstract away many technical details.

While the balance between simplicity and flexibility, as well as between low-level model understanding and model communication, remains debatable, there are opportunities to further simplify implementing hybrid models. Brailsford et al. [1] suggest developing tools that facilitate communication between different specialised software packages (e.g., ABM and SD ones). They also recommend further research on software packages that require minimal or no coding [1]. The emergence of generative artificial intelligence tools, such as ChatGPT, may present an opportunity to address these recommendations. With proper model specification, ChatGPT can generate code in various programming languages and interfaces between them. Although, at this point, the code produced by ChatGPT often contains errors, these AI models are continuously evolving, and this may improve future code reliability.

9.6.7 Areas for Future Research

This section has not only discussed some challenges associated with using hybrid SD-ABM, but also hinted at some opportunities and areas for further research. These include:

- Studying the implications of the use of stochastic v deterministic versions of SD and ABM on the outputs of hybrid models.
- Developing and improving consistent use of more transparent communication and documentation tools for hybrid models could improve learning for future hybrid modelling and support their transparency and validation. Adopting consistent tools would be beneficial for the community to avoid disjointed approaches.
- Developing new methods and a framework for holistic verifying and validating hybrid models.

- Providing more detailed guidance on the use of specific designs.
- Further developing software to support hybrid modelling and the recognition that the emergence of generative artificial intelligence tools may present an opportunity to support this development.

As discussed in Sect. 9.3, some work has been done to propose a practical framework for combining simulation methods [79] and describing linkages between SD and ABM modules in detail [13, 81]. However, this work has only had limited application so far and therefore further application will support understanding its generalisability. Specifically, Nguyen's framework has been developed based on the reflections and insights of a single modeller/researcher within a single healthcare context. Therefore, further testing and application of the framework in diverse contexts would help to assess its usefulness and generalisability. In addition, it may be worth exploring the combination of three simulation methods (SD, ABM, and DES) within a hybrid simulation model and examining how other OR methods can be combined using variations of this framework.

When using the framework described in Sect. 9.3.3 or when building any hybrid model, modellers have choices. Understanding how individual modellers or modelling teams make decisions regarding the adoption of hybrid simulation and the key factors influencing their choices is an area for research. In-depth interviews or observations could be used to gain deeper insights into the decision-making process involved in selecting hybrid simulations. Understanding this decision-making process would be valuable in deriving generalisable insights that can assist in the appropriate selection of methods. Understanding how modellers develop hybrid SD-ABM models can also be gained from sharing models and open-source modelling can help in this regard.

The use of SD and ABM in a hybrid simulation is becoming increasingly popular and, therefore, this chapter has sought to provide a comprehensive overview of methodological and practical considerations to provide insight and support to modellers wishing to use this combination of methods. It is a fruitful area for future research and one where new advances will be of benefit to both the modelling community and the wide range of decision-makers that these models support.

References

1. Brailsford SC, Eldabi T, Kunc M, Mustafee N, Osorio AF (2019) Hybrid simulation modelling in operational research: a state-of-the-art review. Eur J Oper Res 278(3):721–737
2. Howick S, Ackermann F (2011) Mixing OR methods in practice: past, present and future directions. Eur J Oper Res 215(3):503–511
3. Swinerd C, McNaught KR (2012) Design classes for hybrid simulations involving agent-based and system dynamics models. Simul Model Pract Theory 25:118–133
4. Villa F, Costanza R (2000) Design of multi-paradigm integrating modelling tools for ecological research. Environ Model Softw 15(2):169–177

5. Borshchev A, Filippov A (2004) Anylogic—multi-paradigm simulation for business, engineering and research. In: The 6th IIE annual simulation solutions conference, vol 150, p 45
6. Nguyen L, Megiddo I, Howick S (2022) Hybrid simulation modelling networks of heterogeneous care homes and the inter-facility spread of covid-19 by sharing staff. PLOS Comput Biol 18(1)
7. Djanatliev A, German R, Kolominsky-Rabas P, Hofmann BM (2012) Hybrid simulation with loosely coupled system dynamics and agent-based models for prospective health technology assessments. In: Proceeding of winter simulation conference, pp 69:1–69:12
8. Kazakov R, Howick S, Morton A (2021) Managing complex adaptive systems: a resource/agent qualitative modelling perspective. Eur J Oper Res 290(1):386–400
9. Schieritz N (2002) Integrating system dynamics and agent-based modeling. In: Proceedings of the 20th international conference of the system dynamics society, July 2002.
10. Shafiei E, Stefansson H, Asgeirsson EI, Davidsdottir B, Raberto M (2016) Integrated agent-based and system dynamics modelling for simulation of sustainable mobility. In: Handbook of applied system science. Routledge, pp 341–366
11. Kieckhäfer K, Walther G, Axmann J, Spengler T (2009) Integrating agent-based simulation and system dynamics to support product strategy decisions in the automotive industry. In: Proceedings of the 2009 winter simulation conference. IEEE, pp 1433–1443
12. Jo H, Lee H, Suh Y, Kim J, Park Y (2015) A dynamic feasibility analysis of public investment projects: an integrated approach using system dynamics and agent-based modeling. Int J Project Manag 33(8):1863–1876
13. Wurth B (2020) Simulating academic entrepreneurship and inter-organisational collaboration in university ecoystems, a hybrid system dynamics agent-based simulation. PhD Thesis, University of Strathclyde, UK
14. Forrester JW (1958) Industrial dynamics: a major breakthrough for decision makers. Harv Bus Rev 36(4):37–66
15. Forrester JW (2007) System dynamics—a personal view of the first fifty years. Syst Dynam Rev: J Syst Dynam Soc 23(2–3):345–358
16. Sterman JD (2000) Business dynamics: systems thinking and modelling for a complex world. Irwin/McGraw-Hill
17. Morecroft JDW (2007) Strategic modelling and business dynamics: a feedback systems approach. Wiley
18. Lane DC (2001) *Rerum cognoscere causas*: Part I—How do the ideas of system dynamics relate to traditional social theories and the voluntarism/determinism debate? Syst Dynam Rev: J Syst Dynam Soc 17(2):97–118
19. Lane DC (2008) The emergence and use of diagramming in system dynamics: a critical account. Syst Res Behav Sci: Off J Int Fed Syst Res 25(1):3–23
20. Randers J (1980) Guidelines for model conceptualization. In: Randers J (ed) Elements of the system dynamics method. Pegasus Communications, Waltham, MA
21. Richardson GP, Pugh AL (1989) Introduction to system dynamics modeling. Pegasus Communication, Waltham, MA
22. Davahli MR, Karwowski W, Taiar R (2020) A system dynamics simulation applied to healthcare: a systematic review. Int J Environ Res Public Health 17(16):5741
23. Lyneis J, Ford D (2007) System dynamics applied to project management: a survey, assessment, and directions for future research. Syst Dyn Rev 23(2–3):157–189
24. Shepherd SP (2014) A review of system dynamics models applied in transportation. Transport B: Transport Dynam 2(2):83–105
25. Akkermans H, Dellaert N (2005) The rediscovery of industrial dynamics: the contribution of system dynamics to supply chain management in a dynamic and fragmented world. Syst Dyn Rev 21(3):173–186
26. Andersen DF, Rich E, MacDonald R (2009) Public policy, system dynamics applications to. In: Meyers R (ed) Complex systems in finance and econometrics. Springer, New York

27. Ford A (2018) System dynamics models of environment, energy, and climate change. In: Meyers R (ed) Encyclopedia of complexity and systems science. Springer, Berlin, Heidelberg
28. Radzicki MJ (2009) System dynamics and its contribution to economics and economic modeling. In: Meyers R (ed) Encyclopedia of complexity and systems science. Springer, New York
29. Holland JH (1992) Complex adaptive systems. Daedalus 121:17–30
30. Kauffman SA (1995) At home in the universe: the search for laws of self-organization and complexity. Oxford University Press, USA
31. Brown DG, Page S, Riolo R, Zellner M, Rand W (2005) Path dependence and the validation of agent-based spatial models of land use. Int J Geogr Inf Sci 19:153–174
32. Macal CM, North MJ (2009) Agent-based modelling and simulation. In: Proceedings of the 2009 winter simulation conference, Austin, USA, pp 86–98
33. Simon HA, Feldman J (1959) Theories of decision-making in economics and behavioural science. Am Econ Rev 49(3):253–283
34. Kahneman D, Tversky A (1979) Prospect theory: an analysis of decision under risk. Econometrica 47(2):263–292
35. Jennings NR, Wooldridge M (1998) Applications of intelligent agents. Agent technology: foundations, applications, and markets, pp 3–28
36. Grimm V, Berger U, Bastiansen F, Eliassen S, Ginot V, Giske J, Goss-Custard J, Grand T, Heinz SK, Huse G, Andreas H, Jepsen JU, Joegensen C, Mooij WM, Muller B, Pe'er G, Piou C, Railsback SF, Robbins AM, Robbins MM, Rossmanith E, Ruger N, Strand E, Souissi S, Stillman RA, Vabo R, Visser U, DeAngelis DL (2006) A standard protocol for describing individual-based and agent-based models. Ecol Model 198:115–126
37. Grimm V, Berger U, DeAngelis DL, Polhill JG, Giske J, Railsback SF (2010) The ODD protocol: a review and first update. Ecol Model 221:2760–2768
38. Grimm V, Polhill G, Touza J (2017) Documenting social simulation models: the ODD protocol as a standard. In: Edmonds B, Meyer R (eds) Simulating social complexity. Springer
39. DeAngelis DL, Gross LJ (1991) Individual-BASED MODELS AND Approaches in ecology. CRC Press
40. Tracy M, Cerda M, Keyes K (2018) Agent-based modelling in public health: current applications and future directions. Annu Rev Public Health 39:77–94
41. Cassidy R, Singh NS, Schiratti P, Semwanga A, Binyaruka P, Sachingongu N, Chama-Chiliba CM, Chalabi Z, Borghu J, Blanchet K (2019) Mathematical modelling for health systems research: a systematic review of system dynamics and agent-based models. BMC Health Serv Res 19:845
42. Hamill L, Gilbert N (2016) Agent based modelling in economics. Wiley, Chichester
43. Lättilä L, Hilletofth P, Lin B (2010) Hybrid simulation models—when, why, how? Expert Syst Appl 37(12):7969–7975
44. Scheidegger AP, Pereira TF, de Oliveira ML, Banerjee A, Montevechi JA (2018) An introductory guide for hybrid simulation modelers on the primary simulation methods in industrial engineering identified through a systematic review of the literature. Comput Ind Eng 124:474–492
45. Nguyen L, Megiddo I, Howick S (2020) Simulation models for transmission of health care-associated infection: a systematic review. Am J Infect Control 48(7):810–821
46. Pfeffer J, Salancik GR (1978) The external control of organizations: a resource dependence perspective. Harper & Row, New York
47. Hillman AJ, Withers MC, Collins BJ (2009) Resource dependence theory: a review. J Manag 35(6):1404–1427
48. Barney J (1991) Firm resources and sustained competitive advantage. J Manag 17(1):99–120
49. Wernerfelt B (1984) A resource-based view of the firm. Strateg Manag J 5(2):171–180
50. Peteraf MA (1993) The cornerstones of competitive advantage: a resource-based view. Strateg Manag J 14(3):179–191
51. Tversky A, Kahneman D (1974) Judgment under uncertainty: heuristics and biases. Science 185(4157):1124–1131

52. Kahneman D (2003) A perspective on judgment and choice: mapping bounded rationality. Am Psychol 58(9):697
53. Rosen R (1985) Anticipatory systems: philosophical, mathematical and methodological foundations Pergamon Press, Oxford
54. Pezzulo G (2008) Coordinating with the future: the anticipatory nature of representation. Mind Mach 18:179–225
55. Gary MS, Kunc M, Morecroft JD, Rockart SF (2008) System dynamics and strategy. Syst Dynam Rev: J Syst Dynam Soc 24(4):407–429
56. Dierickx I, Cool K (1989) Asset stock accumulation and sustainability of competitive advantage. Manag Sci 35(12):1504–1511
57. Kahneman D, Tversky A (1982) The simulation heuristic. Judgment under uncertainty: heuristics and biases. Cambridge University Press, New York, pp 201–208
58. Butz MV, Sigaud O, Pezzulo G, Baldassarre G (2006) Anticipations, brains, individual and social behavior: an introduction to anticipatory systems. In: Workshop on anticipatory behavior in adaptive learning systems. Springer, Berlin, Heidelberg, pp 1–18
59. Kim DH, Juhn JH (eds) (1997) System dynamics as a modeling platform for multi-agent systems. In: The 15th international conference of the system dynamics society, Istanbul, Turkey
60. Parunak HVD, Savit R, Riolo RL (eds) (1998) Agent-based modeling vs. equation-based modeling: a case study and users' guide. In: International workshop on multi-agent systems and agent-based simulation. Springer
61. Akkermans H (2001) Renga: a systems approach to facilitating inter-organizational network development. Syst Dyn Rev 17(3):179–193
62. Schieritz N, Größler A (2009) Emergent structures in supply chains - a study integrating agent-based and system dynamics modeling. In: 36th annual Hawaii international conference on system sciences, 6–9 Jan 2003, p 9
63. Borshchev A, Filippov A (2004) From system dynamics and discrete event to practical agent based modeling: reasons, techniques, tools. In: Proceedings of the 22nd international conference of the system dynamics society. Citeseer, Oxford
64. Lorenz T, Jost A (2006) Towards an orientation framework in multi-paradigm modeling. In: Proceedings of the 24th international conference of the system dynamics society. System Dynamics Society, Albany, NY
65. Bobashev GV, Goedecke DM, Feng Y, Epstein JM (eds) (2007) A hybrid epidemic model: combining the advantages of agent-based and equation-based approaches. In: 2007 winter simulation conference, Dec 2007
66. Martinez-Moyano I, Sallach D, Bragen M, Thimmapuram PR (2007) Design for a multilayer model of financial stability: exploring the integration of system dynamics and agent-based models
67. Brailsford S, Desai S, Viana J (2010) Towards the holy grail: combining system dynamics and discrete-event simulation in healthcare. In: Proceedings of the 2010 winter simulation conference. IEEE
68. Vincenot CE, Giannino F, Rietkerk M, Moriya K, Mazzoleni S (2011) Theoretical considerations on the combined use of system dynamics and individual-based modeling in ecology. Ecol Model 222(1):210–218
69. Swinerd C, McNaught KR (2014) Simulating the diffusion of technological innovation with an integrated hybrid agent-based system dynamics model. J Simul 8(3):231–240
70. Onggo BS (2014) Elements of a hybrid simulation model: a case study of the blood supply chain in low- and middle-income countries. In: Proceedings of the winter simulation conference
71. Djanatliev A, German R (2015) Towards a guide to domain-specific hybrid simulation. In: Winter simulation conference
72. Mustafee N, Brailsford S, Djanatliev A, Eldabi T, Kunc M, Tolk A (eds) (2017) Purpose and benefits of hybrid simulation: contributing to the convergence of its definition. In: Winter simulation conference

73. Wallentin G, Neuwirth C (2017) Dynamic hybrid modelling: switching between AB and SD designs of a predator-prey model. Ecol Model 345:165–175
74. Shanthikumar JG, Sargent RG (1983) A unifying view of hybrid simulation/analytic models and modeling. Oper Res 31(6):1030–1052
75. Bennett PG (1985) On linking approaches to decision-aiding: issues and prospects. J Oper Res Soc 36(8):659–669
76. Chahal K, Eldabi T (eds) (2008) Applicability of hybrid simulation to different modes of governance in UK healthcare. In: Winter simulation conference, 7–10 Dec 2008
77. Chahal K, Eldabi T, Young T (2013) A conceptual framework for hybrid system dynamics and discrete event simulation for healthcare. J Enterp Inf Manag 26(1/2):50–74
78. Morgan JS, Howick S, Belton V (2017) A toolkit of designs for mixing discrete event simulation and system dynamics. Eur J Oper Res 257(3):907–918
79. Nguyen LKN (2022) Hybrid health systems simulation modelling: controlling Covid-19 infections in care homes: PhD Thesis, University of Strathclyde, UK
80. Jones W, Kotiadis K, O'Hanley JR, Robinson S (2022) Aiding the development of the conceptual model for hybrid simulation: representing the modelling frame. J Oper Res Soc 73(12):2775–2793
81. Nguyen LKN, Howick S, Megiddo I (2024) A framework for mixing system dynamics and agent based modelling in a conceptual hybrid simulation model. Eur J Oper Res
82. Caudill L, Lawson B (eds) (2013) A hybrid agent-based and differential equations model for simulating antibiotic resistance in a hospital ward. In: Proceedings of the 2013 winter simulation conference—simulation: making decisions in a complex world
83. Djanatliev A, German R (2013) Prospective healthcare decision-making by combined system dynamics, discrete-event and agent-based simulation. In: Proceedings of the 2013 winter simulation conference—simulation: making decisions in a complex world
84. Kolominsky-Rabas PL, Djanatliev A, Wahlster P, Gantner-Bär M, Hofmann B, German R et al (2015) Technology foresight for medical device development through hybrid simulation: The ProHTA Project. Technol Forecast Soc Change 97:105–114
85. Jo H, Lee H, Suh Y, Kim J, Park Y (2015) A dynamic feasibility analysis of public investment projects: an integrated approach using system dynamics and agent-based modelling. Int J Project Manag 33(8):1863–1876
86. Cernohorsky P, Voracek J (2012) Towards public health policy formulation. Proc IFKAD-ISSN 2280:787X
87. Verburg PH, Overmars KP (2009) Combining top-down and bottom-up dynamics in land use modelling: exploring the future of abandoned farmlands in Europe with the Dyna-CLUE model. Landsc Ecol 24(9):1167
88. Robledo LF, Sepulveda J, Archer S (2013) Hybrid simulation decision support system for university management. In: Proceedings of the 2013 winter simulation conference, pp 2066–2075
89. Mazhari EM, Zhao J, Celik N, Lee S, Son Y-J, Head L (eds) (2009) Hybrid simulation and optimization-based capacity planner for integrated photovoltaic generation with storage units. In: Proceedings of the 2009 winter simulation conference (WSC). IEEE
90. Chen S, Desiderio S (2020) Job duration and inequality. Economics: Open-Access Open-Assess E-J 14:1–27
91. Swinerd C, McNaught KR (2015) Comparing a simulation model with various analytic models of the international diffusion of consumer technology. Technol Forecast Soc Change 100:330–343
92. Vincenot CE, Moriya K (2011) Impact of the topology of metapopulations on the resurgence of epidemics rendered by a new multiscale hybrid modelling approach. Ecol Inform 6(3–4):177–186
93. Mathieu P, Morvan G, Picault S (2018) Multi-level agent-based simulations: four design patterns. Simul Model Pract Theory 83:51–64
94. Chen CC, Hardoon DR (2010) Learning from multi-level behaviours in agent-based simulations: a systems biology application. J Simul 4(3):196–203

95. Mingers J, Brocklesby J (1997) Multimethodology: towards a framework for mixing methodologies. Omega 25(5):489–509
96. Brailsford SC, Viana J, Rossiter S, Channon AA, Lotery AJ (2013) Hybrid simulation for health and social care: the way forward, or more trouble than it's worth? In: 2013 winter simulations conference (WSC), 8 Dec 2013. IEEE, pp 258–269
97. Rahmandad H, Sterman J (2008) Heterogeneity and network structure in the dynamics of diffusion: comparing agent-based and differential equation models. Manag Sci 54(5):998–1014
98. Harper A, Mustafee N, Yearworth M (2021) Facets of trust in simulation studies. Eur J Oper Res 289(1):197–213
99. Anderson RM, May RM (1991) Infectious diseases of humans: dynamics and control. Oxford University Press
100. Daley DJ, Gani J (2001) Epidemic modelling: an introduction. Cambridge University Press
101. Djanatliev A (2015) Hybrid simulation for prospective healthcare decision-support: system dynamics, discrete-event and agent-based simulation. PhD Thesis, University of Erlangen-Nuremberg
102. Kotiadis K, Mingers J (2006) Combining PSMs with hard OR methods: the philosophical and practical challenges. J Oper Res Soc 57:856–867
103. Mingers J (2001) Combining IS research methods: towards a pluralist methodology. Inf Syst Res 12(3):240–259
104. Roberts M, Andreasen V, Lloyd A, Pellis L (2015) Nine challenges for deterministic epidemic models. Epidemics 10:49–53
105. Rock K, Brand S, Moir J, Keeling MJ (2014) Dynamics of infectious diseases. Rep Prog Phys 77(2):026602
106. Monks T, Currie CSM, Onggo BS, Robinson S, Kunc M, Taylor SJE (2019) Strengthening the reporting of empirical simulation studies: introducing the STRESS guidelines. J Simul 13(1):55–67
107. Grimm V, Railsback SF, Vincenot CE, Berger U, Gallagher C, DeAngelis DL, Edmonds B, Ge J, Giske J, Groeneveld J, Johnston AS (2020) The ODD protocol for describing agent-based and other simulation models: a second update to improve clarity, replication, and structural realism. J Artif Soc Soc Simul 23(2)
108. Vincenot CE, Mazzoleni S, Moriya K, Cartenì F, Giannino F (2015) How spatial resource distribution and memory impact foraging success: a hybrid model and mechanistic index. Ecol Complex 22:139–151
109. Martin R, Schlüter M (2015) Combining system dynamics and agent-based modeling to analyze social-ecological interactions—an example from modeling restoration of a shallow lake. Front Environ Sci 3:66
110. Müller B, Bohn F, Dreßler G, Groeneveld J, Klassert C, Martin R, Schlüter M, Schulze J, Weise H, Schwarz N (2013) Describing human decisions in agent-based models—ODD+ D, an extension of the ODD protocol. Environ Model Softw 48:37–48
111. An L, Grimm V, Sullivan A, Turner Ii BL, Malleson N, Heppenstall A, Vincenot C, Robinson D, Ye X, Liu J, Lindkvist E (2021) Challenges, tasks, and opportunities in modeling agent-based complex systems. Ecol Model 457:109685

Chapter 10
Deployable Healthcare Simulations: A Hybrid Method for Combining Simulation with Containerisation and Continuous Integration

Alison Harper, Thomas Monks, and Sean Manzi

Abstract Methods or approaches from disciplines outside of OR modelling and Simulation (M&S) can potentially increase the functionality of simulation models through hybrid modelling. In healthcare research, where simulation models are commonly used, we see few applications of models that can easily be deployed by other researchers or by healthcare stakeholders. Models are treated as disposable artefacts, developed to deliver a set of results, and rarely address deployment for model reuse. We propose one potential solution to deploying free and open-source simulations using containerisation with continuous integration. A container provides a self-contained environment that encapsulates the model and all its required dependencies including the operating system, software, and packages. This overcomes a significant barrier to sharing models developed in open-source software, which is dependency management. Isolating the environment in a container ensures that the simulation model behaves the same way across different computing environments. It also means that other users can interact with the model without installing software, supporting both use/reuse, and reproducibility of results. We illustrate the approach using a model for orthopaedic elective recovery planning, developed with a user-friendly interface in Python, including a step-by-step approach to support M&S researchers to deploy their own models using our hybrid framework.

Keywords Simulation · Deployment · Containerisation · Continuous integration · Healthcare

---

A. Harper (✉)
Centre for Simulation, Analytics and Modelling, University of Exeter Business School, Rennes Drive, Exeter, UK
e-mail: a.l.harper@exeter.ac.uk

A. Harper · T. Monks · S. Manzi
The National Institute for Health and Care Research Applied Research Collaboration West (NIHR PenARC), University of Exeter St Luke's Campus, Heavitree Road, Exeter, UK
e-mail: t.m.w.monks@exeter.ac.uk

S. Manzi
e-mail: s.s.manzi@exeter.ac.uk

10.1 Introduction

Hybrid modelling in the field of modelling and simulation (M&S) refers to the combined use of simulation techniques with frameworks, methods, tools, and approaches from academic disciplines distinct from M&S [1]. While hybrid simulation mainly focuses on model implementation, hybrid modelling leverages knowledge artefacts from diverse disciplines. *Type E Hybrid Models* [2] combine simulation models with established research approaches in fields such as software engineering and applied computing at different stages of the M&S lifecycle [1, 3, 4]. By utilising research approaches developed in these fields of study, researchers can deploy hybrid methods that complement, rather than supplement, the techniques traditionally used within the M&S knowledge domain. This can enhance accessibility, functionality, or usability of simulation models for end-users and other M&S researchers.

In health care, simulation is used extensively to understand, analyse, and optimise complex systems, contributing to enhanced efficiency, effectiveness, and preparedness [5, 6]. Frequently, complex models are developed for a single problem in a single organisation, results are generated, a report is produced, and the study is published. This means that decision-makers no longer have access to the model should they wish to investigate additional scenarios. Further, across health and care services, similar problems are seen in similar service configurations. These situations contribute to research waste, as models are not effectively integrated into decision-making processes, and subsequently, duplicated effort is required as researchers continue to develop similar models for similar problems. By building on previous work, model reuse can both reduce inefficiency and cost, and improve model quality and reliability [7].

A secondary issue with simulation modelling studies is reproducibility of simulation results. The ability to reproduce published results is central to the scientific method [8–10]. The reproducible research movement in data science and M&S is encouraging researchers to make their code or model available, along with a description of the hardware and software environments [11–13], but even this level of detail may not be enough to replicate results.

Developing models for deployment by healthcare staff or for results replication ensures that they are ready for use in a specific operational environment. However, there are significant challenges to deploying models for these purposes [10]. The use of free and open-source software overcomes cost barriers associated with license fees but can create new challenges. For example, simulation models require software to be downloaded and installed, integration with existing systems, computational resources, and model maintenance and updates. For reproducing results, the correct version of the operating system, software, allocation of memory resources, etc., may be required. When using free and open-source software, package versions also need to be specified. While formal dependency management tools can be used, further problems can occur when a project or application relies on many dependencies or software components, and these dependencies have complex and conflicting requirements or dependencies of their own. This is further complicated when the model needs to be deployed across different operating systems, for example, across both Windows and MacOS.

Deploying models as a web-based application overcomes a number of these barriers [14–17]. One way of achieving this using free and open-source software is through a hybrid application of simulation with containerisation and continuous integration.

Using this approach, *simulation model* development can be enhanced by group collaboration via an online repository for development, monitoring and de-bugging. It can use any simulation method, although should not use a commercial software package which is dependent upon proprietary licenses. *Continuous integration* involves frequently merging code changes—which may be from multiple collaborating modellers—into a central *remote repository*. Modellers can use version control systems, such as Git, to manage their code changes. These can be done in development branches, which can be merged into the main branch as work on a specific feature or bug fix is completed, and once tests or other quality checks are passed. Once the build and tests pass successfully, a new release can be created and made available for deployment via *containerisation*. The packaged artefact, a *container image*, is stored in a *container registry*, which acts as a centralised repository for storing container images. Users can pull the image to their own machine using a containerisation tool, creating a container.

Containers provide a self-contained environment that encapsulates the model, and all the dependencies required by it. This isolation ensures that the dependencies within the container do not conflict with the host system or other applications running on it and ensure consistency across different computing environments. *Continuous deployment* tools monitor the container registry for new images. When a new image is available, it is automatically deployed, replacing the previous version of the application, and ensuring efficient and consistent updates. Figure 10.1 illustrates the relationship between these concepts. Together, these approaches enable model deployment through a hybrid ecosystem of research artefacts. According to the definition outlined by Mustafee and Powell [1], the hybridisation takes place at the solution understanding/informing practice stage of the M&S development lifecycle, where model accessibility directly supports decision-making. However, the approach also has wider applicability for M&S researchers, enabling collaboration, reproducibility, and research continuity.

The next section looks at existing literature where studies have used software engineering and applied computing with simulation. Section 10.3 describes the deployment concepts introduced here in more detail, and Sect. 10.4 provides a case study of our hybrid approach using a free and open-source discrete-event simulation (DES) model developed for orthopaedic planning. In Sect. 10.5, we illustrate its deployment using Docker, a widely used containerisation tool. We additionally provide a clear set of steps to support M&S researchers interested in this approach to apply our hybrid deployment framework. Finally, we conclude with a discussion of the strengths and challenges of this approach in healthcare simulation, the limitations of this work, and future work.

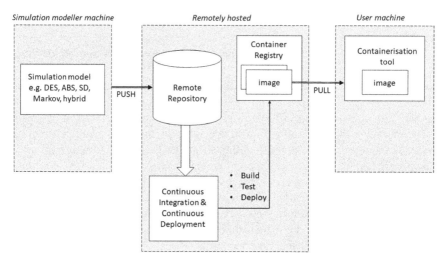

Fig. 10.1 Hybrid model deployment framework

10.2 Previous Work

While the benefits of model use and reuse in health care are clear [18–20], the use of commercial off-the-shelf software (COTS) dominates in healthcare simulation research [11]. Dagkakis and Heavey [21] outlined a number of disadvantages to the use of COTS for simulation reuse, including difficulty customising the model, the high license cost of software packages, and additional costs such as training and maintenance. Free and open-source software overcomes many of these issues, but can bring additional challenges, such as dependency management, providing a functional user interface, and addressing the needs of different users. Deploying the model as a hybrid application removes the need to manually download software, code dependencies, and version control; enables reuse including licensing; can offer an enhanced user experience; supports collaboration; offers the ability to control access; and supports customisation and model maintenance [7, 10, 16, 17].

Web-based simulation research is one approach to hybrid simulation deployment, and a 2010 review of web-based simulation and tools [16] outlined a range of methods for deploying simulation applications on the web. The benefits they described continue to be relevant and have been used extensively to advantage in domains such as environmental modelling [22] and hydrology [23]. Gan et al. [23] described enhanced collaboration, computer platform independence, reproducibility of modelling workflows and results, and the ability of users to create, describe, share, repeat, modify, and analyse model outputs. In geoscience, Qin et al. [24] developed a web-based modelling framework to access model results, and share and reuse models.

Cloud-based simulation has seen a rise in popularity in recent years [17, 25]. As containerisation has gained popularity, it has become a major deployment method in cloud environments [26]. Containerisation has been used in complex biomedical simulation studies such as event-based in silico trials to support collaboration, portability, reproducibility, and scaling using multiple containers for individual model components [27–29]. Containers are useful where a researcher needs to run multiple concurrent experiments, as each can be run in its own container, and can be easily started and stopped without affecting the other experiments. These studies emphasised portability within a research group, while Moreau et al. [10] discussed the applicability of containerisation to open research.

In healthcare M&S, there are few examples of interactive deployed simulation models focusing on reuse. Anagnostou et al. [30] developed a hybrid DES/agent-based simulation using a user-friendly dashboard to enable end-users to input parameters for COVID-19 regional management, deployed on a demo web server. With a focus on model use and reuse in health care, Tyler et al. [31] developed a tool in R, with an R Shiny user interface. The model is available to download and install, but potentially faces the issues previously discussed, namely ongoing dependency management. Capocchi et al. [32] described an approach for facilitating discrete-event system specification (DEVS) model simulation on the web through a web app, which can be hosted in a cloud solution through Docker, and proposed a healthcare case study for COVID-19 surveillance. However, across published applied healthcare M&S research, we know of no examples of hybrid simulation models deployed using containerisation [11]. This approach offers value, as it can be used to support model reuse in health care, and overcomes both the challenges of local implementation, and the costs of deploying on a web server.

To increase uptake of these methods, we provide an example of a simulation model deployed with Docker, with a step-by-step method to enable other M&S researchers to reproduce our method for their own simulation models and improve the accessibility and reusability of their work.

10.3 Deployment Concepts

This section details the concepts introduced in Sect. 10.1 and Fig. 10.1. We first introduce the concept of free and open-source software (FOSS). This is important, as container images are typically built using open-source technology. Additionally, FOSS eliminates license fees and provides flexibility in terms of customisation and adaptation to specific needs. The subsequent sections describe containerisation tools, remote code repositories, and continuous integration and deployment, which alongside the model itself, creates a set of research artefacts used to inform practice.

10.3.1 Free and Open-Source Software (FOSS)

Free and open-source software tools are software applications or programs released under a license that allows users to freely use, modify, and distribute the software's source code [21]. While containerisation does not inherently rely on FOSS, many containerisation solutions such as Docker are built on top of FOSS technologies, and its core components are open source. Typically, a FOSS operating system is based on a customisation of Linux operating system. The licenses used with Linux grant users the freedom to redistribute (possibly adapted) copies under the same terms and conditions, it can be used for any form of computing, and no license cost is involved. Before sharing code, it is essential that authors select an appropriate FOSS license for the code and other research artefacts to ensure appropriate use of the model and remove liability of the authors. A common approach in data science is to adopt a permissive license such as MIT (see Monks et al. [33] for more details on license types).

Python is a popular FOSS programming language known for its simplicity and versatility. For DES, Python packages include SimPy [34], Salabim [35], Ciw [36], and De-sim [37]. These have been used in several Operations Research relevant publications in healthcare [30, 38–40].

For agent-based modelling, Python libraries include Mesa [41] and for System Dynamics models, PySD [42]. PySD is also available in R via the PySD2R package, while DES simulation libraries available for R include R Simmer [43] and for Julia, SimJulia [44]. Packages also exist for providing user-friendly, interactive front-ends to open-source models. These include Streamlit for Python [45] and Shiny for R and Python [46]. These are browser-based technologies that can be deployed on the web using (for example) streamlit.io or shinyapps.io, although these come with resource limits and cannot handle high user load.

10.3.2 Containerisation Tools

In computing, *containers* refer to lightweight, portable, self-contained software units that package together an application, such as a simulation model, and all its software dependencies, including libraries, frameworks, data, and system tools. Containers provide a consistent and isolated runtime environment for running applications, ensuring that they behave the same way across different computing environments [47, 48]. Each container operates as a separate entity, with its own file system, networking, and resources, while sharing the host operating system's kernel. Containers enable efficient deployment, scaling, and management of applications, as they can be easily moved between different computing environments without compatibility issues. Commonly used open-source tools for developing, managing, and running containers include Docker [48] (these may be specialised images, e.g. Rocker for R [49]), Singularity/Apptainer [50], and Podman [51]. As well as encapsulating all dependencies within the container to avoid conflicts, containers support efficient, collaborative research, and reproducibility, as the exact same environment is used across systems [10, 52].

A container registry is a centralised repository used to store and distribute container images. An image is a blueprint for what you want to build, including the operating system, software, and packages. A container is an instantiation of an image: multiple copies of the same image can run simultaneously. When a container image is built, it is pushed to a container registry, where it can be pulled by different teams, systems or cloud platforms. Popular container registries include DockerHub, Google Artefact Registry, and Azure Container Registry. These enable efficient collaboration and deployment in container-based environments. The container is portable, in that once created, it can run on any system that supports containerisation technology without modification, and supports scalability as the application can be easily replicated and deployed across multiple instances. It provides an excellent solution for deploying interactive applications such as Streamlit and Shiny. While sharing dependencies may be an appropriate approach, wrapping everything in a container such as Docker and posting it to a public (or private) container registry such as DockerHub, reduces the need to recreate the environment and makes the model more accessible for users.

10.3.3 Code Repositories

Central to deployment using containerisation is a remote code repository. The most popular solutions are GitHub, GitLab, and BitBucket. All code artefacts (including licence, metadata files, a citation file, and README) should be committed to the remote repository to provide long-lasting version control.

GitHub and other online code repositorys' primary purpose is version control of code. That is, modellers incrementally commit updates to their code and models during development. These tools allow for easy *rollback,* inspection, or use of older versions of the code. Another major feature of version control is *branching.* The main (sometimes called master) branch of a repository contains the *production model*, i.e. a simple, but functional simulation model that can be deployed to users. Incremental developments are typically made on a development or feature branch, tested, and then merged into the main branch when a modeller is confident it will not break the production model. GitHub is an online tool. In traditional workflows, all of the model coding is done offline in a local code repository that is managed by a local version control system such as *Git*. A modeller codes their model on their own machine and *pushes* their commits to GitHub (or a similar repository). A new user of the code would *clone* the GitHub repository to their own local machine. A user who is already using the code might *pull* updates from the remote repository and merge changes into their own branch.

As a simple example, consider a model of an emergency department (ED) developed in Python DES package SimPy. The model logic is fully coded and it can be run from the command line. The working SimPy model is currently committed to the main branch. A National Health Service (NHS) analyst can clone the main branch of the repository and execute the model to analyse their own ED. Simultaneously, a simulation modeller can develop a graphical interface for the model on a separate feature branch (e.g. called *interface*). When the interface is complete the modeller can issue a pull request to the main branch. A pull request is effectively a request to merge code from one branch into another, e.g. from *interface* to main. It provides a safety net for the changes to be reviewed before merging (and potentially breaking a production model). Once merged the NHS analyst can simply pull the new code from the main branch, although there are no guarantees the new model interface will install easily on the NHS analyst's machine.

A secondary use of tools like GitHub is for collaboration on scientific studies and model development. GitHub provides many mechanisms for its users to share, review, and control code. One simple example is *Issues*. This is a discussion log that might report bugs, potential improvement, new ideas, and general queries. Another example is the concept of a *release*. A release tags a snapshot of the code. In our ED model example, the SimPy command line model might be tagged as version 1.0.0, while the ED model with a graphical user interface might be tagged as version 2.0.0. A release is an easy way for modellers to talk to their users about the correct version of the model: it supports a shared language between modellers and users.

10.3.4 Continuous Integration and Continuous Deployment

Continuous practices include continuous integration and continuous deployment. These are software development industry practices that enable frequent and reliable releases of new features and products [53].

Continuous integration involves regularly integrating code changes from multiple developers into a shared code repository such as GitHub. The main goal of continuous integration is to detect and address integration issues early by automatically building, testing, and validating the code changes. Continuous deployment is the ability to make releases quickly and reliably through automation support.

An example is GitHub Actions, a feature provided by GitHub enabling automation of various tasks and workflows within a GitHub repository, including continuous integration and deployment [54]. With GitHub Actions, custom workflows can be defined using Yet another Markup Language (YML) files that describe the steps to be executed when specific events occur, such as push events (code changes) or pull request events. These workflows can include building and testing code, running linters (which flag syntax errors), performing code reviews, deploying applications, and more [55].

The next section briefly describes our case study DES model and its user interface before we step through the stages required to build a container for the model.

10.4 Case Study: Elective Orthopaedic Capacity Planning Model

10.4.1 Background

This case study describes an application of DES for use as a generic planning tool for regional post-COVID-19 orthopaedic waiting list recovery. Elective joint replacement surgeries are among the most common elective procedures. In the aftermath of the pandemic, elective waiting lists were at record levels in the UK. To address this challenge, the government allocated funds for healthcare organisations to expand capacity and meet new interim elective targets, with rapid planning occurring on a large scale across the UK.

The DES model has been described in detail elsewhere [56]. It enables various configurations of bed numbers, theatre capacity, and productivity measures such as theatre scheduling, hospital lengths-of-stay and rates of delayed discharges to estimate resultant surgical throughput and capacity requirements, developed for a proposed orthopaedic facility. It is developed to be used by healthcare planners, managers, and clinicians, can be reused for similar applications, and can be readily adapted for other specialties or scaled for multiple specialties.

10.4.2 Overview of the DES Model

The FOSS DES model was developed in Python 3.8 using SimPy, with Streamlit to create a user-friendly interface, in collaboration with a large NHS Trust in England.

In the model, patients are categorised by surgical procedure and enter deterministically according to a baseline operating theatre schedule. If a bed is not available within a tolerance of 0.5–1 day, the theatre slot is lost. A conceptual view of the model is in Fig. 10.2. Lengths-of-stay are sampled by surgical type, and a proportion of patients have their discharge delayed, often for downstream reasons such as lack of availability of community health or social care.

An optional user-defined schedule can vary, per day of week, the numbers of theatres available, the daily number of theatre sessions, and the allocation of procedure types to each session. Parameters can be changed in the Streamlit interface using sliders and buttons; these additionally include numbers of beds, patient mean lengths-of stay per procedure type, length-of-stay of patients with a delayed discharge, and proportion of patients with a delayed discharge. This enables comparing scenarios including resource configurations, adding evening or weekend theatre activity, and changing the way surgical procedures are scheduled, for example, scheduling complex surgeries with longer, more variable lengths-of-stay earlier in the week. An illustrative screenshot of the Streamlit landing page is in Fig. 10.3, which provides information about the project and the model to users.

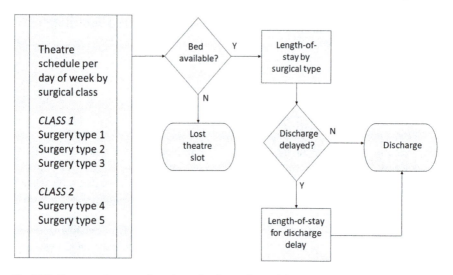

Fig. 10.2 Conceptual process flow chart of orthopaedic model

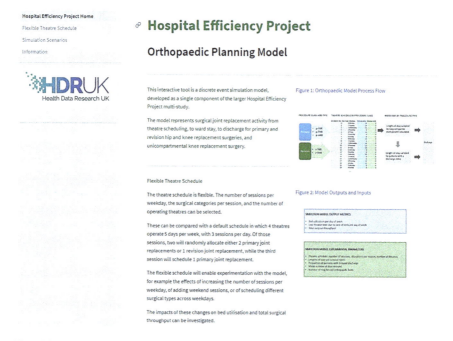

Fig. 10.3 Screenshot of the streamlit landing page

Experimental scenarios are therefore a combination of resourcing and efficiency measures. For decision-making, different NHS user bases are supported, for example, clinicians and service managers can influence how patient pathways are configured and standardised to support reducing lengths-of-stay and delayed discharges towards national benchmarks [57]. Finance and departmental managers will have control over resourcing decisions, and clinical, department, and service managers can make decisions regarding workforce capacity, for example, for weekend or evening activity.

Outputs can be used to provide quantitative evidence in support of regional business cases to secure core capital funding to support NHS elective surgical recovery. Outputs by weekday include (i) total surgical throughput per procedure, with the aim of identifying the configuration which best achieves this within other constraints; (ii) bed utilisation to identify days of the week where the system is under pressure; (iii) lost theatre slots for system reasons, which represents a mismatch between the balance of bed utilisation to theatre activity.

Given the requirement for NHS use across a wide, non-technical user base, and the generic nature of the model for modification and reuse, NHS deployment was integral to its design and development. We show one way that this has been achieved through hybridising the simulation model using a container, with continuous integration and deployment. As this approach is novel in applied healthcare M&S research, we have additionally outlined the steps taken to achieve the hybrid integration. The purpose of this is to support M&S researchers interested in deploying applied healthcare simulation models to be used by NHS staff, with the aim of increasing the potential to make a real impact on healthcare service delivery.

10.5 Deployed Implementation

We illustrate our DES model deployment method using Docker for containerisation, DockerHub for container hosting, and GitHub actions for continuous integration and deployment. All simulation code, build scripts, and tests are available on GitHub and archived at Zenodo [58]. We chose to use Docker for several reasons. It is the most popular containerisation platform, with a large and engaged community of developers and users. It is also relatively easy to use. It also supports multiple operating systems and provides seamless integration with other tools [59]. All links are provided in the 'Code Sharing' section at the end of this chapter.

The Docker container can be found on DockerHub. Figure 10.4 illustrates the application of our hybrid deployment framework. Each component is clearly explained using a stepwise approach with code listings.

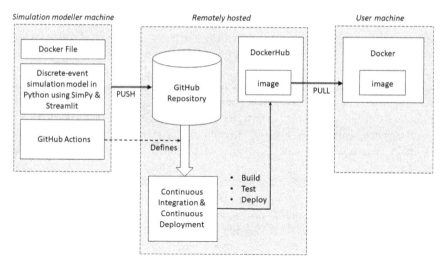

Fig. 10.4 Application of hybrid model deployment framework

10.5.1 The Docker Container

The build for a Docker container is specified within a Dockerfile. Listing 1 details the Dockerfile used within our example. It is comprised of the following major steps.

Step 1: The Initial Image

Docker works in a modular way by building a new image based on other published images. Here we start by using *Python 3.8-slim*. This provides a light weight (smaller size) Debian Linux image with Python 3.8 pre-installed. An alternative would be to use a full version of Linux such as *Ubuntu:latest*.

Step 2: Install Any Extra Software Needed

The slim build of Linux and Python is useful to keep an image to a relatively small size. The downside is that it may miss some essential low-level software needed to run a simulation model, and Streamlit web apps. It will also exclude version control software such as Git that that we need to use to clone the GitHub repository during the build. Our second stage installs these essential tools including Git using a Linux package management tool. Any cached files are removed at the end of the installation.

Step 3: Clone the Remote Code Repository

We intend this script to be run remotely as part of our continuous integration and deployment pipeline. We therefore use *git clone* to copy the most recent version of the code and software dependency list to our working directory on the image.

```
# Step 1: the initial image
FROM python:3.8-slim

# streamlit working dir is /app
RUN mkdir /app
WORKDIR /app

# Step 2: Install extra software
# Git means we can clone from a repo instead of local
copy if needed.
# Clean up after.
RUN apt-get update && apt-get install -y \
build-essential \
git \
software-properties-common \
&& rm -rf /var/lib/apt/lists/*

# Step 3: clone the repository to the app directory
RUN git clone https://github.com/TomMonks/hep-deploy /app

# Step 4: pip install the python software dependencies
e.g. simpy 4.0
RUN pip3 install -r requirements.txt

# Step 4: Container EntryPoint
# EXPOSE PORT 8501 - standard for streamlit.
EXPOSE 8501
# run app on 0.0.0.0:8501
# This means that the app will run when the image is
launched.
ENTRYPOINT ["streamlit", "run", "Hospital_Effi-
ciency_Project_Home.py", "--server.port=8501", "--
server.address=0.0.0.0"]
```

Listing 1 Script used to build Dockerfile for model and streamlit app

Step 4: Install the Simulation Model Dependencies

In order for the simulation model to run, we need to install the correct software dependencies (e.g. SimPy and Streamlit). In our example, we are installing our dependencies with the Python Package Index (PyPI). We are using the pip package manager for Python to install our dependencies on our image. These are held in the *requirements.txt* file stored in our git repository. You can think of this as a simple shopping list. It is a list of package names and versions.

Step 5: Create a Container Entry Point

The last step is included to immediately launch the Streamlit front end to our simulation model when the container is run. Streamlit provide examples and instructions online. In simple terms we are opening port 8501 as this is the port used by the Streamlit web server. We then specify that the container will run Streamlit and point it at the Python module Hospital_Efficiency_Project.py. This contains our Streamlit script.

10.5.2 Continuous Integration of New Code

Continuous integration allows us to quickly update our simulation models and their interfaces. An important part of this process is testing that our changes do not break the simulation model. The full details of testing simulation models are out of scope for this paper. Here we illustrate testing by embedding a simple test that checks the simulation model will run and return results after an update is made. We do this using two automation tools: *pytest* and GitHub actions.

Testing the Model Still Runs

In our repository, we have created a sub directory called *tests/*. The directory contains a Python model with a single Python function called *test_model_can_run*(). The function simply creates an instance of the model and performs a single run. A test is successful if it returns results with non-zero length. Prefixing the name of this function means that it can automatically be discovered by the Python package *pytest*. This package will run all tests it finds and returns the results to the user.

Automating Pytest on New Code Commits

Our simple test will execute automatically each time a change is made to the code and committed to the main branch of the GitHub repository. Github actions are written in Yet another Markup Language (YML). Listing 2 details the code used in test_model.yml.

In summary, Listing 2 will be run if new code is pushed to the main branch of the GitHub repository or if a pull request is created (i.e. a suggested update to the main branch that is checked before merging). The tests will be run on Ubuntu Linux using Python versions 3.8 and 3.9. For each version of Python, pip is used to install the

```
name: Test model

on:
  push:
branches: [ main ]
  pull_request:
branches: [ main ]

jobs:
  build:

runs-on: ubuntu-latest
strategy:
  matrix:
    python-version: ["3.8", "3.9"]

steps:
  - uses: actions/checkout@v3
  - name: Set up Python ${{ matrix.python-version }}
    uses: actions/setup-python@v4
    with:
      python-version: ${{ matrix.python-version }}
  - name: Install dependencies
    run: |
      python -m pip install --upgrade pip
      pip install pytest
      if [ -f requirements.txt ]; then pip install -r requirements.txt; fi
  - name: Test with pytest
    run: |
      pytest
```

Listing 2 GitHub action to automate code testing during continuous integration

dependencies and pytest is called to automatically discover and execute tests. Users can navigate to the GitHub 'actions' tab to view all workflows and browse the results of the tests. If an error has occurred, the pull request can be rejected.

10.5.3 Automated Build and Deployment to DockerHub

Our final step builds the Docker image remotely and pushes to DockerHub. We again use GitHub actions to automate this process. We aim to minimise the number of Docker builds we produce. Therefore, we limit this operation to when we issue a GitHub release. A release is when we decide that a major, minor, or patch (bug fix) has been made to our simulation model code. A release might have a name or more usefully a clear version number. For example, version 0.1.1. Listing 3 details the YML instructions.

Listing 3 can be used as a template for other implementations of our continuous integration and deployment pipeline. There are two key aspects the user must adapt: credentials and the image name.

Users must first create two GitHub repository secrets *DOCKER_USERNAME* and *DOCKER_PASSWORD*. These are the account details for DockerHub. They are entered via GitHub settings and are stored in an encrypted format.

Listing 3 specifies the image used in our applied example. When adapted, users must specify their own image name. This is in the format *my-docker-hub-namespace/my-docker-hub-repository*.

We again note that the action will only be initiated on a new release of the code. A release is created through GitHub itself. This is intuitive (achieved by clicking on the new release link on the main page of the GitHub repository), but we recommend users consult up-to-date GitHub documentation first.

Depending on the complexity of the software environment, the image build will take a few minutes to several hours. Our experience is that typical simulation environments take minutes rather than hours to build. Once the action has successfully run, a new Docker image will be available to pull from DockerHub. The script above will tag the image twice. It will automatically add the tags *latest* (overwriting the current *latest* version) and <release>. Taking our previous example this would be *tommonks01/hep-deploy:latest* and *tommonks01/hep-deploy:v0.1.1*.

10.5.4 Pulling the Updated Image to a Local Machine

Users of the simulation model can now pull the updated model from DockerHub. A pre-requisite is that they must have container software installed on their machine. Docker, for example, will work on Windows, Mac, and Linux. Our DockerHub page has up-to-date instructions on running the Docker container once downloaded. Listing 4 details the commands in order.

```yaml
name: Publish Docker image

on:
  release:
types: [published]

jobs:
  push_to_registry:
name: Push Docker image to Docker Hub
runs-on: ubuntu-latest
steps:
  - name: Check out the repo
    uses: actions/checkout@v3

  - name: Log in to Docker Hub
    uses: docker/login-action@f4ef78c080cd8ba55a85445d5b36e214a81df20a
    with:
      username: ${{ secrets.DOCKER_USERNAME }}
      password: ${{ secrets.DOCKER_PASSWORD }}

  - name: Extract metadata (tags, labels) for Docker
    id: meta
    uses: docker/metadata-action@9ec57ed1fcdbf14dcef7dfbe97b2010124a938b7
    with:
      images: tommonks01/hep-sim

  - name: Build and push Docker image
    uses: docker/build-push-action@3b5e8027fcad23fda98b2e3ac259d8d67585f671
    with:
      context: .
      file: ./Dockerfile
      push: true
      tags: ${{ steps.meta.outputs.tags }}
      labels: ${{ steps.meta.outputs.labels }}
```

Listing 3 GitHub action to automate docker build and push to DockerHub (Listing 3)

```
docker pull tommonks01/hep-sim
docker run -p 8501:8501 tommonks01/hep-sim:latest
```

Listing 4 Commands to run the container

10.6 Discussion and Conclusion

10.6.1 Summary and Strengths

Containers are being used in a growing number of computational fields, enabling efficient, collaborative, shared research. We have presented an example of a hybrid modelling approach for applied healthcare research using discrete-event simulation. Our application used a DES model developed for orthopaedic elective planning, combined with methods from software engineering for continuous integration and containerisation. By enabling deployment, this solution increases the functionality of our free and open-source simulation model, which has been developed for use/reuse by NHS users. The approach overcomes a significant barrier to sharing and re-using models developed in free and open-source software, which is compatibility across systems and dependency management. Isolating the model and its environment, including software and all required packages, in a container ensures that the simulation model can be used across different computing environments by any user, with a consistent runtime environment. This means that healthcare staff and other researchers can use the model without installing software and packages, supporting maintenance, and use/reuse. As well as making it easier to share work, containers also allow more efficient collaboration, address issues of reproducibility, and support scientific progress by reducing duplication of effort and allowing others to build upon existing work [10].

In healthcare M&S, there are few examples in the literature of shared, executable simulation models, and fewer again that are deployable by end-users [11]. As this is a novel hybrid application in healthcare M&S research, we have outlined in detail the steps required. In doing so, our aim is to support M&S researchers to apply our hybrid deployment framework to convert simulation models from disposable research artefacts to reusable solutions across health and social care. Our example application uses Docker, as it is the most common platform used to build and share containers. By allowing modellers to abstract away the complexities of the underlying infrastructure, Docker allows users to focus on writing model code, enabling testing of applications on their own machines and then deploying them to any environment running Docker. The model is runnable on Windows, MacOS, or Linux distribution; users simply need to install the correct version of Docker.

Other methods for sharing simulation models are available, each with advantages and disadvantages [60]. For example, a simple approach is the use of a package manager (such as pip or conda) to create a *virtual environment*. The package manager is installed directly on the users' machine and will attempt to install the correct operating system-specific version of software packages directly. However, conflicts between packages can still arise, as different packages may have incompatible dependencies or require specific versions of shared libraries, and potentially requires more maintenance and testing. Containers eliminate the need to manually install and configure dependencies on each hosting platform. By providing a consistent environment, containerisation ensures that the application can be reproduced exactly even when deployed on different systems.

For healthcare M&S researchers, a long-standing challenge is translating model results into real-world practice. Using the method we have described overcomes some of the barriers of model use/reuse in healthcare by enabling decision-makers to access and use the model. It forms a *Type E Hybrid Model* [2, 4], with the main objective of deploying the model outside of the research team, supporting the final stage of the M&S development lifecycle. We have additionally focused on model reuse to support application in other health and care organisations, and adaption to similar problem areas by M&S researchers. Despite many examples of successful healthcare M&S applications, there remains little evidence of long-term, widespread use/reuse [31]. While many factors contribute to model uptake and reuse, we have presented a hybrid deployment framework and an application demonstrating one approach to deploying simulation models within a healthcare setting for this purpose.

10.6.2 Challenges for the M&S Community

Using containers naturally presents challenges for M&S researchers. To start, Docker and other containerisation platforms require the use of FOSS software such as R or Python, as COTS simulation software have proprietary licenses and each container may require a paid software license (that may be time-limited in nature, such as an annual subscription). COTS packages may also be Microsoft Windows based: an operating system that is typically not used in containerisation solutions due to licensing. There are costs and benefits to be considered. For instance, researchers may prefer not to programme using FOSS, for example, to take advantage of COTS visual interactive simulation.

A further challenge is that FOSS deployment comes at the cost of time and effort [61]. There is likely to be a learning curve for M&S researchers, as Docker and other tools come with their own set of concepts, commands, and workflows. There may also be adjustments for FOSS users who are used to using IDEs or notebooks to adopt new tools. Ensuring that everything works seamlessly within a container can require additional effort and troubleshooting. However, the advantages can outweigh these challenges, and more example applications such as ours can help researchers to adopt the approach.

10.6.3 Limitations

Our example application only demonstrates the use of a DES developed in Python using SimPy and Streamlit, hosted in a GitHub repository, using Docker for containerisation, DockerHub for container hosting, and GitHub actions for continuous integration and deployment. There are many other containerisation approaches and platforms.

One limitation of our hybrid approach is persistence of model state and results within the Docker container. When the container is shut down both the simulation model state and results vanish. This means that users must save results or inputs to a file on their own machine and load them again each time the model is opened. If a persistence feature is required, a natural extension of our hybrid approach is to include Docker volumes that can persist data long term—these allow Docker to use a local hard drive for storing data.

10.6.4 Future Work

In future work, we aim to develop a framework to test and implement methods to support model deployment for M&S healthcare researchers to increase transparency, longevity, and reuse of models while reducing research waste. This includes helping researchers to overcome common challenges such as time, licensing, technology, and training, and aims to increase the quality and reproducibility of published simulation models.

Code Availability All simulation code, build scripts, and tests are available on GitHub: https://github.com/TomMonks/hep-deploy and archived at Zenodo: https://zenodo.org/record/8011462. The Docker container can be found here: https://hub.docker.com/r/tommonks01/hep-sim

Acknowledgements The original study was funded by the Health Data Research (HDR) UK South West Better Care Partnership (#6.12).

The authors are supported by the National Institute for Health Research Applied Research Collaboration Southwest Peninsula. The views expressed in this publication are those of the author(s) and not necessarily those of the National Institute for Health Research or the Department of Health and Social Care.

References

1. Mustafee N, Powell JH (2018) From hybrid simulation to hybrid systems modelling. In: Johansson B, Jain S, Rose O, Rabe M, Skoogh A, Mustafee N, Juan AA (eds) Proceedings of the 2018 winter simulation conference. IEEE, Piscataway, NJ, 2495–2506
2. Mustafee N, Harper A, Onggo BS (2020) Hybrid modelling and simulation (M&S): driving innovation in the theory and practice of M&S. In: Bae KG, Feng B, Kim S, Lazarova-Molnar

S, Zheng Z, Roeder T, Thiesing R (eds) Proceedings of the 2020 winter simulation conference. IEEE, Piscataway, NJ, 3140–3151
3. Tolk A, Harper A, Mustafee N (2021) Hybrid models as transdisciplinary research enablers. Eur J Oper Res 291(3):1075–1090
4. Mustafee N, Harper A, Fakhimi M (2022) From conceptualization of hybrid modelling & simulation to empirical studies in hybrid modelling. In: Feng B, Pedrielli G, Peng Y, Shashaani S, Song E, Gunes Corlu C, Lee LH, Chew EP, Roeder T, Lendermann P (eds) Winter simulation conference. IEEE, Piscataway, NJ, 1199–1210
5. Philip AM, Prasannavenkatesan S, Mustafee N (2022) Simulation modelling of hospital outpatient department: a review of the literature and bibliometric analysis. Simulation, p 00375497221139282
6. Salmon A, Rachuba S, Briscoe S, Pitt M (2018) A structured literature review of simulation modelling applied to emergency departments: current patterns and emerging trends. Oper Res Health Care 19:1–13
7. Liu Y, Zhang L, Liu Y, Laili Y, Zhang W (2021) Model maturity-based model service composition in cloud environments. Simul Model Pract Theory 113:102389
8. Auer S et al (2021) Science forum: a community-led initiative for training in reproducible research. eLife. https://doi.org/10.7554/eLife.64719
9. Baker M (2016) Reproducibility: seek out stronger science. Nature 537:703–704
10. Moreau D, Wiebels K, Boettiger C (2023) Containers for computational reproducibility. Nat Rev Methods Primers 3:50
11. Monks T, Harper A (2023) Computer model and code sharing practices in healthcare discrete-event simulation: a systematic scoping review. J Simul. https://doi.org/10.1080/17477778.2023.2260772
12. Monks T, Currie C, Onggo BS, Robinson S, Kunc M, Taylor S (2019) Strengthening the reporting of empirical simulation studies: introducing the STRESS guidelines. J Simul 13(1):55–67
13. Taylor SJ, Anagnostou A, Fabiyi A, Currie C, Monks T, Barbera R, Becker B (2017) Open science: approaches and benefits for modeling & simulation. In: Chan VWK, D'Ambrogio A, Zacharewicz G, Mustafee N, Wainer G, Page EH (eds) 2017 winter simulation conference. IEEE, Piscataway, NJ, pp 535–549
14. Monks T, Harper A (2023) Improving the usability of open health service delivery simulation models using python and web apps. NIHR Open Res 3:48
15. Monks T, Harper A (2023) SimPy and StreamLit tutorial materials for healthcare discrete-event simulation. Zenodo. https://doi.org/10.5281/zenodo.8193001
16. Byrne J, Heavey C, Byrne PJ (2010) A review of web-based simulation and supporting tools. Simul Model Pract Theory 18(3):253–276
17. Zhang L, Wang F, Li F (2019) Cloud-based simulation. In: Sokolowski J, Durak U, Mustafee N, Tolk A (eds) Summer of simulation. Simulation foundations, methods and applications. Springer, Cham. https://doi.org/10.1007/978-3-030-17164-3_6
18. Crowe S, Grieco L, Monks T, Keogh B, Penn M, Clancy M, Elkhodair S, Vindrola-Padros C, Fulop NJ, Utley M (2023) Here's something we prepared earlier: development, use and reuse of a configurable, inter-disciplinary approach for tackling overcrowding in NHS hospitals. J Oper Res Soc: 1–16.https://doi.org/10.1080/01605682.2023.2199094
19. Monks T, Robinson S, Kotiadis K (2014) Learning from discrete-event simulation: exploring the high involvement hypothesis. Eur J Oper Res 235(1):195–205
20. Robinson S, Nance RE, Paul RJ, Pidd M, Taylor SJE (2004) Simulation model reuse: definitions, benefits and obstacles. Simul Model Pract Theory 12(7–8):479–494
21. Dagkakis G, Heavey C (2016) A review of open source discrete event simulation software for operations research. J Simul 10(3):193–206
22. Zeng Z, Yuan X, Liang J, Li Y (2021) Designing and implementing an SWMM-based web service framework to provide decision support for real-time urban stormwater management. Environ Model Softw 135:104887

23. Gan T, Tarboton DG, Dash P, Gichamo TZ, Horsburgh JS (2020) Integrating hydrologic modelling web services with online data sharing to prepare, store, and execute hydrologic models. Environ Model Softw 130:104731
24. Qin R, Yang S, Xu Z, Hong T (2023) Development of a web-based modelling framework for harmful algal blooms transport simulation using open-source technologies. J Environ Manage 325:116616
25. Onggo BS, Taylor S, Tulegenov A (2014) The need for cloud-based simulation from the perspective of simulation practitioners. In: Proceedings of the operational research society simulation workshop, 103–112. https://www.theorsociety.com/media/3591/sw14-proceedings-book-final-hw_20042016101209.pdf#page=108
26. Piraghaj SF, Dastjerdi AV, Calheiros RN, Buyya R (2017) ContainerCloudSim: an environment for modeling and simulation of containers in cloud data centers. Softw Pract Exper 47(4):505–521
27. Rudyy O, Garcia-Gasulla M, Mantovani F, Santiago A, Sirvent R, Vázquez M (2019) Containers in HPC: a scalability and portability study in production biological simulations. In: 2019 IEEE international parallel and distributed processing symposium (IPDPS). IEEE, 567–577
28. Miller C, Padmos RM, van der Kolk M, Józsa TI, Samuels N, Xue Y, Payne SJ, Hoekstra AG (2021) In silico trials for treatment of acute ischemic stroke: design and implementation. Comput Biol Medi 137:104802
29. van der Kolk M, Miller C, Padmos R, Azizi V, Hoekstra A (2021) Des-ist: a simulation framework to streamline event-based in silico trials. Computational science–ICCS 2021: 21st international conference, Krakow, Poland, June 16–18, 2021, proceedings, part III. Springer International Publishing, Cham, pp 648–654
30. Anagnostou A, Groen D, Taylor SJ, Suleimenova D, Abubakar N, Saha A, Mintram K, Ghorbani M, Daroge H, Islam T, Xue Y, Okine E, Anokye N (2022) FACS-CHARM: a hybrid agent-based and discrete-event simulation approach for covid-19 management at regional level. In: Feng B, Pedrielli G, Peng Y, Shashaani S, Song E, Gunes Corlu C, Lee LH, Chew EP, Roeder T, Lendermann P (eds) Winter simulation conference. IEEE, Piscataway, NJ, pp 1223–1234
31. Tyler JM, Murch BJ, Vasilakis C, Wood RM (2022) Improving uptake of simulation in healthcare: user-driven development of an open-source tool for modelling patient flow. J Simul 1–18. https://doi.org/10.1080/17477778.2022.2081521
32. Capocchi L, Santucci JF, Fericean J, Zeigler BP (2022) DEVS model design for simulation web app deployment. In: Feng B, Pedrielli G, Peng Y, Shashaani S, Song E, Gunes Corlu C, Lee LH, Chew EP, Roeder T, Lendermann P (eds) Winter simulation conference. IEEE, Piscataway, NJ, 2154–2165
33. Monks T, Harper A, Anagnostou A, Taylor S (2022) Open science for computer. SIMULATION. https://doi.org/10.31219/osf.io/zpxtm
34. Team Simply (2020) SimPy. https://simpy.readthedocs.io/en/latest/about/index.html
35. van der Ham R (2018) Salabim: discrete event simulation and animation in python. J Open Sour Softw 3(27):767
36. Palmer GI, Knight VA, Harper PR, Hawa AL (2019) CIW: an open-source discrete event simulation library. J Simul 13(1):68–82
37. Goldberg AP, Karr JR (2020) DE-Sim: an object-oriented, discrete-event simulation tool for data-intensive modelling of complex systems in python. J Open Sour Softw 5(55):2685
38. Allen M, Bhanji A, Willemsen J, Dudfield S, Logan S, Monks T (2020) A simulation modelling toolkit for organising outpatient dialysis services during the COVID-19 pandemic. PLoS ONE 15(8):e0237628–e0237628
39. Chalk D, Robbins S, Kandasamy R, Rush K, Aggarwal A, Sullivan R, Chamberlain C (2021) Modelling palliative and end-of-life resource requirements during COVID-19: implications for quality care. BMJ Open 11(5):e043795
40. Shoaib M, Mustafee N, Madan K, Ramamohan V (2021) Leveraging healthcare facility network simulations for capacity planning and facility location in a pandemic. Socio-Econ Plann Serv. https://doi.org/10.1016/j.seps.2023.101660

41. Kazil J, Masad D, Crooks A (2020) Utilizing python for agent-based modelling: the mesa framework. In: Social, cultural, and behavioral modelling: 13th international conference, SBP-BRiMS 2020, Washington, DC, USA, 18–21 Oct 2020, proceedings 13(308–317). Springer International Publishing
42. Martin-Martinez E, Samsó R, Houghton J, Solé Ollé J (2022) PySD: system dynamics modelling in python. J Open Sour Softw 7:78(4329)
43. Ucar I, Smeets B, Azcorra A (2019) Simmer: discrete-event simulation for R. J Stat Softw 90(2):1–30
44. Lauwens (2021) SimJulia. https://simjuliajl.readthedocs.io/en/stable/welcome.html
45. Streamlit (2023) Streamlit. https://docs.streamlit.io/
46. Shiny (2023) Share your shiny applications online. https://www.shinyapps.io/
47. Shasha LU, Haili X, Xiaoning W (2022) Application of container technology in high performance computing environment. Front Data Comput 3(6):118–126
48. Merkel D (2014) Docker: lightweight Linux containers for consistent development and deployment. Linux J 239(2):2
49. Boettiger C, Eddelbuettel D (2017) An introduction to rocker: docker containers for R. R J 9:527
50. Kurtzer GM, Sochat V, Bauer MW (2017) Singularity: Scientific containers for mobility of compute. PLoS One 12(5):e0177459
51. Podman (2023) Available at https://docs.podman.io/en/latest/Introduction.html
52. Krafczyk M, Shi A, Bhaskar A, Marinov D, Stodden V (2019) Scientific tests and continuous integration strategies to enhance reproducibility in the scientific software context. In: Proceedings of the 2nd international workshop on practical reproducible evaluation of computer systems, pp 23–28
53. Shahin M, Babar MA, Zhu L (2017) Continuous integration, delivery and deployment: a systematic review on approaches, tools, challenges and practices. IEEE Access 5:3909–3943
54. Kim AY et al (2022) Implementing GitHub actions continuous integration to reduce error rates in ecological data collection. Methods Ecol Evol 13:2572–2585
55. GitHub (2023) GiHub actions. Available at https://docs.github.com/en/actions
56. Harper A, Monks T, Wilson R, Redaniel MT, Eyles E, Jones T, Penfold C, Elliott A, Keen T, Pitt M, Blom A (2023) Development and application of simulation modelling for orthopaedic elective resource planning in England. BMJ Open 13(12)
57. Wall J, Ray S, Briggs TW (2022) Delivery of elective care in the future. Future Healthcare J 9(2):144
58. Harper A, Monks T (2023) Supplementary material: code for the deployment of a simpy and streamlit model to dockerhub. https://zenodo.org/record/8011462
59. Boettiger C (2015) An introduction to docker for reproducible research. ACM SIGOPS Oper Syst Rev 49(1):71–79
60. Harper A, Monks T (2023) A framework to share healthcare simulations on the web using free and open source tools and python. In: Proceedings of the operational research society simulation workshop 2023 (SW23). https://doi.org/10.36819/SW23.030
61. Cadwallader L, Hrynaszkiewicz I (2022) A survey of researchers' code sharing and code reuse practices, and assessment of interactive notebook prototypes. PeerJ 10:e13933

Part III
Applications

Chapter 11
Hybrid Simulation in Healthcare Applications

Anastasia Anagnostou and Simon J. E. Taylor

Abstract Healthcare systems are traditionally characterised by complexity and heterogeneity. With the continuous increase in demand and shrinkage of available resources, the healthcare sector faces the challenge of delivering high quality services with fewer resources. Decision-makers typically are advised by experts in order to inform their policies and management strategies. One of the tools that is widely used to support evidence-based decisions is modelling and simulation (M&S). However, the size and complexity of healthcare systems constitute the conventional M&S techniques inadequate to model the system as a whole. Hybrid simulation, that combines one or more of the agent-based Simulation (ABS), discrete-event Simulation (DES) and System Dynamics (SD) techniques, is employed to enable holistic and detailed analysis of a system and also support model re-use. In this chapter, we discuss the use of hybrid simulation for emergency medical services and pandemic crisis management.

Keywords Agent-based simulation · Discrete-event simulation · Distributed simulation · Emergency medical services · Pandemic crisis management

11.1 Introduction

The health and social care sector face unprecedented challenges. The ageing population, the modern lifestyle and the emerging and re-emerging infectious diseases put a strain on healthcare services worldwide. At the same time, there is scarcity of the necessary resources to tackle these challenges. Healthcare professionals and policymakers are called to make difficult decisions such as which interventions to fund,

---

A. Anagnostou (✉) · S. J. E. Taylor
Modelling and Simulation Group, Department of Computer Science, Brunel University London, Kingston Lane, Uxbridge UB8 3PH, Middx, UK
e-mail: anastasia.anagnostou@brunel.ac.uk

S. J. E. Taylor
e-mail: simon.taylor@brunel.ac.uk

© The Author(s), under exclusive license to Springer Nature Switzerland AG 2024
M. Fakhimi and N. Mustafee (eds.), *Hybrid Modeling and Simulation*, Simulation Foundations, Methods and Applications, https://doi.org/10.1007/978-3-031-59999-6_11

how to allocate resources, what treatment is most effective and/or cost-effective, how to distribute supplies, among others.

One of the tools that is widely used to support evidence-based decisions in healthcare is modelling and simulation (M&S). In recent years, M&S of healthcare applications has gained momentum. Salleh et al. [1], in a review of systematic literature reviews on simulation in healthcare, found that 30 reviews were published since 2010 while there were only two published prior to 2005 and five between 2005 and 2009. The power of M&S is its ability to represent a real system and to study the impact of changes without actually altering the physical system. It provides insight into operational processes and can be used for dynamic forecasting, what-if analysis and optimisation at strategic, tactical and operational levels planning. Some of the most common M&S techniques in healthcare applications are system dynamics (SD), discrete-event simulation (DES) and agent-based simulation (ABS).

However, the size and complexity of healthcare systems constitute the conventional M&S techniques inadequate to model the system as a whole. M&S practices have some limitations, especially when large-scale systems are being modelled. For example, an M&S technique might not be able to capture the wider system due to its heterogeneity or compromises in the level of detail of a model and/or the amount of experimentation performed in a project might be necessary due to lack of time or compute resources. Large-scale systems usually involve subsystems either within the same organisation or in autonomous organisations with interdependencies. It may be convenient to create a hybrid model that combines one or more of ABS, DES and SD techniques to represent different systems [2]. According to Bell et al. [3] combining M&S methods as part of a hybrid simulation (HS) is better able to support diverse stakeholder perspectives, provide better understanding of the wider system and offer system insights and robust assumptions across models.

There is significant interest in HS for healthcare applications. This led to substantial advances in HS theory, especially with regard to hybridisation approaches and model connectivity. In a state-of-the-art review of the literature on hybrid simulation, Brailsford et al. [4] identified four types of hybridisation. These are as follows: (a) sequential, where two or more distinct single-method models that are executed sequentially, so that the output of one becomes the input to another; (b) enriching, where there is one dominant method with limited use of other methods; (c) interaction, where distinct single-method submodels interact cyclically at runtime; and (d) integration, where there is one seamless model in which it is impossible distinguish the boundaries of the simulation methods. There is also discussion in the literature about the connectivity approaches of HS. For example, Mustafee et al. [5] identified three main connectivity approaches. That is, to manually execute the models and transfer the relevant variables between them (non-synchronised manual execution), to automate the variable transfer but in a non-synchronised way (non-synchronised automated execution), and to use simulation packages that support multiple simulation techniques (synchronised CSP-driven execution) such as Anylogic.[1] Nonetheless, there is a risk of expertise bias, as identified by Kar et al. [6] in their recent review of HS in healthcare.

Despite the advances in the theoretical and methodological aspects of HS in healthcare, dos Santos et al. [7] highlight the lack of studies in the literature that include practical implementation of HS in healthcare applications. Mustafee et al. [8] give some indicative empirical studies where HS has been applied in practice in different fields.

So, how can we use hybrid simulation to study large and complex systems such as healthcare? To shed light on this question, we provide two practical examples of HS in two different healthcare applications. In both examples, we used hybrid DES and ABS techniques with different hybridisation and connectivity approaches. The first example is an interaction-based HS of emergency medical services with synchronised automated execution. The type of connectivity is achieved by developing the HS as a distributed simulation (DS) federation. The second example is a sequential-based HS of pandemic crisis management with non-synchronised manual execution. The chapter is structured as follows. In Sect. 11.2 discusses DES and ABS briefly and provides a review of the advances in M&S studies in the related application areas. Next, in Sects. 11.3 and 11.4, we describe our model development approach for each of the cases. Finally, in Sect. 11.5, we discuss our vision for future HS applications.

11.2 Literature Review

DES is a simulation technique that models a system's behaviour as discrete events. That is an instant of time at which an entity enters or leaves an activity. An activity changes the state of an entity [9]. For example, if the entity is a customer awaiting some service, while in the queue, the state of the customer could be "awaiting to be served". Once the server is available, the service will start, this event will change the state of the customer to "being served". DES is being used in various industries to analyse process behaviours within a system [10, 11]. DES is a popular technique, second to Monte Carlo simulation (MCS), in healthcare applications [12]. It is being employed to analyse patient flow, processes, disease progression, and health economic policies, among others [13, 14].

A more contemporary simulation techniques is ABS. It is characterised by the agents as individuals that have certain properties. The agents interact with other agents and the environment of the system. As a result, these interactions change the agents' properties, which define the agents' behaviour. ABS is being used mainly to analyse individual behaviours within a system. Arguably, ABS is continuously gaining in popularity within the simulation community. One of the reasons is its similarities with object oriented paradigm [15] and its ability to model people dynamics

---

[1] https://www.anylogic.com.

and social interactions. ABS has been employed in several healthcare and public health applications [16], including lifestyle, health behaviours [17] as well as communicable and non-communicable diseases, epidemiology [18] and socioeconomic inequalities in health [19].

Regardless the fundamentally different philosophies of the two simulation paradigms, there are significant similarities, too. For example, in both DES and ABS, the simulation time is progressing in discrete time steps [20] and their modelling view is a bottom-up approach of a system's behaviour.

11.2.1 ABS and DES for Emergency Medical Services

The literature on HS for EMS is limited. Ambulance management policies were analysed by Aringhieri [21] using a combination of ABS and DES in a single simulation package. Anagnostou et al. [22] employed DS to develop an interoperating hybrid ABS-DES model for holistic analysis of EMS considering emergency departments as well as ambulance services models. Olave-Rojas and Nickel [23] used ABS and DES in combination with machine learning (ML). They developed their HS model in a single package.

There are however many existing models of ambulance services, most of them in the context of emergency services and disaster planning. Jotshi et al. [24] study emergency vehicles dispatching and routing policies after an earthquake. Wu et al. [25] use ABS, DES, and geographical information systems (GIS) to introduce a decision support system for real-time disaster management. Campbell and Schroder [26] make use of simulation to present a training tool for emergency response. Henderson and Mason [27] use DES to model the Auckland, New Zealand ambulance service and GIS for geospatial visualisation. Their tool, BartSim, includes a travel model which produces deterministic computations of travel time when the travel time for the same distance varies during the day. DES is selected by Silva and Pinto [28] to analyse ambulance services in a Brazil city. They use optimisation techniques for ambulance deployment when testing different improvement scenarios. Ridler et al. [29] also used DES for their EMS model built using the JEMSS package implemented in Julia programming language.

Similarly, operations of the A&E departments have received much attention in the M&S community. Various simulation techniques have been deployed for A&E analysis, typically using DES that seems to be the most popular technique in operational research (OR) [30]. Eatock et al. [31] use DES to study the different A&E strategies used to meet the four-hour length of stay operational standard imposed by the UK government. They reproduced accurately the different strategies used to meet the operational standards. DES is used by Komashie and Mousavi [32] to analyse various key performance indicators in emergency departments. Alternatively, Escudero-Marin and Pidd [33] employ ABS to model emergency departments, where the agents can change behaviour according to some system variables. Ordu et al. [34] developed a DES decision support system to study demand and capacity in A&E.

they developed their model in the simulation package Simul8.[2] Patient waiting times in emergency departments is studied using DES and optimisation by Sasanfar et al. [35]. The authors developed their model in the simulation package ARENA[3] and they conducted a case study in an Iranian hospital. Gul et al. [36] employed ML and DES to study the preparedness of emergency departments after physical disasters such as earthquake.

11.2.2 ABS and DES for Pandemic Crisis Management

There are many HS studies that modelled various aspects of the Coronavirus pandemic. For example, Cimini et al. [37] used HS to assess the impact of containment measured for COVID-19. Their HS consists of an ABS virus contagion model and a DES people flow model in indoor environments. The authors conducted a case study at the University of Bergamo and their experiments shown the results of different class size examples. They reported that the use of HS allowed for more flexibility and insight gaining. Bouchnitaa and Jebranea [38] used multi-agent systems with social force modelling to study how the movement of people affect the transmission of COVID-19 and tested different non-pharmaceutical interventions in a case study in Morocco.

Arguably, DES is the most popular simulation method in the area of hospital bed capacity management and other pandemic related operations such as testing. Several studies have been conducted to estimate ICU bed capacity [39], to balance hospital resources for COVID-19 [40]. Saidani et al. [41] developed a DES to study COVID-19 testing capacity on a University campus.

ABS has been used widely to model the spread of infectious diseases. Examples include Kerr et al. [42] where the authors used ABS to model disease dynamics within social networks and the effect of preventive measures. The model includes hospitalisations and hospital bed capacity as input parameters. Similar study was conducted by Hinch et al. [43]. The authors developed their ABS in Python and with R interfaces. Wang et al. [44] used the Unity game engine[4] to develop an ABS of COVID-19. They reported that the visualisation of their simulation contributed to a better user experience.

Most of the ABS disease spread models consider bed capacities in an aggregated level but, usually, they do not include the operations of a hospital. Arguably, a hybrid approach where hospitals are explicitly modelled as well as the population interactions in the community is beneficial in gaining insight of the progress of a pandemic and the impact of public health preventive measures on the community and the healthcare system. This is particularly interesting at a regional level where the character of an area can be captured in the model.

---

[2] https://www.simul8.com/.

[3] https://www.rockwellautomation.com/en-us/products/software/arena-simulation.html.

[4] https://unity.com/.

11.3 Interaction-Based Hybrid Simulation of Emergency Medical Services

11.3.1 Background and Motivation

EMS are complex and multidimensional systems that provide immediate care to patients with acute illnesses or serious injuries. The role of EMS is to offer transport to those patients that were unable to get to hospital by their own means and to offer pre-hospital care on the site of an incident and during transport to the hospital.

EMS are accessible to the public by an emergency telephone number that put them through a control centre. The control centre personnel initially assess the incidents and find and dispatch the appropriate emergency vehicle and crew. EMS can be public, private or voluntary organisations. Usually, the offered services are classified into two categories, the basic life support (BLS) and the advanced life support (ALS). BLS deals with less serious illnesses and injuries and the crew does not have medical training, whereas ALS deals with serious illnesses and injuries, where the flashing blue lights are on, and the crew has medical skills, i.e. paramedics and emergency medical technicians.

A timely response and transfer to the regional hospitals' accident and emergency (A&E) departments, more than often, has saved the life of a patient. It is clear that there is close collaboration between the ambulance service and A&E departments. A global view of an emergency service system therefore includes the ambulance service and the A&E departments, usually situated in the hospitals of the coverage area. The fact that two different organisations involved raises the question, whether a single simulation technique is capable to accommodate the functionalities of those two organisations.

The ambulance service model, in a rough outline, includes the emergency call centre, or centres, the vehicles and the crews. The call operators have to respond to the emergency call, assess the incident severity in order to send the appropriate vehicle and crew, find the closest available vehicle to the site of the incident and send the vehicle to the patient. The crew, in turn, apart from the medical treatment on site, has to decide whether the patient needs to be transferred to a hospital or released after the on-scene treatment. If the patient has to be transferred, the closest available hospital and the fastest route should be found (see Fig. 11.1). All the above indicate a high degree of interaction among the system's objects. Both ABS and DES can be used to model such a system, however ABS simplifies the conceptualisation and interaction elements.

On the other hand, the A&E departments in a hospital are highly process oriented organisations. When a patient arrives at the A&E, generally and regardless of the mean of arrival, a treatment decision is made according to the patient's condition. If the required resources are available (e.g. cubicle, bed, nurse or doctor, etc.), the patient progresses through the system at the next activity. Otherwise, the patient enters a queue until the resources become available. Similarly, in DES, an entity passes through a system's processes, being processed when there are available resources

11 Hybrid Simulation in Healthcare Applications

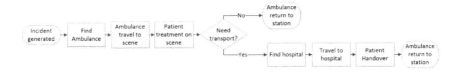

Fig. 11.1 Ambulance service process flow

or waiting in a queue as appropriate. An entity's state changes in accordance with the model's activities. However, the entity itself is not able to take decisions or interact with the objects of the simulation, but rather it is driven by the system's processes until, eventually, exits the simulation. DES therefore appears to be the most appropriate for an A&E patient flow model.

The majority of EMS simulation projects consider only the ambulance service up to the point of patient handover. However, the performance one system hugely affect the performance of the other. For example, lack of resources in an A&E will lead to hospital handover delays and as a consequence increase in ambulance response times. In this work, a comprehensive perspective of EMS is taken in order to perceive a global view of the emergency systems. Thus, in order to model the whole EMS, the A&E departments of the hospitals are included in the simulation model and operate in collaboration with the ambulance service.

At an EMS simulation, the interacting models exchange information frequently and often this information is more complex than a single variable. For example, an A&E model must be able to transfer its availability to the ambulance service model as soon as there is a change, the ambulance service model should be able to transfer patient information to a hospital model when a patient is transferred to it (usually this is in a form of an object that holds all of the patient data), etc. Therefore, the appropriate hybridisation method for this system is thought to be the interaction-based hybrid simulation with automated and synchronous connectivity. To implement this approach, we employed high level architecture (HLA)-based distributed simulation [45].

The hybrid distributed ABS-DES model can be used to explore the impact of emergency services operational and strategic decisions. Example questions are as follows:

- What would be the impact of changes in A&E discharge policies on ambulance response times?
- How changes in hospital handover protocols would affect the ambulance service performance?
- How would a new A&E impact the ambulance service?
- What would be the impact of A&E and ambulance services strategies on patient's health?

11.3.2 Agent-Based Simulation of Ambulance Services

The first element of EMS is the ambulance service. The ambulance services can be either air- or land-based. Air ambulance services respond to serious injuries when timing is extremely critical or the landscape is inaccessible by road. The land ambulance services coordinate the emergency calls and decide which vehicle and crew to be sent to an incident. Generally, the ambulance service fleet consists of fast response cars, ambulances and two-wheel vehicles. Furthermore, there are two types of crews: the BLS crew that deal with non-life threatening incidents, and the ALS crew that are able to provide medical care and deal with life-threatening injuries. Depending on the incident, ambulances transfer patients to hospitals or just treat them at the scene.

The timeline of the ambulance service is characterised by three distinguished periods: The *waiting period* is the time from the emergency call received to the time that an ambulance is found. The *service period* spans from the starting of a journey to the scene to the ending of the journey to the hospital; in the case that no hospitalisation is needed, the service time ends at the completion of on-scene treatment. The last period is the *response time* period that starts when receiving an emergency call and ends when the ambulance reaches the scene of the incident. The above three time periods are considered to be some of the performance measures of the ambulance service organisations.

At the conceptualisation stage, a decision should be taken about the level of detail that the simulation models will include. One of the objectives of this study is to demonstrate the HS approach. Thus, it was decided to model the participating simulations with just enough details to validate the models. The ambulance service model therefore includes the land ambulance service and car ambulance vehicles. The emergency call centre is a simplified form of the real system and the ambulance crew is not modelled.

Simulation agents are of two types, passive agents, which are part of the environment, and active agents, which interact with each other and the environment. Passive agents are the hospitals and the ambulance stations which have predetermined attributes such as location coordinates and capacity. The active agents are the emergency calls, the patients and the ambulances. Emergency calls are generated according to an arrival probability distribution. A generated emergency call carries location and incident information. Simultaneously with a call generation, a patient is generated that adopts the location and incident information from the call. Incident information is the patient condition and whether this patient needs transfer to an A&E. In real ambulance services, when a call arrives at the dispatch centre, an assessment is carried out and then the appropriate ambulance crew is sent to the incident scene. An assumption of this model is that all calls will be attended, thus the emergency call searches for the nearest available ambulance. Once an ambulance is found the call is removed from the simulation. When an ambulance is found, it is flagged as unavailable and starts the journey towards the incident scene. At the scene, there is a delay for on-site treatment. If the patient is flagged as one that needs transfer to an A&E, the ambulance searches for the nearest available A&E and starts

11 Hybrid Simulation in Healthcare Applications

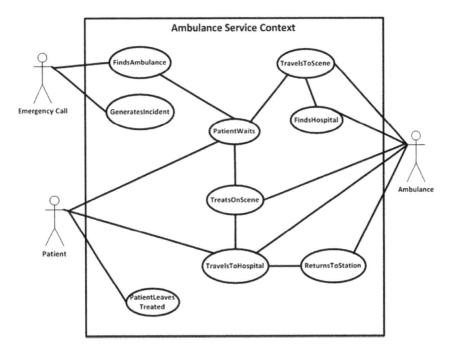

Fig. 11.2 Ambulance service model agents interaction

the journey to the hospital. After the patient's handover, the ambulance goes back to station and is flagged as available. The agent's interactions are shown in Fig. 11.2.

Travel times are calculated assuming an average ambulance speed and the Euclidean distance between the current location and the destination location. According to a study by Jones et al. [46], Euclidean distance with a corrective factor can be used in research with a high degree of confidence to represent real driving distance in an urban setting. Silva and Pinto [28] used a corrective coefficient to realistically represent the relationship between the actual distance and the Euclidean distance of two points in an urban environment. Each location point has an X and a Y Cartesian coordinate, so the Euclidean distance s between two points is calculated by Eq. 11.1. The travel time t is calculated by Eq. 11.2.

$$s = \frac{\sqrt{(X_A - X_B)^2 + (Y_A - Y_B)^2}}{(X_A - X_B)^2 + (Y_A - Y_B)^2}, \quad (11.1)$$

$$t = c \times s/v, \quad (11.2)$$

where

t is the travel time in hours,
s is the distance in miles,

v is the average speed in miles per hour, and
c is a corrective coefficient calculated as 1.32 for large urban settings.

11.3.3 Discrete-Event Simulation of Accident and Emergency Departments

The second element of EMS is the A&E departments in the region. For the purpose of this work, a general A&E department is considered with high level of abstraction. The normal A&E flow includes an initial triage, the minors units and the majors units. The minors units treat minor injuries and illnesses while the major units deal with seriously ill patients. Also, in a general A&E department, there are two streams of patient input: the ambulance arrivals and the walk-in arrivals.

The patient flow of a general A&E is shown in Fig. 11.3. Patients arrive at the hospital, either by ambulance or by their own means (walk-in). Ambulance arrivals are directed to the appropriate section (i.e. minors or majors) according to their condition. Walk-in arrivals are directed to the above sections or leave the hospital after an initial assessment (i.e. triage service). Walk-in patients enter a queue for triage. The triage staff performs the initial assessment and directs patients accordingly. Then the patients join a queue for the minors or majors department according to their condition. When the required resources become available, patients receive treatment in the respective unit. If no treatment is required, they leave the system after the triage service. Ambulance patients go directly to the minors or majors department, based on the previous communication with the ambulance model, and do not enter the subsequent queues. This is due to the fact that the hospital has been notified of the ambulance patient arrival and has reserved the appropriate resources. After the end of the service, patients exit the A&E and the resources are released. Clinical staff is modelled as one type of resource. Therefore, there is no differentiation among nursing, medical or technical personnel. Another simplification of our model is that imaging or lab tests processes are not included.

The hospital availability dependents on the clinical staff resources and the particular type of A&E observation beds/cubicles (from now on are mentioned as beds), namely minors and majors. This can be calculated for each particular bed type as per Eq. 11.3.

$$H_a = \min(A_{cs}, A_{b_{\{type\}}}) \quad (11.3)$$

where

H_a is the is the hospital availability,
A_{cs} is the available clinical staff,
$A_{b_{type}}$ is the available beds of a particular type, and
type {minors, majors}

11 Hybrid Simulation in Healthcare Applications

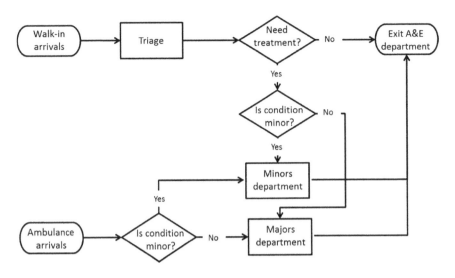

Fig. 11.3 A%E department process flow

The available clinical staff and available beds are calculated by Eqs. 11.4 and 11.5, respectively.

$$A_{cs} = CS_c - CS_o, \tag{11.4}$$

$$A_{b_{\{type\}}} = BC_{\{type\}} - BO_{\{type\}} \tag{11.5}$$

where

CS_c	clinical staff capacity,
CS_o	is clinical staff occupancy,
$BC_{\{type\}}$	is the bed capacity of a particular type, and
$BO_{\{type\}}$	is the bed occupancy of a particular type.

The hospital availability is updated in the DES A&E department model. This information is passed to the ABS ambulance service model in order to locate the suitable hospital when a patient transfer is needed.

11.3.4 Hybrid Distributed ABS-DES of Emergency Medical Services

The hybrid EMS model is an interaction-based hybrid simulation with automated synchronous connectivity. To achieve this, we implemented an HLA-based DS. A typical DS is composed of number of simulation models which are known as federates

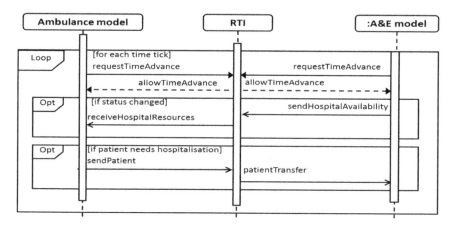

Fig. 11.4 Communication via RTI

and all these federates run under one federation, connected through a middleware, as if they were a single simulation. In an HLA DS, this middleware is called a run time infrastructure (RTI) and follows the IEEE 1516 standard [47].

Two aspects of requirements were considered in order to integrate the ABS ambulance model with the various DES models of the A&E departments. First, we had to ensure that the models can run as independent simulations. Secondly, we had to enable these independent simulations to behave as components of a DS system that interoperate with the RTI middleware. Being able to run as part of a DS, each federate model must provide an interface to connect to another federate model. Standardisation of data representation is achieved through the RTI middleware in order to enable data exchange among the different federates. Another very important role of the RTI interface is the implementation of time synchronisation mechanism among all federate models of a federation. Figure 11.4 shows the interactions between the federates and the RTI middleware.

The message passing and time management agreement among the federates via the RTI are described in Fig. 11.4. The distributed model initialises when all federates register to the RTI. A time synchronisation is agreed between federates through RTI and time advance request is initiated by each federate at every simulation time unit in accordance to the tenth rule of HLA specification [47]. There are two optional messages sent and received by the federates on change of state. These messages update the local status of the federates. These are data structures, variables or objects that update shared data. When there is a need for hospital transfer, a patient object, or entity, is passing from the ambulance model to the A&E model and it time-stamped by both models. When there is a change in the A&E availability, this information is being known to the ABS ambulance model by updating a variable via the RTI middleware. The ambulance model, in turn, is using this information to find the most suitable hospital for the patient on transfer. Also, when a patient is on the way to an A&E, information is sent by the ambulance model in order to update the hospital availability when the decision for transfer is taken and not when the patient arrives at the hospital. This can prevent a conflict at the hospital availability variable update.

11.4 Sequential-Based Hybrid Simulation of Pandemic Crisis Management

11.4.1 Background and Motivation

The Coronavirus pandemic is arguably the worst pandemic event in the 21st century. It has posed unparallel challenges to almost all sectors in all countries across the whole world. At the date of writing (October 2023) the World Health Organisation (WHO) recorded more that 771 million cases worldwide and more than 6.96 million recorded deaths [48]. The impact of COVID-19, a disease caused by SARS-CoV-2, however is estimated to be higher. In a recent report, WHO estimated that the excess mortality associated with COVID-19 during the first two years of the pandemic (January 2020 to December 2021) has reached almost 15 million worldwide [49].

One of the interesting aspects of the response to COVID-19 is its local nature. National and local governments have implemented various measures at various points in time according to the severity of the disease in that location, the economic constraints and inputs from experts. Moreover, similar sets of measures can vary in their ability to contain the spread of infection in different locations. Arguably, this is due to the differences among the regions in terms of demographics, the economy, the density and behaviour of the population, as well as the degree of compliance to public health preventive measures. Further, healthcare resources as well as health needs vary significantly between districts. Therefore, it is essential to evaluate the efficiency of these measures for a given region.

M&S has been used widely to support decision-making for COVID-19. Early on in the pandemic Currie et al. [50] discussed the use of M&S in the context of reducing the impact of COVID-19. They identified key decisions that M&S can support and they suggested M&S methods for each key decision based on the focus of the study. For example, ABS is identified as an appropriate method when individual behaviour is considered important. On the other hand, DES is most suited for studying operational matters. HS is also identified as the method that is used when a single approach cannot capture the complexity of the system.

Arguably, dealing with a crisis of that scale is complex. It is unlikely that a single model can encompass the multiple dimensions of a pandemic. Some of the reasons behind that maybe that attempting to include all of the aspects of such a crisis can lead to unmanageably complex models or that different modelling methods are better suited to model different dimensions of the crisis, among others [51]. Many modelling efforts emerged in the past two years. There are many models that focus on the epidemic dynamics of the disease. Some examples are discussed in Kerr et al. [42], Hinch et al. [43] and Wang et al. [44]. There are also numerous models with emphasis on resources allocation. Examples include Zhu et al. [39], Melman et al. [40] and Saidani et al. [41]. However, few studies attempt to tackle both and investigate the impact of prevention measures at a local scale.

In this work we present a HS ABS-DES simulation approach that aspires to provide a comprehensive manner for managing COVID-19 at a regional level. Our hybrid model consists of the Flu And Coronavirus Simulator (FACS), an ABS model that predicts the spread of infections in specified area, and the dynamiC Hospital wARd Management (CHARM) model, a DES mode of intensive care unit (ICU) patient flow that caters for reconfiguration of ICU wards at runtime. Details of both models are described in Sects. 11.4.2 and 11.4.3.

The Hybrid FACS-CHARM model can be used to explore questions regarding public health interventions and their impact on the local hospital facilities. Example questions are:

- What would be the impact of public health intervention on the case numbers?
- What would be the impact of public health intervention on the local hospitals?
- When would COVID-19 bed surge be needed?
- What would be the impact of elective cancellation strategies on bed capacities?

11.4.2 Agent-Based Simulation of Spread of a COVID-19

The Flu and Coronavirus Simulator (FACS) is an ABS model that uses geospatial, demographic and disease-specific data to model the temporal and spatial progression of COVID-19 in a specified geographical location (e.g. city or borough). It identifies the amenities and houses in the location and creates a local spatial network of people (agents) and amenities. It then simulates the movement of population across the amenities according to their needs which results in the spread of infection. The needs are specified based on an agent's age. For example, school age population go to school, working age population go to offices, etc. All agents spend most of their day at home and a proportion of their time in shopping, outdoor and indoor leisure facilities. Infected agents self-isolate and if the disease is severe, they are hospitalised. FACS has the capability to simulate the effects of various preventive measures such as lock-downs and vaccination strategies as well as the impact of emergence of new variants of the virus and the impact of distribution of amenities in the region. FACS can be easily adapted to model other similar air-borne diseases. FACS consists of three main modules, these are the *FACS pre-processing module*, the *FACS core module* and the *FACS post-processing module*.

FACS Pre-processing Module. This module is used for the preparation of input data using CSV and YAML files, the parsing of different buildings types from the Open-StreetMap[5] files and preparing the geospatial location graphs. The input data can be categorised into three types that is population data, disease data and location data.

---

[5] https://www.openstreetmap.org/.

Population data include the age distribution of agents in the area and is obtained from openly published demographic information, it also includes data that determine the behaviour of the population, i.e. it needs based on age. Disease data contain the disease parameters and prevention measures data, including pharmaceutical (e.g. vaccination) and non-pharmaceutical (e.g. social distancing, lockdowns, etc.) interventions. The disease parameters are the infection rate, the mortality rate, hospitalisation probability, incubation period, time to hospitalisation, length of stay in ICU, and time to recovery. Prevention measures data are in the form of a schedule. For each time period, there are several pharmaceutical and non-pharmaceutical measures related parameters that are defined in a YAML file. With regard to vaccinations, these are the number of vaccinations administered per day, eligible age group, efficacy in terms of disease severity and ability to transmit the virus while vaccinated and booster doses administered per day. With regard to non-pharmaceutical measures, the parameters are the facilities under lockdown, mask wearing and uptake, track and trace efficiency and social distance advice. Location data includes the type of amenities and their coordinates as well as their size in square metres and all facilities of a specific type, i.e. a school type amenity includes schools, colleges and universities, an office type amenity includes all work spaces, i.e. office, industrial area, construction site, etc.

FACS Core Module. This module contains the rules that govern the behaviour of the population and the disease as well as the impact the prevention measures have on the disease outcome. The population follows daily routines, e.g. they go to school or work, they stay in their household, they visit shopping and leisure amenities, etc. Based on these routines, they spend time in outdoor or indoor facilities of a certain size where they come into contact with other people. The size of the area, the air quality in this area, the time spent within as well as their immunity, the infection rate and the number of infected people they come into contact are the factors that affect the probability P_{inf} of getting infected. The infection probability is calculated by Eq. 11.6.

$$P_{inf} = M \times \frac{IR}{SF} \times LS_{inf} \frac{NI_{avg} \times A_{per}}{A_{loc} \times Max_{per}} \times \frac{LS_{inf}}{OD_{loc}} \qquad (11.6)$$

where

IR	is the infection rate,
SF	is a scaling factor,
LS_s	is the length of stay of a susceptible person in an area in minutes,
NI_{avg}	is the average number of infectious contacts per day,
A_{per}	is the physical area of a single standing person,
A_{loc}	is the size of the location in m²,
Max_{per}	is the maximum number of persons that can fit in 4 m²,
LS_{inf}	is the length of stay of an infectious person in the area in minutes, and
OD_{loc}	is the opening time of a location on a given day.

In the core module, the attributes of the agents are updated. These attributes indicate whether the agents are susceptible, infected, recovered, or dead. If infected, agent attributes indicate whether the symptoms are mild or severe and in the latter case whether they are admitted to hospital. Other attributes include an agent's immunity, vaccination status, geolocation at a given time its needs and household (size and geolocation).

The prevention measures rules are essentially the implementation of the measures as these are described in the input parameters. Each agent, based on its age and history follows the respective prevention measures with a given uptake probability.

FACS Post-processing Module. This module processes the raw simulation outputs. It is used to collate all replication outputs, plot the output graphs and visualise the disease progression on an interactive map. Output results include the daily number of agents in each compartment, i.e. susceptible, infected, recovered and dead, daily hospitalisations, and a history log of the behaviour of each agent, e.g. places visited, where an infection occurred, etc. Example of a FACS output is shown in Fig. 11.5, where the number of daily cases (b) and hospitalisations (b) for the London borough of Brent are plotted. The graph shows experiments ran in the first year of the COVID-19 pandemic. The shaded area indicates 95% confidence intervals. Actual validation data are plotted for the first wave and the vertical red lines denote key lockdown interventions. FACS is described in detail in Mahmood et al. [52].

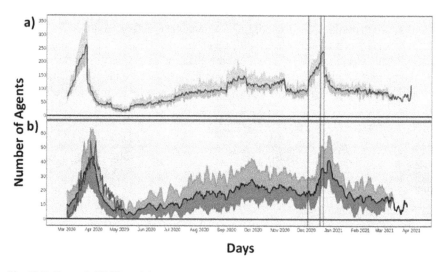

Fig. 11.5 Example FACS model output

11.4.3 Discrete-Event Simulation for ICU Bed Capacity Management

Hospitals across the globe face capacity challenges due to the Coronavirus pandemic. Large numbers of COVID-19 patients require admission to ICU and very often for a long period of time. Another big challenge is the highly infectious nature of the virus. Hospitals make huge ward rearrangements in order to prevent nosocomial infections as well as create surge bed capacity for the anticipated COVID-19 admissions. During the first wave, hospitals cancelled all elective surgeries and were able to deal with only a small number of emergency incidents. When the first crisis started to ease, hospitals main question was how to manage future COVID-19 outbreaks, or other highly infectious diseases, and at the same time continue normal operation? Arguably, cancellation of all scheduled surgeries is not a viable strategy. The backlog of elective operations has a massive impact on the care that healthcare systems can provide to the population. Consequently, this has a negative impact on the quality of life and the economy.

In an attempt to support hospitals in planning their ICU bed capacity, we developed the dynamiC Hospital wARd Management (CHARM) model. CHARM is a DES model of the ICU patient flow process that allows for dynamic reconfiguration of hospital wards. It considers three types of admissions for emergency, elective and COVID-19 arrivals. A routing logic allocates the patients to the respective wards. COVID-19 capacity can be pooled from elective and emergency capacity when there is a surge of COVID-19 admissions. The resources are reversed to their original configuration when the surge eases. Hospital wards in CHARM are allocated in zones that represent six types, i.e. COVID-19 (C) and COVID-19 recovery (CR), emergency (EM) and emergency recovery (EMR), and elective (EL) and elective recovery (ELR). These zones are the pool of bed resources for each type. Each ward is dynamically recharacterised as of different zone type according to bed occupancy. The model logic is shown in Fig. 11.6 and applies on every patient that enters the model. There are three types of arrivals for C, EM and EL patients. When there is a new arrival, the model checks whether there is bed availability in the respective wards. If there is no bed available, the patient is moved out of the specific facility. If a bed is available, the patient stays in the ICU ward for the designated length of stay (LoS). Mortality check takes place at the end of this stay. If the patient does not die, the patient will attempt to move to a respective recovery ward. Similarly, to ICU, if there is no recovery bed available, the patient will move to another facility. If there is recovery bed available, the patient will occupy this resource and stay for a designated LoS. Mortality check takes place at the end of the recovery stay, too. If the patient does not die, the patient will be discharged.

At every simulation time unit, the model performs some calculations and checks. The 7-day average of bed occupancy for type C beds is calculated to account for ward change setup time. Based on the bed occupancy, ward re-configuration occurs. If COVID-19 bed occupancy is greater than a set threshold, emergency and elective zones are converted to COVID-19 zones, respectively. If COVID-19 bed occupancy

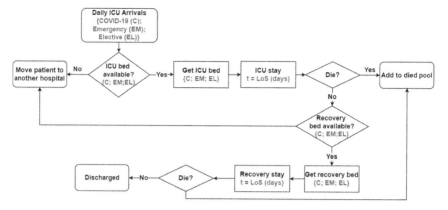

Fig. 11.6 CHARM model patient flow

falls below a set threshold, elective and emergency zones converted back to their original type.

CHARM is built in Python using the SimPy libraries.[6] It takes two types of input data, that is daily COVID-19 patient arrivals and input parameters data.

COVID-19 arrivals, for the time horizon of the simulation, are read from a CSV file. This file in our hybrid FACS-CHARM is generated by a FACS simulation. However, CHARM can run independently too using retrospective primary COVID-19 ICU arrival data or predicted from other models. The input parameters data are also read from CSV files. These include daily emergency and elective arrivals, LoS for each zone type, mortality probability, the upper and lower thresholds of COVID-19 ICU bed occupancy and the number of replications that we wish to run the simulation for. Input parameters also include the initialisation of zones with their types and capacities.

11.4.4 Hybrid ABS-DES of Pandemic Crisis Management

Based on Brailsford's [4] hybridisation classification, the FACS-CHARM hybrid model can be described as sequential-based hybridisation with manual (or automated) interconnection. Our model has a single connecting variable. This is the daily COVID-19 ICU hospitalisations.

FACS predicts the daily COVID-19 ICU hospitalisations in the coverage area, among others. It also outputs geolocation information of the admitted cases. Using this feature, we can identify the total cases as well as the number of cases arriving at the individual hospitals serving the modelled region. CHARM in turn uses this information that is the number of cases that arrive at the specific hospital, to predict ICU bed occupancy and whether surge capacity is likely to be needed.

---

[6] https://gitlab.com/team-simpy/simpy.

11 Hybrid Simulation in Healthcare Applications

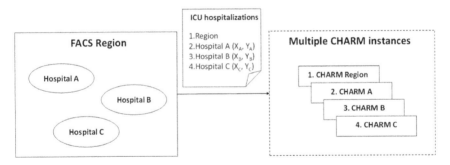

Fig. 11.7 Hybrid FACS-CHARM data exchange

The combined capabilities of the hybrid FACS-CHARM model allow for experimentation with different public health interventions, e.g. lockdowns, vaccinations, etc., and ICU bed capacity strategies for more efficient management of a pandemic. FACS can predict the case numbers and the ICU hospitalisations in a region under different preventive measures while at the same time CHARM can predict the impact of these on the local hospitals and more accurately plan for bed surge capacity and elective cancellations.

FACS-CHARM can be used to study the COVID-19 impact on ICU capacity in the region as a whole and/or on individual hospitals within the region. For example, let's assume that the area modelled with FACS has three hospitals with ICU beds—hospitals A, B and C as shown in Fig. 11.7. FACS records the temporal and spatial ICU hospitalisations. We therefore have information of the daily admissions in each of the hospitals in the area (Long = $X\{A, B, C\}$, Lat = $Y\{A, B, C\}$) and in the whole region. In this example, four CHARM instances can be run for analysing ICU bed strategies in the region and/or in hospitals A, B and/or C. Input data for each case are daily ICU hospitalisations for the region, and/or for hospitals A, B, and/or C, respectively.

The CHARM multiple instance option is denoted with a dotted line to indicate that this is optional. Each of the instances is independent and the choice to execute one or more depends on the scope of the study.

11.5 Discussion and Conclusions

M&S is a powerful tool for evidence-based decision-making. But simulation models are expensive to build, both in terms of time and expertise. It is also difficult for non-expert end-users to understand and use the models, especially when the systems are large and complex and require a combination of modelling techniques, such as the use of HS. Understandability, reusability, and openness, therefore, are important properties that could help HS to achieve its potential.

Commercial simulation packages provide visualisation capabilities to make the models easily understandable and usable by the end-users. For models that are developed in a programming language, modellers typically do not provide user-friendly interfaces. However, this could support easier acceptance and by end-users. To support easy interaction with the model, we developed web interfaces. Examples of these are shown in Fig. 11.8. A dashboard application for the CHARM model is shown in (a) where users can enter input parameters in a user-friendly manner and visualise and interact with the output plots. We also developed a dashboard for FACS that enables end-users to enter certain pre-processed locations in a user-friendly manner and visualise the spread of a disease on a map interactive plots. The code of the CHARM[7] and FACS[8] dashboards are available on open access repositories.

With regard to reusability and openness, in a panel discussion where Aleman et al. [53] debated different modelling approaches for COVID-19, it was highlighted that availability and reuse of existing models could have resulted in faster modelling efforts and therefore better support to decision-makers, especially at the early stages of the Coronavirus pandemic. To support this effort, it is important to have well-documented models using standard reporting methods such as the ODD protocol [54] and the STRESS guidelines [55]. Also, Lather and Eldabi [56] discussed the benefits of an HS hub for pandemic situations where models can be easily found and accessed. In that respect, the code of both FACS and CHARM models is available on public repositories.[9,10] Arguably, DS also supports reusability by designing composable component models (federates) that can be recombined with each other or with different interoperable models and form different DS systems (federations) [57].

To conclude, this article discussed the use of HS in healthcare applications. The benefits of using a combination of M&S techniques were discussed in the context of simulation studies of large-scale and complex systems. Two indicative application examples were presented with different hybridisation implementations. The first exemplar case demonstrated an interaction-based HS of EMS with synchronised automated execution. The EMS HS was comprised of an ABS ambulance service model and DES A&E department models, as many as the coverage area. The synchronised automated execution was achieved by implementing the model as an HLA-based DS. The second exemplar case was an HS of pandemic crisis management. This was a sequential-based HS with manual execution that can easily be automated. Two models constituted the system, an infectious disease ABS model and an ICU bed capacity management DES model. We also discussed our vision for easily understandable, reusable, and open models that could potentially support rapid HS development in the future.

---

[7] CHARM dashboard code repository: https://gitlab.com/anabrunel/charm-app.
[8] FACS dashboard code repository: https://github.com/arindamsaha1507/FADE_Dashboard.
[9] FACS code repository: https://github.com/djgroen/facs.
[10] CHARM code repository: https://gitlab.com/anabrunel/charm.

11 Hybrid Simulation in Healthcare Applications

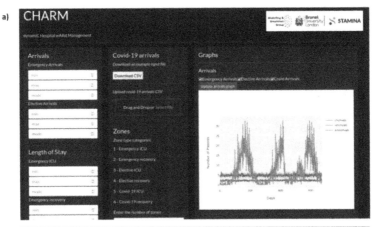

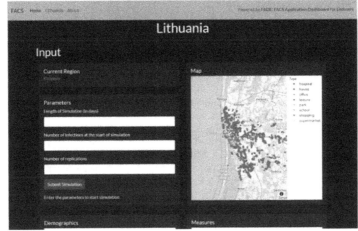

Fig. 11.8 CHARM (**a**) and FACS (**b**) dashboards

Glossary

A&E Accident and emergency
ABS Agent-based Simulation
ALS Advanced life support
BLS Basic life support
CHARM dynamiC Hospital wARd Management
CSV Comma separated values
DES Discrete-event Simulation
EMS Emergency Medical Services
FACS Flu and coronavirus simulator
GIS Geographic information system
HLA High-level architecture

HS Hybrid Simulation
ICU Intensive care unit
LoS Length of stay
ML Machine Learning
MCS Monte Carlo Simulation
M&S Modelling and Simulation
OR Operational Research
RTI Run time infrastructure
SD System Dynamics
WHO World Health Organization

References

1. Salleh S, Thokala P, Brennan A, Hughes R, Booth A (2017) Simulation modelling in healthcare: an umbrella review of systematic literature reviews. Pharmacoeconomics 35:937–949
2. Anagnostou A, Taylor SJE (2017) A distributed simulation methodological framework for OR/MS applications. Simul Model Pract Theory 70:101–119
3. Bell D, Cordeaux C, Stephenson T, Dawe H, Lacey P, O'Leary L (2016) Designing effective hybridization for whole system modeling and simulation in healthcare. In: 2016 winter simulation conference (WSC). IEEE, pp 1511–1522
4. Brailsford SC, Eldabi T, Kunc M, Mustafee N, Osorio AF (2019) Hybrid simulation modelling in operational research: a state-of-the-art review. J Oper Res 278(3):721–737
5. Mustafee N, Sahnoun M, Smart A, Godsiff P, Baudry D, Louis A (2015) Investigating execution strategies for hybrid models developed using multiple M&S methodologies. In: 48th annual simulation symposium. Alexandria, VA, pp 78–85
6. Kar E, Eldabi T, Fakhimi M (2022) Hybrid simulation in healthcare: a review of the literature. In: 2022 winter simulation conference (WSC). IEEE, pp 1211–1222
7. dos Santos VH, Kotiadis K, Scaparra MP (2020) A review of hybrid simulation in healthcare. In: 2020 winter simulation conference (WSC). IEEE, pp 1004–1015
8. Mustafee N, Harper A, Fakhimi M (2022) From conceptualization of hybrid modelling & simulation to empirical studies in hybrid modelling. In: 2022 winter simulation conference (WSC). IEEE, pp 1199–1210
9. Pidd M (2004) Computer simulation in management science, 5th edn. Wiley, Chichester, UK (2004)
10. Brailsford SC, Bolt T, Connell C, Klein JH, Patel B (2009) Stakeholder engagement in health care simulation. In: 2009 winter simulation conference (WSC). IEEE, pp 1840–1849
11. Robinson S (2005) Discrete-event simulation: from the pioneers to the present, what next? J Oper Res 56(6):619–629
12. Katsaliaki K, Mustafee N (2011) Applications of simulation within the healthcare context. J Oper Res Soc 62(8):1431–1451
13. Mustafee N, Katsaliaki K, Taylor SJE (2010) Profiling literature in healthcare simulation. SIMULATION 86(8–9):543–558
14. Forbus JJ, Berleant D (2022) Discrete-event simulation in healthcare settings: a review. Modelling 3(4):417–433
15. North MJ, Macal CM (2007) Managing business complexity. Oxford University Press Inc., New York
16. Barnes S, Golden B, Price S (2013) Applications of agent-based modeling and simulation to healthcare operations management. In: Handbook of healthcare operations management: methods and applications. Springer, New York, pp 45–74

17. Yang Y (2019) A narrative review of the use of agent-based modeling in health behavior and behavior intervention. Transl Behav Med 9(6):1065–1075
18. Tracy M, Cerda M, Keyes KM (2018) Agent-based modeling in public health: current applications and future directions. Annu Rev Public Health 39:77–94
19. Boyd J, Wilson R, Elsenbroich C, Heppenstall A, Meier P (2022) Agent-based modelling of health inequalities following the complexity turn in public health: a systematic review. Int J Environ Res Public Health 19(24):16807
20. Pawlaszczyk D, Strassburger S (2009) Scalability in distributed simulations of agent-based models. In: 2009 winter simulation conference (WSC). IEEE, pp 1189–1200
21. Aringhieri R An integrated DE and AB simulation model for EMS management. In: 2010 IEEE workshop on health care management (WHCM). IEEE, pp 1–6
22. Anagnostou A, Nouman A, Taylor SJE (2013) Distributed hybrid agent-based discrete event emergency medical services simulation. In: 2013 winter simulations conference (WSC). IEEE, pp 1625–1636
23. Olave-Rojas D, Nickel S (2021) Modeling a pre-hospital emergency medical service using hybrid simulation and a machine learning approach. Simul Model Pract Theory 109:102302
24. Jotshi A, Gong Q, Batta R (2009) Dispatching and routing of emergency vehicles in disaster mitigation using data fusion. Socioecon Plann Sci 43(1):1–24
25. Wu S, Shuman L, Bidanda B, Prokopyev O, Kelley M, Sochats K, Balaban C (2008) Simulation-based decision support system for real-time disaster response management. In: IIE annual conference. IISE, p 58
26. Campbell BD, Schroder KA (2009) Training for emergency response with RimSim: response! In: Modeling and simulation for military operations IV, vol 7348. SPIE, pp 136–143
27. Henderson SG, Mason AJ (2004) Ambulance service planning: simulation and data visualisation. Operations research and health care: a handbook of methods and applications, pp 77–102
28. Silva PMS, Pinto LR (2010) Emergency medical systems analysis by simulation and optimization. In: 2010 winter simulation conference (WSC). IEEE, pp 2422–2432
29. Ridler S, Mason AJ, Raith A (2022) A simulation and optimisation package for emergency medical services. Eur J Oper Res 298(3):1101–1113
30. Brailsford SC, Hilton NA (2001) A comparison of discrete event simulation and system dynamics for modelling healthcare systems. In: 2000 ORAHS, pp 18–39
31. Eatock J, Clarke M, Picton C, Young T (2011) Meeting the four-hour deadline in an A&E department. J Health Organ Manag 25(6):606–624
32. Komashie A, Mousavi A (2005) Modeling emergency departments using discrete event simulation techniques. In: 2005 winter simulation conference (WSC). IEEE, pp 2681–2685
33. Escudero-Marin P, Pidd M (2011) Using ABMS to simulate emergency departments. In: 2011 winter simulation conference (WSC). IEEE, pp 1239–1250
34. Ordu M, Demir E, Tofallis C (2020) A decision support system for demand and capacity modelling of an accident and emergency department. Health Systems 9(1):31–56
35. Sasanfar S, Bagherpour M, Moatari-Kazerouni A (2021) Improving emergency departments: Simulation-based optimization of patients waiting time and staff allocation in an Iranian hospital. Int J Healthcare Manag 14(4):1449–1456
36. Gul M, Fuat Guneri A, Gunal MM (2020) Emergency department network under disaster conditions: the case of possible major Istanbul earthquake. J Oper Res Soc 71(5):733–747
37. Cimini C, Pezzotta G, Lagorio A, Pirola F, Cavalieri S (2021) How can hybrid simulation support organizations in assessing COVID-19 containment measures? Healthcare 9(11):1412
38. Bouchnitaa A, Jebranea A (2020) A hybrid multi-scale model of COVID-19 transmission dynamics to assess the potential of non-pharmaceutical interventions. Chaos Sol Fractals 138:109941
39. Zhu Z, Hen BH, Teow KL (2012) Estimating ICU bed capacity using discrete event simulation. Int J Health Care Qual Assur 25(2):134–144
40. Melman GJ, Parlikad AK, Cameron EAB (2021) Balancing scarce hospital resources during the COVID-19 pandemic using discrete-event simulation. Health Care Manage Sci 24(2):356–374

41. Saidani M, Kim H, Kim J (2021) Designing optimal COVID-19 testing stations locally: a discrete event simulation model applied on a University campus. PLoS ONE 16(6):e0253869
42. Kerr CC, Stuart RM, Mistry D, Abeysuriya RG, Rosenfeld K, Hart GR, Nunez RC, Cohen JA, Selvaraj, P, Hagedorn B, George L (2021) Covasim: An agent-based model of COVID-19 dynamics and interventions. PLoS Comput Biol 17(7):e1009149
43. Hinch R, Probert WJ, Nurtay A, Kendall M, Wymant C, Hall M, Lythgoe K, Bulas Cruz A, Zhao L, Stewart A, Ferretti L (2021) OpenABM-covid19—an agent-based model for non-pharmaceutical interventions against COVID-19 including contact tracing. PLoS Comput Biol 17(7):e1009146
44. Wang Y, Xiong H, Liu S, Jung A, Stone T, Chukoskie L (2021) Simulation agent-based model to demonstrate the transmission of COVID-19 and effectiveness of different public health strategies. Front Comput Sci 3(642321):1–8
45. Nouman A, Anagnostou A, Taylor SJE (2013) Developing a distributed agent-based and des simulation using portico and repast. In: 2013 IEEE/ACM international symposium on distributed simulation and real time applications (DS-RT). IEEE, pp 97–104
46. Jones SG, Ashby AJ, Momin SR, Naidoo A (2010) Spatial implications associated with using Euclidean distance measurements and geographic centroid imputation in health care research. Health Serv Res 45(1):316–327
47. IEEE standard for modeling and simulation (M&S) high level architecture (HLA)—frame-work and rules. IEEE Standard 1516 (2000)
48. World Health Organization (2023) WHO coronavirus (COVID-19) dashboard. https://covid19.who.int/Cited06July2023
49. World Health Organization (2022) 14.9 million excess deaths associated with the COVID-19 pandemic in 2020 and 2021. https://www.who.int/news/item/05-05-2022-14.9-million-excess-deaths-were-associated-with-the-covid-19-pandemic-in-2020-and-2021. Cited 06 July 2023
50. Currie CS, Fowler JW, Kotiadis K, Monks T, Onggo BS, Robertson DA, Tako AA (2020) How simulation modelling can help reduce the impact of COVID-19. J Simul 14(2):83–97
51. Mustafee N, Brailsford S, Djanatliev A, Eldabi T, Kunc M, Tolk A (2017) Purpose and benefits of hybrid simulation: contributing to the convergence of its definition. In: 2017 winter simulation conference (WSC). IEEE, pp 1631–1645
52. Mahmood I, Arabnejad H, Suleimenova D, Sassoon I, Marshan A, Serrano-Rico A, Louvieris P, Anagnostou A, Taylor SJE, Bell D, Groen D (2020) FACS: a geospatial agent-based simulator for analysing COVID-19 spread and public health measures on local regions. J Simul 16(4):355–373
53. Aleman DM, Anagnostou A, Currie CS, Fowler JW, Gel ES, Rutherford AR (2021) Panel on simulation modeling for COVID-19. In: 2021 winter simulation conference (WSC). IEEE, pp 1–12
54. Grimm V, Railsback SF, Vincenot CE, Berger U, Gallagher G, DeAngelis DL, Edmonds B, Ge J, Giske J, Groeneveld J, Johnston AS (2020) The ODD protocol for describing agent-based and other simulation models: a second update to improve clarity, replication, and structural realism. J Artif Soc Soc Simul 23(2)
55. Monks T, Currie CS, Onggo BS, Robinson S, Kunc M, Taylor SJE (2019) Strengthening the reporting of empirical simulation studies: introducing the STRESS guidelines. J Simul 13(1):55–67
56. Lather JI, Eldabi T (2020) The benefits of a hybrid simulation hub to deal with pandemics. In: 2020 winter simulations conference (WSC). IEEE, pp 992–1003
57. Anagnostou A, Taylor SJE, Onggo BS, Heavey C, Monks T (2014) Towards a methodology for building large-scale distributed hybrid agent-based and discrete-event simulations: the case of emergency medical services. In: 8TH simulation workshop (SW14). ORS, pp 14–25

Chapter 12
Hybrid Modelling Approach Using Reinforcement Learning in Conjunction with Simulation: A Case Study of an Emergency Department

Vishnunarayan Girishan Prabhu and **Kevin M. Taaffe**

Abstract The last several years have seen a significant increase in adopting hybrid modelling and simulation to model, understand, analyze, and enhance various healthcare aspects, including patient flows, resource allocation, scheduling, policy evaluation, etc. The idea of hybrid simulation is to develop simulation models combining at least two methods: discrete-event simulation, system dynamics, and agent-based simulation. In contrast, hybrid modelling combines various simulation approaches with modelling methods other than simulation from the broader operations research and management sciences disciplines. The ability of hybrid modelling and simulation to capture various intricacies and represent complex systems has encouraged researchers and practitioners to apply these techniques in healthcare. While their application in healthcare has increased exponentially, there are further opportunities to integrate advanced modelling approaches. This chapter discusses a case study of an emergency department (ED) using a hybrid modelling and simulation approach combining a forecasting, hybrid simulation, and mixed-integer linear programming model to improve physician shift scheduling, patient flow and safety. Compared to the current practices, the proposed model reduced handoffs and patient time in the ED by 5.6% and 9.2% without a significant increase in budget. Finally, we propose our future work for integrating reinforcement learning to further enhance the model.

Keywords Hybrid simulation · Hybrid modelling · Healthcare systems modelling · Mixed-integer programming · Reinforcement learning

---

V. Girishan Prabhu (✉)
Industrial and Systems Engineering, University of North Carolina at Charlotte, Smith 314, 9201 University City Blvd, Charlotte, NC, USA
e-mail: vgirisha@charlotte.edu

K. M. Taaffe
Department of Industrial Engineering, Clemson University, Clemson, USA
e-mail: taaffe@clemson.edu

© The Author(s), under exclusive license to Springer Nature Switzerland AG 2024
M. Fakhimi and N. Mustafee (eds.), *Hybrid Modeling and Simulation*, Simulation Foundations, Methods and Applications, https://doi.org/10.1007/978-3-031-59999-6_12

12.1 Introduction

Healthcare operations involve managing various administrative, financial, legal, clinical, and quality improvement activities necessary to support the core functions of providing medical care, enhancing patient experience, and satisfying organizational goals. These activities are often influenced by various factors, including patient volume fluctuations, diverse treatment modalities, regulatory compliance, resource availability, technology adoption, etc., making the effective functioning of health systems challenging. Additionally, the interplay of multiple stakeholders, including physicians, nurses, administrators, insurance providers, and patients, further adds to the intricacies and complexity of efficient healthcare operations. Effective healthcare operations and management require strategic planning, efficient resource allocation, streamlined communication, and constant adaptation to the evolving healthcare landscape. Given the inherent uncertainty and complexity of health systems, it is critical that strategic planning and resource allocation strategies are based on evidence-based tools that can account for multifaceted goals and details for efficient healthcare operations. Over the last several decades, various operations research (OR) and management sciences (MS) approaches have been implemented to address different aspects of healthcare operations. Specifically, to improve patient flow, resource allocation, staff scheduling, appointment scheduling, testing new policies and change management, researchers have utilized methods including simulation modelling, mathematical optimization, heuristics, queueing theory, and other stochastic models (Markov decision processes, Monte Carlo simulation) [1–3]. While each method has its benefits and drawbacks, the unique capability of simulation models to test "what-if scenarios," capture micro and macro details, and identify the impact of each factor has made it one of the most popular tools used by researchers and practitioners in the healthcare sector [4, 5].

Among the widely used simulation modelling approaches (discrete-event simulation, system dynamics, and agent-based simulation) in healthcare, discrete-event simulation (DES) is the most prevalent and established approach, with the first application dating back six decades [6–8]. In terms of application areas pertinent to healthcare, DES has been used to model patient flow in emergency departments (EDs), outpatient scheduling for primary care clinics, resource allocation in EDs, process improvement of cardiac catheterization labs, and impact of radiology department on ED process flow to name a few applications [7–9]. While not as popular as DES, system dynamics (SD) has existed for a similar length of time, and its capability to incorporate feedback loops, stocks and flow function, and time delay has led to some specific applications in public health, such as chronic disease management, patient behavior and adherence to treatment plans, policy analysis, etc. [10–13]. Finally, agent-based simulation (ABS) is comparatively newer than the other two approaches in the OR/MS community and healthcare operations. However, the last few decades have seen significant adoption of ABS for healthcare operations. Specifically, the capability of ABS to represent micro-details and agent-level characteristics, interactions, and environment specifications has led to its adoption for infectious disease modelling, public health policy analysis, disaster management planning, etc. [14, 15]. The aim of this chapter is not to discuss in depth the capabilities of each modelling method or compare them; hence, we assume that readers are familiar with these, and for those who are not, please refer to the prior research that provides an introduction on all three methods [16–18]. While literature shows that each simulation approach has advantages and capabilities, relying on a single approach could fail to capture the system details, which are often incorporated into simulations as assumptions [19, 20].

Hybrid simulation (HS), defined as "one single conceptual simulation model that, when implemented in computer software, uses more than one simulation paradigm," can address the limitations of using a single approach by leveraging the capabilities of multiple approaches to better represent a system [21, 22]. Additionally, researchers have categorized simulation models within HS based on interactions between the different simulation approaches used [23]. Further, a recent study provides a much more comprehensive classification of HS, accounting for the historical terminology, current landscape, and potential future evolution [24]. In this chapter, we will use the terminology "simulation models combining at least two of the following methods: DES, SD, and ABS" to define HS. The prevalence of HS is comparatively lesser because of the challenge of coding HS models from scratch and the limited number of off-the-shelf software tools available [20, 25]. However, in terms of application, a 2019 literature review reported that HS applications were most popular in healthcare, contributing to 22% of peer-reviewed articles using HS [20]. As mentioned earlier, the inherent complexities and variabilities of healthcare are one of the main motivators driving the popularity and adoption of HS in healthcare. Finally, in terms of common modelling combinations used for HS, researchers and practitioners have combined all three simulation approaches, but DES + SD is the most common HS approach used in healthcare applications [20, 26].

As discussed, HS allows for capturing and better representing complex systems, including healthcare. However, hybrid modelling (HM), integrating various simulation approaches discussed above with OR/MS modelling approaches, such as mathematical optimization, forecasting, heuristics, graph theory, etc., allows for broadening the scope of the project by leveraging the capabilities of each method and deriving better solutions [27]. Specifically, for this case study, by integrating a forecasting model, a mixed-integer linear programming model with an HS model allows for developing a robust long-term plan by predicting the patient arrivals, creating an optimal resource allocation plan and validating them before implementation. We assume that readers are aware of various popular OR/MS methods, and for a detailed classification of OR/MS methods and HM, we recommend readers read these recent papers that discuss these topics in great detail, including classifications and terminologies [1, 27]. Pertinent to healthcare applications, prior studies have explored HM methods for several decades to address various issues, including staffing and scheduling, resource allocation, appointment scheduling, etc., with simulation optimization being the most common approach [28–30]. There is a developing literature on using HS with HM. However, it is not as well-established as using individual simulation methods in HM because of the challenges discussed earlier. Similarly, from modelling approaches, not a lot of HS models have been integrated with methods such as reinforcement learning or deep learning for improving healthcare operations.

In this chapter, we discuss a case study using HS and HM to improve patient flow and patient safety in an ED while considering the staffing budget. Specifically, we first develop forecasting models for predicting the patient arrivals to the ED, including their severity levels, which is inputted into a linear programming model for developing the staffing schedule. The objective of the mathematical model is to identify optimal shift schedules that minimize the combined cost of patient wait times, handoffs: transfer of a patient's care, and responsibility from one physician to another, and physician shifts, thus considering patient flow, patient safety, and staffing budget. Furthermore, to investigate the impact of generated schedules on ED performance, we test these in the validated HS model representative of the ED. Finally, we discuss a potential scenario where the optimization algorithm can be replaced by a reinforcement learning model, which will be integrated with a simulation model for active policy testing.

12.2 Background

The ED is a critical division in a health system where patients receive care for various conditions, including life-threatening emergencies, chronic ailments, and non-emergent situations. This diverse nature of ED to care for patients and the Emergency Medical Treatment and Labor Act (EMTALA), which mandates ED to provide screening and stabilizing care to all patients regardless of their ability to pay, makes ED a primary access point for patients seeking care [31, 32]. Over the last several years, patient arrivals to EDs in the USA have increased from 96.5 million annual visits in 1995 to 115.3 million in 2005 and 151 million in 2019 [33, 34]. At the same time, the number of EDs in the USA has decreased by over 15% in the last decade [35]. The ever-increasing patient volume and the decreasing number of EDs lead to mismatch, predisposing EDs to crowding [36–38]. The American College of Emergency Physicians (ACEP) defines ED crowding as the situation that "occurs when the identified need for emergency services exceeds available resources for patient care in the ED, hospital, or both" [39]. Crowding in ED is a global concern, and studies have often linked this as a factor leading to suboptimal care, delays in care, and higher chances of medical errors [36, 40].

A few leading causes of ED crowding include high patient census (patient arrivals), inadequate resources (beds, medical devices, etc.), inadequate planning, and poor ED design [41, 42]. Some of the most commonly adopted solutions to avoid ED crowding include expanding ED capacity, stopping boarding admitted patients in the ED, adding hallway beds, increasing on-call providers, and adding temporary resources [43]. While these solutions are effective temporary fixes, they can often be costly and negatively impact patient safety and physician well-being. A recent study investigating ED crowding identified that access to future patient demands (arrivals to ED) during shift scheduling and resource allocation can improve ED planning and potentially avoid crowding [38]. Most EDs, including our partner ED, build their clinician schedules about one quarter (3 months, 90 days) ahead. Hence, it is critical to have robust 90-day forecasts to assist ED administrators in planning clinician schedules to improve ED performance.

Researchers have used various forecasting methods to predict patient arrivals to the ED for different horizons [44–56]. Additionally, a few studies have focused on forecasting specific types of patient arrivals to the ED (primarily patients with respiratory diseases) [57, 58]. Regarding the methodology for forecasting patient arrivals to the ED, autoregressive moving average (ARMA), vector autoregressive moving average (VARMA), Holt-Winters, linear regression, multiple linear regression (MLR), autoregressive integrated moving average (ARIMA), seasonal ARIMA (SARIMA), and machine learning models (ANNs and RNNs) have been used extensively. In terms of forecasting horizon, researchers have forecasted hourly, daily, and monthly patient arrivals. Finally, regarding the data used, some studies have reported better performance when using multiple input variables. However, these often require expert manipulation, making them difficult for ED administrators and stakeholders to use. A significant gap among prior studies that forecasted patient arrivals to the ED is that none included the emergency severity index (ESI) levels of the patients while generating the forecasts. ESI is a standardized index assigned to each patient presenting to the ED across North America that varies from 1 to 5, representing the patient's severity and the expected resources the patient would require [59]. Hence, including ESI levels in the patient forecasts is critical for better planning the ED resource allocation.

Specific to ED, a variety of OR/MS approaches, including mathematical optimization models, queuing theory models, simulation modelling, and probabilistic models have been used to address various challenges observed in the ED, including resource allocation [2, 60–62]. While simulation is popular for addressing operational issues, its outcome is a realization and not an optimal solution. However, by identifying a specific objective, researchers have developed mathematical models to identify optimal staffing levels, generate schedules, determine optimal beds or other resource requirements, etc. One study used a mixed-integer linear programming (MILP) model to minimize understaffing with respect to patient volumes, which resulted in significant improvements in median length of stay, door-to-provider time, and door-to-bed time [63]. Researchers have also used HM approaches combining simulation and optimization models to identify and test optimal solutions before their implementation [64]. Most of the studies in ED have focused on identifying solutions that reduce patient waiting time, reduce length of stay, or improve ED throughput. While a few studies have used patient wait times as surrogates for patient safety, we introduce a new metric—handoffs, directly quantifying patient safety [65].

Patient safety is a crucial part of the ED as continuous patient flow and interactions with multiple departments and providers make it prone to errors. Additionally, researchers have observed ED as one of the hospital departments with high error rates. Among different issues that lead to medical errors, studies have identified handoffs, transfer of a patient's care, and responsibility from one physician to another as major patient safety issues [65, 66]. Specifically, studies investigating ED handoffs observed that the vital signs were not communicated for approximately 75% of the patients, and errors were observed in about 60% of cases [66]. Hence, while developing the ED physician shift schedules, it is crucial to consider patient safety metrics such as handoffs.

Our literature review shows that researchers have used time-series forecasting, mathematical models, simulations (including HS), and HM approaches to improve ED operations. However, to our knowledge, these studies have not combined these approaches, specifically an HS simulation in an HM approach, and included patient severity in their forecast. Moreover, as noted earlier, few have considered a direct patient safety metric in the mathematical model. In this chapter, we integrate HS (DES + ABS) with a mixed-integer linear programming model and forecasting model to form an HM approach to improve patient flow and patient safety in an ED while considering the staffing budget. By integrating a forecasting model, a mixed-integer linear programming model with an HS model allows for developing a robust long-term plan by predicting the patient arrivals, creating an optimal resource allocation plan and validating them before implementation. To our knowledge, this is the first study that combines these approaches to leverage the capabilities of various modelling approaches to develop a robust staff scheduling plan.

12.3 Methods

12.3.1 Data

Input data for developing the forecasting model included two specific data points: the time of the day and the ESI levels assigned to the patient presenting to the ED. Additionally, we gathered data points to develop the optimization and simulation model, including the number of beds, allowable physician shifts, patient arrivals, patient time in the ED, and the number of interactions between physicians and patients. All data points utilized for developing the models were collected from the PRISMA Health Greenville Memorial Hospital (GMH), Greenville, South Carolina, USA. Furthermore, the research team included ED physicians working in GMH for guidance and addressing any other physician-dependent activities in the ED to be included in the model. PRISMA Health is the largest healthcare provider in South Carolina and serves as a tertiary referral center. The flagship GMH academic ED is an Adult Level 1 Trauma Center, seeing over 106,000 patients annually.

Figure 12.1, represents the patient arrivals to the GMH ED averaged for 2017–2019 used for our forecasting model. One of the first patterns to notice is the impact of hour of the day. Patient arrivals are low during the early hours, steadily increase from 7:00 am until 12:00 pm, and then remain at this level until 7:00 pm. This patient arrival trend aligns with some prior studies [54, 67]. Another pattern that

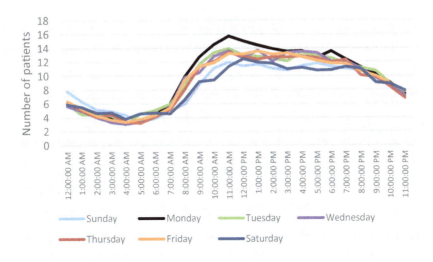

Fig. 12.1 Patient arrivals to the partner ED

can be noticed is the difference between weekdays and weekends, where weekdays have higher patient arrivals compared to weekends, and Mondays have the highest patient arrivals. These observations clearly suggest the need for long-term planning as each day and hour of the day require different staffing levels. In terms of ESI levels, we observed that about 50% of patient arrivals to the ED are ESI-3 patients, followed by ESI 2 and ESI 4, which contributed 25% and 20% of the patient arrivals. Finally, ESI 1 and 5 each contributed only 2–3% of the total arrivals. ESI 1 refers to severely unstable patients who need immediate intervention, and ESI 5 patients are the most stable patients and may be treated non-urgently and mostly require the least resources. ESI levels are critical during forecasting as each ESI demands different resources.

For training the forecasting model, we used 2017, 2018, and the first six months of 2019 data, and the predictions were made on the last six months divided into clusters of three months. Rather than using an entire year of patient arrival and using it for physician scheduling, we created clusters of three months and used the cluster with the highest patient arrivals for this research (July 2019 September 2019). We created clusters of three months because the partner ED generated their staffing schedules 90 days in advance (1 cluster) for their longitudinal planning. For example, staffing plans for the second quarter were generated during the end of the first quarter of a year and so on. Further, we decided to use the cluster with the most patient arrivals to comprehend the worst-case scenario and highest patient load experienced by the ED. Additionally, using the whole year's data was impossible because of the variability and cannot help create a robust plan for each quarter. Further, relying on daily or weekly data will fail to account for operational biases such as leave of absence, vacations, etc., which could impact patient times in the ED. Clustering the data by quarters allowed us to address these issues. Additionally, based on expert opinions from the ED physicians, we wanted to use the pre-COVID-19 data as the patient arrivals varied significantly during 2020. Another reason for using this specific period was to test the optimal schedule in our validated simulation model that used the same patient arrivals. However, both models were developed such that any patient arrivals can be used to generate a weekly schedule.

Next, we introduce Table 12.1, which represents the time a patient spends in the ED based on their ESI level. We split the data into two parts: "Bed to Disposition" and "Disposition to ED Departure." Bed to disposition represents the time a patient occupies an ED bed and is provided care by physicians and other medical providers, including performing tests, providing medicines, blood draws, etc. Although patients will be waiting in their beds during this period without receiving direct care, all these delays are due to waiting for their test results, medicines, etc. In general, this represents the period a patient first occupies a bed in the ED until the physicians make a disposition decision (admit, discharge, or transfer). The second part, "Disposition to Departure," is the period for which a patient occupies the ED bed from the time the physician makes a disposition decision until they are physically moved from the ED (discharged, admitted, or transferred). Hence, these are logistical delays where a patient can be either waiting until a bed is available in the hospital (admission) or waiting for transportation (discharged or transfer). While we primarily focus on the bed-to-departure time for this study, our model still accounts for delays before assigning a bed in the ED, where the patients wait in a waiting room until the beds are available, similar to an actual setting. As mentioned earlier, the entire bed-to-disposition time of a patient is not spent with a physician as it includes other activities. Based on literature and discussions with ED physicians, we used between 15 and 30% of total time as the care time where a patient would be cared for by a physician [68]. The percentages were assigned based on severity such that the total time spent with an ESI-1 patient was the highest and that with an ESI-5 patient was the lowest. This approach was used mainly because of the lack of detailed visit-by-visit data available to support detailed modelling.

Further, to build a model representative of ED operations where a physician visits patients multiple times based on their severity (ESI level), we split the care time into multiple smaller windows. Based on our past observational studies and discussions with ED faculties and physicians, on average, an ES1-1 patient was visited four times by a physician, ESI-2 and 3 were visited three times, and ES1-4 and 5 were visited

Table 12.1 Patient time in the ED

Severity	Bed to disposition (mins)	Disposition to ED departure (mins)	Total time (mins)
ESI 1	115 ± 9.6	121 ± 12.1	236 ± 16.8
ESI 2	186 ± 6.2	86 ± 9.4	272 ± 12.4
ESI 3	175 ± 3.6	54 ± 6.2	229 ± 9.6
ESI 4	90 ± 3.2	24 ± 4.9	114 ± 8.9
ESI 5	107 ± 3.7	15 ± 2.6	122 ± 7.1

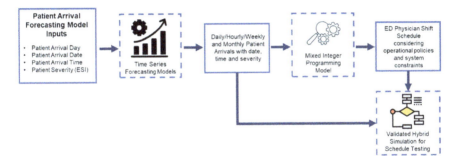

Fig. 12.2 Conceptual framework of the hybrid modelling and simulation approach implemented in the partner ED

two times [69]. The physician's time with a patient for each visit was a constant time block of 15 min, as the MILP modelling approach considers time as a discrete block of events.

12.3.2 Hybrid Modelling

Figure 12.2 below provides the conceptual framework of the hybrid modelling and simulation approach implemented in the partner ED. Here, we first develop forecasting models for predicting the patient arrivals to the ED, including their severity levels, where input data included the patient arrivals data with day, date, time and severity levels for the specific quarter (90 days). Patient arrivals for the next 90 days predicted using the best forecasting model are then inputted into a linear programming model to develop the staffing schedule. The objective of the mixed-integer linear programming model is to identify optimal shift schedules that minimize the combined cost of patient wait times, handoffs: transfer of a patient's care and responsibility from one physician to another, and physician shifts, thus considering patient flow, patient safety, and staffing budget. Finally, to investigate the impact of generated schedules on ED performance, we imported the schedules and the forecasted patient arrivals in the validated HS model representative of the ED.

12.3.2.1 Forecasting Model

We used the moving average naive model as our benchmark model to compare the forecasts from other models. Based on literature and data visualizations, we decided to develop both ARIMA and SARIMA models as these models are effective in forecasting time-series data, especially SARIMA when the data is considered to have seasonality [46, 47, 49, 51, 53, 54]. Additionally, we developed a Holt-Winters forecasting model as this approach can account for the level, trend, and seasonality component in the time-series data. Finally, we also developed two machine learning models: Extreme gradient boosting (XGBoost) and random forest regression model. Both are decision tree machine learning algorithms and require a supervised learning approach where each input requires an output pair within the training model for the model to learn and predict. However, the foundation of each algorithm is different; where random forest regression uses a bagging technique, whereas the XGBoost uses a boosting technique for learning.

To avoid the potential issue of overfitting with the machine learning models and tuning the hyperparameters, we used a blocked-crossed validation approach. This approach was preferred as it can avoid memorizing the patterns and leakage from future data, which is unavoidable when using k-fold cross-validation with time-series data. Specifically, the blocked-cross-validation approach accounts for this issue by adding margins at two positions: (i) between the training and validation folds and (ii) between the folds used at each iteration. Finally, to evaluate the performance of each model, we used root mean square error (RMSE), mean absolute error (MAE), and mean absolute percent error (MAPE).

12.3.2.2 Optimization Model

Before formulating the problem, we first list the key ED operational activities to replicate the partner ED. The first was accounting for the varying patient arrivals to the ED, including patient ESI levels. The second was modelling multiple patient–physician interactions based on the patient's ESI, accounting for minimum delays between patient–physician interactions for secondary care (imaging, blood draws, etc.), ensuring that the same physician provides care for the patient unless the physician ends their shift (handoff), physician shift length is limited to 8 h. Next, we define the notation used in the MILP model. The model included four sets and corresponding indices as follows: (i) I represents the set of patient arrivals to the GMH ED indexed by i, (ii) K represents the set of possible physicians that can be staffed for a day indexed by k, (iii) T represents the set of time slots considered for staff scheduling indexed by t, and (iv) M represents the set of physician visits required by a patient indexed by m.

Here, set I includes all the unique patient arrivals to the GMH ED for a week, which totals more than 1500. Set K consists of the unique physician identification number that can start an ED shift for a day with an upper threshold of 25 physicians per day. Further, T represents timeslots for an entire week (which varies based on slot length). Finally, set M includes values from 1 through 4, representing the patient interaction with a physician. Next, we introduce the parameters considered in the model. Most of the parameters represent various patient characteristics, including severity, arrival time, physician visits, and fixed time slots that should be avoided for calculating patient wait time as these delays are inherent and one parameter defining the ED bed capacity.

- α_i represents the time slot of arrival for patient i.
- β_i represents the severity level of patient i.
- γ_i represents the total number of visits required by patient i.

w_i represents the total time slots for patient i that should not be considered for waiting cost.

- C represents the total bed capacity of the GMH ED.

Finally, we introduce the decision variables in the model:

- U_{ik} $\begin{cases} 1, & \text{If patient } i \text{ served by physician } k \\ 0, & \text{otherwise} \end{cases}$
- $Y\text{start}_{kt}$ $\begin{cases} 1, & \text{If physician } k \text{ starts their shift at time slot } t \\ 0, & \text{otherwise} \end{cases}$
- Y_{kt} $\begin{cases} 1, & \text{If physician } k \text{ is available for service at time slot } t \\ 0, & \text{otherwise} \end{cases}$
- X_{iktm} $\begin{cases} 1, & \text{If patient } i \text{ is served by physician } k \text{ at time slot } t \text{ for their visit } m \\ 0, & \text{otherwise} \end{cases}$.

Minimize:

$$SC * \sum_{kt} Y\text{start}_{kt} + OC * \sum_{ikt} t * X_{ikt1} - \alpha_i$$
$$+ OC * F * \sum_{ikt} (t * X_{ikt\gamma_i} - t * X_{ikt1} - w_i)$$
$$+ HC * \sum_{ik} U_{ik},$$

subject to:

$$\sum_{kt} t * X_{ikt1} \geq \alpha_i \quad \forall i \in I$$

$$\sum_{ktm} X_{iktm} = \gamma_i \quad \forall i \in I$$

$$\sum_{km} X_{iktm} \leq 2 \quad \forall i \in I, \forall t \in T$$

$$\sum_{kt} X_{iktm} = 1 \quad \forall i \in I, \quad \forall m \in M$$

$$\sum_{ikm} X_{iktm} \leq C \quad \forall t \in T$$

$$\sum_{kt} t * X_{iktm} \leq \sum_{kt} t * X_{iktm+1} \quad \forall i \in I$$

$$\sum_{mt} X_{iktm} \leq 4 * U_{ik} \quad \forall i \in I, \ \forall k \in K$$

$$\sum_{im} X_{iktm} \leq 4 * Y_{kt} \quad \forall k \in K, \ \forall t \in T$$

$$\sum_{t} Y\text{strt}_{kt} \leq 1 \quad \forall k \in K$$

$$\sum_{kt} Y\text{strt}_{kt} \leq K$$

$$8 * Y\text{strt}_{kt} \leq \sum_{q=t}^{\text{Min}(168, t+7)} Y_{kq} \quad \forall k \in K, \ \forall t \in T$$

$$U_{ik}, \ Y\text{start}_{kt}, \ Y_{kt}, \ X_{iktm} \in \{0, 1\}.$$

In the formulation, the objective function minimizes the cost of staffing the ED physicians, handoffs, patient onboarding, and patient waiting time in the ED. The cost of staffing an ED physician (SC) was based on the national average rate for ED physicians, and the onboarding cost (OC) for patients based on their ESI level was derived from the literature [70, 71]. However, because of the lack of data on the cost of patient waiting once admitted, we used a factor value (F) between 0 and 1 and multiplied it by the OC to calculate the waiting cost. Finally, for the handoff cost (HC), we used high values ($1000) to avoid any possible handoffs.

The first constraint ensures that a patient is served their first visit ($m = 1$) only after their arrival at the ED. The second constraint ensures that the patient is provided with all their required visits before discharge. As mentioned earlier, each hour represents a time slot, but from observations and discussions with physicians, we assume that a physician can visit four patients in an hour. However, the same patient cannot be visited four times in an hour as that is not realistic as patients wait to get their tests, imaging, radiology, etc., completed. The third constraint ensures that a patient can be visited at most twice by a physician in an hour. The fourth constraint assures that each visit m for a patient cannot exceed 1, making sure that each visit is completed fully during a physician visit. The next constraint ensures that at any given time t, the patients served cannot exceed the ED bed capacity. As patients have multiple interactions with physicians during an ED stay, these visits must be ordered such that a later visit ($m + 1$) follows the prior visit (m) in terms of time slot, and our sixth constraint ensures the visits are ordered. The next two constraints ensure that a patient can be visited a maximum of four times by a physician, and a physician can visit up to four patients during any given time slot (1–h block). The next two constraints ensure that a physician starts the shift only once a day and that the total number of physicians staffed per day does not exceed the maximum number of possible physicians that can work for a day based on health system budget constraints. To ensure that a physician shift, once started, lasts for eight hours, we use the second to the last constraint. Finally, the last constraint defines the variable types, which are all binary in this case.

12.3.2.3 Hybrid Simulation Model

In a traditional DES modelling approach used to model an ED, patients are considered agents that flow in the ED, each with unique attributes cared for by healthcare providers modelled as a resource. This traditional DES approach would suffice to address the issues at a high level, including bed planning and staffing requirements. However, to meet the aim of this research, which focuses on improving patient flow and patient safety by minimizing the number of handoffs and wait times, this approach would not incorporate the impact of the physician's decision-making capabilities based on current conditions in the ED. Hence, in this study, we used a novel approach where patients and physicians were modelled as agents with unique attributes (ABS) and the activities in the ED system as DES. Using this approach allowed replicating a physician's activity as realistic as observed in the ED, unlike the traditional DES approach where a patient would seize a physician resource just once for a particular amount of time and release them to move on to the following process.

To provide further insight into the modelling approach adopted for this study, we introduce Fig. 12.3, which captures the essence and capabilities of various ED physician activities that the model can simulate. In the figure above, dashed lines represent patients, and the solid lines represent the physicians moving in the ED. A patient arriving at the ED undergoes various onboarding processes (discussed in the next paragraph) before being assigned an ED room. Each room in the ED has a single bed that a patient will occupy from room assignment until the physician makes a disposition decision. Each arriving physician has an arrival time, shift end time, and pod assignment in the ED to provide medical care during their shift. Upon arrival to their specific pod, a physician goes to the physician station, and if another physician is leaving the same pod, the patients from the leaving physician are transferred to the new physician—that is, patient handoffs occur. If no physician leaves the same pod, the arriving physician starts assigning themselves new patients who are waiting in the ED without a room assignment. The physician will also spend time in the station reviewing the patient's medical record before visiting each patient. When ready, the physician visits the patient in their room, with the time required depending on the patient's ESI level. Following the patient visit, the physician returns to their station to order tests, labs, and imaging as necessary while the patient waits in the bed for the requested tests. Secondary care, including labs, medicines, imaging, etc., are either performed while patients are on the bed, or in a few instances, patients might be rolled out of the ED, but the bed/room will not be assigned to another patient (based on observations in the ED). After the first visit with a patient, our approach links a physician and the specific patient based on their unique IDs. This ensures that the same physician will provide the subsequent care for the patient unless they are ending their shift and the patients are handed off to another physician. From a modelling standpoint, we have an array where each physician ID can handle multiple patient IDs, but each patient ID can link only to one physician at a time, thus replicating how an actual ED functions. The number of subsequent visits and time spent with the patient during each visit again depends on the ESI level of the patient. Using the ABM approach for both patients and physicians overcomes the limitations of traditional

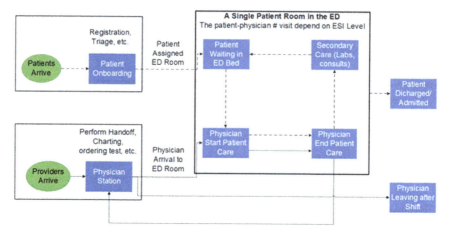

Fig. 12.3 Hybrid simulation approach for physician–patient interaction

DES, where physicians are modelled as resources and cannot make intelligent decisions. By modelling physicians as a resource, the decision-making capabilities are limited to patients where patients seize the physicians for a certain amount of time and release them, making the resource (physicians) available for the next patient. Further, a physician stays idle and doesn't flow in the ED and cannot make intelligent decisions limiting their ability to pick and choose patients based on their workload, time remaining in the shift, and pod limitations. However, modelling both patients and physicians as agents, as discussed in the research, allows both agents to make intelligent decisions based on rules replicating the actual ED activities, including charting, adding orders, handoffs, etc. Finally, our current modelling approach allows for the flexibility of continuous model development, especially when modelling secondary resources as they would act as independent activities.

Before beginning the model development, the first step was to capture the day-to-day activities at GMH ED. Through observations and meetings with ED physicians, we developed a detailed process map using Microsoft Visio. A majority of the patients are triaged, where they are assessed by a triage nurse and assigned an ESI level based on their medical condition. However, a few severe cases (e.g., car crashes, ST-Elevation Myocardial Infarction, etc.) might not be triaged and are provided care in the trauma bay. The triaged patients are then registered into the hospital's electronic health record and then directed to the waiting room, where they are prioritized based on their assigned severity level. When a bed is available in one of the ED pods, an ED nurse takes the patients from the waiting room based on their severity level and the capability of the ED pod. This is because certain pods in the ED do not have medical equipment and other capabilities to handle high-severity patients. Apart from the patient arrivals, rest of the activities are specific for each pod, and each pod was modelled separately. The ED rooms and trauma bays are modelled as resources and divided into four pods where the capacities and capabilities of these resources are the same as in the GMH ED.

Upon a physician's arrival for a shift in a specific pod, the physician who will be leaving the ED will transfer their patients to the arriving physician. As mentioned earlier, this process of transferring the care of a patient is defined as a handoff. In the absence of a physician in the station, the new physician will take a new patient and later meet the leaving physician for handoffs. These are usually rare because physicians leaving the ED do not tend to provide care during the last 15 min of the shift, as they would be focused on completing the patient charts. For the handoff process, we use a delay using a distribution based on the data collected from observations. In case no physician is leaving an ED pod, then there would be no handoffs, and the arriving physician would start taking new patients. Finally, in the case when a physician leaves the ED and a new physician is not arriving at the ED, which happens during night shifts, the leaving physician will handoff their remaining patients to the existing physician in the ED. Post handoff, the physician decides on taking a new patient depending on the current number of cases handled. The model ensures that the physicians working in the same pod simultaneously share the patient load equally. It should be noted that a physician's workload is considered balanced based on the number of patients they are providing care to and not necessarily based on the ESI level of the patient, as that is the practice followed in the ED.

After accepting a new patient, the physician would then meet the patient in the ED room for the first evaluation and then returns to the physician station to document in the medical record, order test, labs, consults, and medicines as necessary. As the physician places the order, the nurse then completes the required documentation, the ordered tasks, medication administration, and runs bedside tests or ordered interventions. Additionally, patient care often includes diagnostic imaging that may require the patient to be moved out of the ED to the radiology suite or samples sent to the lab. Following the drug administration, imaging, and diagnostic testing, the physician returns to the patient for the subsequent evaluation, and the physician provides care until the patient is clinically stabilized. After a subsequent patient visit, the physician might not necessarily return to the physician's station immediately. Hence, based on expert opinion, we used a 40% probability that the physician might visit another patient before returning to his or her station. After the final evaluation, the patient is either discharged or admitted as an inpatient to the hospital, and the physician may take on a new patient. During all the scenarios in the model, whenever a level 1 patient is presented in the ED, irrespective of all the policies and rules, the immediately available physician serves the patient.

While we did not consider nurses, consults, and ancillary resources as specific entities in the simulation model, because of interdependencies of different departments, the simulated time a patient spends in the system was validated against the actual time, which is discussed in detail later. The hybrid simulation approach allowed us to replicate physician activities and daily operations from an ED standpoint and investigate our primary aim of understanding the impact of overlapping shifts on the number of handoffs and patient flow without any restrictions. These activities would have been challenging to replicate using a traditional modelling approach where physicians are modelled as resources. A detailed description of the simulation model is discussed in our prior work [72].

Table 12.2 Simulation model validation

Severity	Actual time in ED (mins)	Simulated time in ED (mins)	Percent difference (%)
ESI 1	236 ± 16.8	218 ± 0.82	−7.6
ESI 2	272 ± 12.4	281 ± 4.4	3.3
ESI 3	229 ± 9.6	216 ± 3.3	−5.7
ESI 4	114 ± 8.9	121 ± 1.5	6.1
ESI 5	122 ± 7.1	122 ± 2.8	0

After developing the simulation model of PRISMA Health ED, the next step was validating the model against actual data. We used the patient time in the ED for each ESI level and the daily number of handoffs as the validation metrics to ensure that the model represents the partner ED. The model was simulated for a three-week schedule with an additional two-day warm-up period for the model to attain equilibrium. A total of 60 replications were performed, such that the margin of error on time in the ED metric was ± 10 min (at $\alpha = 0.05$). We first compared the simulated weekly throughput and the daily number of handoffs with the retrospective data from partner ED. On performing a t-test, we observed that the actual weekly throughput (1508 ± 8.2) and simulated weekly throughput (1505 ± 5.3) values did not vary significantly (p-value = 0.90). Similarly, we observed that the actual daily handoffs (92 ± 4.2) and simulated daily handoffs (93 ± 5.8) values did not vary significantly (p-value = 0.11). Finally, comparing the simulated average time in the ED to the actual data for each ESI, we did not observe any significant differences (p-value > 0.05). Table 12.2 below represents the simulated and actual data for each ESI level.

12.4 Results

Upon developing the model and tuning the parameters on the training data, the next step was to use these models to forecast patient arrivals to the ED. We first discuss the findings from the long-term forecasting model—ED patient arrivals for the next 90 days. Table 12.3 represents the performance metrics score for each model output for the long-term forecasts.

Table 12.3 Model performance for the 90-day forecast

Model	RMSE	MAE	MAPE (%)
MA	30.1	23.6	14.2
ARIMA	27.2	21.6	10.6
SARIMA	25.6	19.2	9.9
Holt-Winters	26.8	19.8	10.0
XGBoost	16.6	14.1	5.9
Random Forest	17.4	14.6	6.4

It is evident that both machine learning models outperformed the naïve model and other traditional time-series models. However, it is interesting to notice that the Holt-Winters approach outperformed the ARIMA model, and this can be primarily attributed to the fact that the Holt-Winters model can account for seasonality. However, comparing the SARIMA model to the Holt-Winters model, SARIMA was slightly better. The most significant improvements were observed with the machine learning models, where the MAPE value was reduced by half compared to the traditional time-series forecasting model. Among the two machine learning models, XGBoost outperformed the random forest model for all the performance metrics. One of the key observations here is the high RMSE values irrespective of the forecasting approach, which could be caused by extreme values (outliers). Even with a significant change in patient demands and arrivals, the machine learning models forecast was robust (based on RMSE, MAE, and MAPE), as models with a MAPE value of 5.0% are considered excellent. However, to avoid bias and over-relying on one value, we look at RMSE (16.6), which is comparatively low given the daily arrivals vary from 150 patients a day to as high as 270 patients.

After identifying the best-performing model, the next step was to look at the ESI predictions for the 90-day forecast. Table 12.4 above represents the performance metrics score for each ESI level from the XGBoost forecast. The first thing to notice is the varying RMSE, MAE, and MAPE values across the ESI levels. Specifically, it can be noticed that MAPE values are high for ESI 1 and 5 and minimum for ESI 3, whereas the RMSE and MAE behave vice versa. This represents the bias associated with each metric where MAPE penalizes heavily when the forecasted values are smaller as it is a percentage value. However, by using a combination of performance metrics, we can identify that the ESI-level forecasts from the model are robust.

The forecasted patient arrivals, along with other data, were inputted into the mathematical model to identify staffing schedules that can minimize patient hand-offs, physician shifts, and patient wait times while considering the staffing budget. Table 12.5 below represents the physician shift start times for the week. While the table below represents the shift start times for the whole week, the mathematical model output provides specific start times and the number of shifts for each day, as some days require more resources. Note that certain times are not included (12:00–6:00 am, 6:00 pm–9:00 pm) in Table 12.5 below as a shift cannot be started at those times. Although the shift start times are very similar between the two policies as certain time frames are restricted to the operation policies, one of the interesting

Table 12.4 XGBoost ESI level 90-day forecast

ESI	RMSE	MAE	MAPE (%)
ESI 1	3.1	2.8	38.0
ESI 2	8.9	7.0	12.5
ESI 3	13.4	10.1	8.4
ESI 4	6.0	5.1	15.5
ESI 5	1.9	1.4	46.1

factors to notice is how the schedule generated by the mathematical model recommends starting a shift in a staggering approach as opposed to starting shifts only at particular time frames (e.g., 7:00, 9:00, etc.) as observed in the current policy. Finally, in terms of total hours staffed for the week, there were no significant differences.

After generating the new schedule, the next step was to compare the new policy to the current (baseline) policy in the validated simulation model. We used three ED performance metrics: the number of handoffs, patient time in the ED, weekly throughput, and change in hours measured in full-time equivalent (FTE). Here, the number of handoffs is the patient safety metric, the next two are patient flow metrics, and the change in FTEs represents budget implications. FTEs are a unit of measurement that calculates the number of full-time hours a physician works in the ED, where 40 h per week translates to 1 FTE. This method allows for adding up the hours of full-time, part-time, and various other types of employees into measurable "full-time" units. Each policy was simulated for a three-week schedule and replicated until the margin of error on time in the ED metric was ± 10 min (at $\alpha = 0.05$).

From Table 12.6 above, one of the first things to notice is the similar weekly throughput observed across both scenarios, irrespective of changes in other performance metrics. This is primarily because the ED is simulated for a fixed three-week duration with the same forecasted patient arrivals, leading to a limited fixed demand. Further, on running an independent t-test, we did not observe a significant difference (p-value $= 0.34$) in weekly throughput across the two policies. However, we observed that both handoffs per day (p-value < 0.01) and patient time in the ED (p-value $= 0.03$) varied significantly between the two policies, where the new policy outperformed the baseline policy. Compared to the baseline policy, the new policy reduced the patient time in the ED by 5.6% and handoffs by 9.6% with a slight non-significant (p-value $= 0.21$) increase in change in hours/week measured as FTEs.

Table 12.5 Weekly physician shift start times

Time	7	8	9	10	11	12	13	14	15	16	17	22	23
Current	22	4	20	5	0	2	0	0	21	4	20	15	21
New	14	14	14	7	7	0	7	7	14	21	0	0	21

Table 12.6 Simulation model results

Policy	Weekly throughput	# Handoffs per day	Time in the ED (mins)	Change in hours/week (FTE)
Baseline	1505	93	213 ± 4.6	0
New	1503	84	201 ± 5.9	$+6$ (+0.15 FTEs)

12.5 Conclusions and Future Work

Protecting the ED from crowding is one of the highest public health priorities to ensure timely patient care and patient safety. Although most EDs across the USA plan in advance to avoid ED crowding, studies have observed that most EDs still rely on quick temporary fixes. Although sometimes ad hoc actions are required because of unexpected issues such as evacuations and natural disasters, most of the time, these are required because of inadequate short and long-term planning. One of the most important inputs required for robust planning is the future patient census (arrivals) to the ED. Over the last decades, several studies have applied numerous approaches for forecasting patient arrivals to the ED and have generated acceptable results. However, most of these studies have focused on predicting daily patient arrivals to the ED, except for two recent studies that have explored hourly patient arrival forecasting [47, 56]. Surprisingly, none of these studies has included ESI levels of forecasted patient arrivals, which significantly influences resource requirements.

The case study presented utilized an HM approach that first developed traditional time-series models and machine learning models to forecast long-term (90 days ahead) patient arrivals to the partner ED with the patient's ESI levels. XGBoost algorithm generated the best long-term forecasts with MAPE values of 5.9%, outperforming prior studies. Moreover, we forecast ESI levels of these arrivals with a maximum RMSE value of 13.4. These findings are promising, given the simple input variables and the long-term forecasts. These forecasts were inputted into a mixed-integer linear programming model to generate an optimal ED physician staffing schedule that minimizes the overall cost of patient onboarding time, patient waiting time after ED admission, number of patient handoffs, and cost of hiring an ED physician. Finally, the generated schedule was tested in our validated HS model representative of the partner ED, and we observed that the new schedule reduced the patient time in the ED by 5.6% and handoffs by 9.6% with a non-significant increase in staffing hours measured in FTEs.

This case study presents how integrating HS models with an HM approach can overcome challenges that cannot be addressed by using a single simulation approach or OR/MS approach. As mentioned in the introduction, integrating these approaches often allows to reduce the number of assumptions and potentially achieve better solutions compared to using a singular modelling approach. However, it should be noted that our current approach still has fundamental and practical limitations that can be further improved using advanced approaches.

From a practical standpoint, although it will be easy to use the developed hybrid model by practitioners and ED administrators without any modelling background, generalizing the model to another ED or other departments in the hospital would require model changes. Specifically, the forecasting and mathematical models would require minimal changes, such as changing input data and constraints within the hybrid model. However, the hybrid simulation model would require significant changes as the current modelling uses a micro-modelling strategy specific to the partner ED. A potential approach to make the simulation model more generalizable is to adopt a macro-modelling strategy without getting into specifics such as pods, number of physician–patient interactions, etc. From a fundamental standpoint, to improve our forecasting model, we will focus on fine-tuning the forecasting model by incorporating other parameters that can be exported from EHR to investigate if the model predictions can be improved. Additionally, from a modelling standpoint, a hierarchical forecasting approach with an optimization function could potentially improve ESI-level forecasting and reduce errors as the ESI-level predictions and total volume predictions are reconciled using the optimization function. Regarding the simulation model, a major limitation is that we are not representing ancillary resources to the ED, including labs and consults, as separate processes. However, the model still accounts for these delays using the retrospective data. In future work, we plan to include the impact of these ancillary resources as specific processes with resources to better represent the partner ED.

The biggest opportunity we identify in this case study is to enhance our optimization model approach with a reinforcement learning approach. Reinforcement learning is a type of machine learning where an agent learns to behave in an environment by taking specific actions. The agent receives rewards for taking actions that lead to desired outcomes and penalties for taking actions that lead to undesired outcomes. Over time, the agent learns to take actions that maximize its rewards. There are various algorithms/methods, such as brute force approach, value function, temporal difference methods, Q-learning algorithm, deep Q-learning, etc., that can be used to explore various actions and help agents take the best action during each step. Pertinent to this case study, we plan to use a multi-agent reinforcement learning (MARL) method where multiple agents interact with each other and in an environment where each agent has its own set of goals, and the agents must learn to coordinate their actions in order to achieve their goals. Here, patients and physicians will be the two sets of agents, and our simulation model will act as the test bed for real-time testing of different policies. This approach can overcome a lot of assumptions in the current optimization model where times are divided into windows and the decision to serve which patient is based on ESI and wait times. Moreover, in the current optimization model, the time between different patient visits is modelled with some hard constraints, which is not an ideal way to capture a stochastic system such as ED. By using a MARL approach integrated with the HS model, we can develop policies that can provide recommendations for patient–physician matching and physician actions (sign up a new patient vs. care for an existing patient, etc.). Overall, this modelling approach can allow for developing policies that minimize the patient length of stay, improve patient safety and patient flow, and also account for physician load.

While the proposed strategy is indeed interesting and can provide policies with much higher fidelity compared to the current approach discussed in the case study, the biggest challenge is integrating the HS with the MARL model. One specific out-of-the-box simulation tool that has explored the idea of integrating HS with reinforcement learning is AnyLogic using the ALPyne, which allows for integration with the local Python environment. Our aim is to use this integration and potentially explore Python for the modelling of the whole system, including the forecasting model, HS model, and reinforcement learning model. Although in this chapter, we discuss a specific case of integrating HS with reinforcement learning, these methods could be applied for other healthcare applications such as (i) optimizing hospital resource allocation, (ii) developing personalized treatment plans, (iii) patient scheduling and appointment booking, (iv) optimizing drug dosage, (v) patient flow and discharge planning to name a few.

References

1. Mustafee N, Katsaliaki K (2020) Classification of the existing knowledge base of OR/MS research and practice (1990–2019) using a proposed classification scheme. Comput Oper Res 118:104920
2. Rais A, Vianaa A (2011) Operations research in healthcare: a survey. Int Trans Oper Res 18(1):1–31. https://doi.org/10.1111/j.1475-3995.2010.00767.x
3. Capan M, Khojandi A, Denton BT, Williams KD, Ayer T, Chhatwal J et al (2017) From data to improved decisions: operations research in healthcare delivery. Med Decis Mak 37(8):849–859. https://doi.org/10.1177/0272989X17705636
4. Brailsford SC, Harper PR, Patel B, Pitt M (2009) An analysis of the academic literature on simulation and modelling in health care. J Simul 3:130–140. https://doi.org/10.1057/jos.2009.10
5. Arisha A, Rashwan W (2016) Modeling of healthcare systems: past, current and future trends. In: Proceedings—winter simulation conference, pp 1523–1534
6. Goldman J, Knappenberger HA, Eller JC (1968) Evaluating bed allocation policy with computer simulation. Health Serv Res 3(2):119
7. Zhang X (2018) Application of discrete event simulation in health care: a systematic review. BMC Health Serv Res 18(1):1–11. https://doi.org/10.1186/s12913-018-3456-4
8. Vázquez-Serrano JI, Peimbert-García RE, Cárdenas-Barrón LE (2021) Discrete-event simulation modeling in healthcare: a comprehensive review. Int J Environ Res Public Health 18(22):12262
9. Jacobson SH, Hall SN, Swisher JR (2013) Discrete-event simulation of health care systems. Int Ser Oper Res Manage Sci 206:273–309. https://doi.org/10.1007/978-1-4614-9512-3_12
10. Brailsford SC (2008) System dynamics: what's in it for healthcare simulation modelers. In: Proceedings—Winter Simulation Conference, pp 1478–1483
11. Homer JB, Hirsch GB (2006) System dynamics modeling for public health: background and opportunities. Am J Public Health 96(3):452–458. https://doi.org/10.2105/AJPH.2005.062059
12. Davahli MR, Karwowski W, Taiar R (2023) A system dynamics simulation applied to healthcare: a systematic review. Int J Environ Res Public Health 17(16):5741
13. Forrester JW (2017) Industrial dynamics. J Oper Res Soc 48(10):1037–1041. https://doi.org/10.1057/palgrave.jors.2600946
14. Tracy M, Cerdá M, Keyes KM (2018) Agent-based modeling in public health: current applications and future directions. Annu Rev Public Health 39:77–94. https://doi.org/10.1146/annurev-publhealth-040617-014317
15. Silverman BG, Hanrahan N, Bharathy G, Gordon K, Johnson D (2015) A systems approach to healthcare: agent-based modeling, community mental health, and population well-being. Artif Intell Med 63(2):61–71
16. Barnes S, Golden B, Price S (2013) Applications of agent-based modeling and simulation to healthcare operations management. Int Ser Oper Res Manage Sci 184:45–74. https://doi.org/10.1007/978-1-4614-5885-2_3
17. Brailsford S, Churilov L, Dangerfield B (2014) Discrete-event simulation and system dynamics for management decision making, vol 9781118349021, pp 1–342. Wiley, London. https://doi.org/10.1002/9781118762745

18. Railsback SF, Lytinen SL, Jackson SK (2006) Agent-based simulation platforms: review and development recommendations. Simulation 82(9):609–623. https://doi.org/10.1177/0037549706073695
19. Sumari S, Ibrahim R, Zakaria NH, Ab Hamid AH (2013) Comparing three simulation model using taxonomy: system dynamic simulation, discrete event simulation and agent based simulation. Int J Manage Excell 1(3):54
20. Brailsford SC, Eldabi T, Kunc M, Mustafee N, Osorio AF (2019) Hybrid simulation modelling in operational research: a state-of-the-art review. Eur J Oper Res 278(3):721–737
21. Lättilä L, Hilletofth P, Lin B (2010) Hybrid simulation models—when, why, how? Expert Syst Appl 37(12):7969–7975
22. Mustafee N, Brailsford S, Djanatliev A, Eldabi T, Kunc M, Tolk A (2017) Purpose and benefits of hybrid simulation: contributing to the convergence of its definition. In: Proceedings—winter simulation conference, pp 1631–1645
23. Morgan JS, Howick S, Belton V (2017) A toolkit of designs for mixing discrete event simulation and system dynamics. Eur J Oper Res 257(3):907–918
24. Mustafee N, Powell JH (2019) From hybrid simulation to hybrid systems modelling. In: Proceedings—Winter Simulation Conference, pp 1430–1439
25. Brailsford SC (2016) Hybrid simulation in healthcare: new concepts and new tools. In: Proceedings—Winter Simulation Conference, pp 1645–1653
26. Dos Santos VH, Kotiadis K, Scaparra MP. A review of hybrid simulation in healthcare. In: Proceedings—Winter Simulation Conference, pp 1004–1015
27. Mustafee N, Harper A, Onggo BS (2020) Hybrid modelling and simulation (MS): driving innovation in the theory and practice of MS. Proc Winter Simul Conf 2020:3140–3151
28. Cabrera E, Taboada M, Iglesias ML, Epelde F, Luque E (2012) Simulation optimization for healthcare emergency departments. Proc Comput Sci 9:1464–1473
29. Ordu M, Demir E, Tofallis C, Gunal MM (2020) A novel healthcare resource allocation decision support tool: a forecasting-simulation-optimization approach. J Oper Res Soc 72(3):485–500. https://doi.org/10.1080/01605682.2019.1700186
30. Wang L, Demeulemeester E (2022) Simulation optimization in healthcare resource planning: a literature review. IISE Trans. https://doi.org/10.1080/24725854.2022.2147606
31. Laxmisan A, Hakimzada F, Sayan OR, Green RA, Zhang J, Patel VL (2007) The multitasking clinician: decision-making and cognitive demand during and after team handoffs in emergency care. Int J Med Inform 76:801–811
32. McDonnell WM, Gee CA, Mecham N, Dahl-Olsen J, Guenther E (2013) Does the emergency medical treatment and labor act affect emergency department use? J Emerg Med 44(1):209–216
33. Centers for Disease Control and Prevention (2010) NCHS pressroom—fact sheet—emergency department visits. https://www.cdc.gov/nchs/pressroom/04facts/emergencydept.htm
34. Cairns C, Ashman JJ, Kang K (2019) Emergency department visit rates by selected characteristics: United States, 2019. Atlanta, Georgia
35. Hsia RY, Kellermann AL, Shen YC (2011) Factors associated with closures of emergency departments in the United States. JAMA J Am Med Assoc 305(19):1978–1985
36. Di Somma S, Paladino L, Vaughan L, Lalle I, Magrini L, Magnanti M (2015) Overcrowding in emergency department: an international issue. Intern Emerg Med 10(2):171–175
37. George F, Evridiki K (2015) The effect of emergency department crowding on patient outcomes results. Health Sci J 9(1):1–6
38. Kelen GD, Wolfe R, D'onofrio G, Mills AM, Diercks D, Stern SA et al (2021) Emergency department crowding: the canary in the health care system. N Engl J Med Catal
39. American College of Emergency Physicians. Crowding (2019)
40. Kulstad EB, Sikka R, Sweis RT, Kelley KM, Rzechula KH (2010) ED overcrowding is associated with an increased frequency of medication errors. Am J Emerg Med 28(3):304–309
41. Morley C, Unwin M, Peterson GM, Stankovich J, Kinsman L (2018) Emergency department crowding: a systematic review of causes, consequences and solutions. PLoS ONE 13(8):1–42. https://doi.org/10.1371/journal.pone.0203316

42. Moskop JC, Sklar DP, Geiderman JM, Schears RM, Bookman KJ (2009) Emergency department crowding, part 1—concept, causes, and moral consequences. Ann Emerg Med 53(5):605–611
43. Derlet RW, Richards JR (2008) Ten solutions for emergency department crowding. Western J Emerg Med 9(1):24
44. Aboagye-Sarfo P, Mai Q, Sanfilippo FM, Preen DB, Stewart LM, Fatovich DM (2015) A comparison of multivariate and univariate time series approaches to modelling and forecasting emergency department demand in Western Australia. J Biomed Inform 57:62–73
45. Batal H, Tench J, McMillan S, Adams J, Mehler PS (2001) Predicting patient visits to an urgent care clinic using calendar variables. Acad Emerg Med 8(1):48–53
46. Carvalho-Silva M, Monteiro MTT, de Sá-Soares F, Dória-Nóbrega S (2018) Assessment of forecasting models for patients arrival at emergency department. Oper Res Health Care. 18:112–118
47. Choudhury A, Urena E (2020) Forecasting hourly emergency department arrival using time series analysis. Br J Health Care Manage 26(1):34–43. https://doi.org/10.12968/bjhc.2019.0067
48. Côté MJ, Smith MA, Eitel DR, Akçali E (2013) Forecasting emergency department arrivals: a tutorial for emergency department directors. Hosp Top 91(1):9–19
49. Hertzum M (2017) Forecasting hourly patient visits in the emergency department to counteract crowding. Ergon Open J 10(1):1–13
50. Jones SS, Thomas A, Evans RS, Welch SJ, Haug PJ, Snow GL (2008) Forecasting daily patient volumes in the emergency department. Acad Emerg Med 15(2):159–170. https://doi.org/10.1111/j.1553-2712.2007.00032.x
51. Kadri F, Harrou F, Chaabane S, Tahon C (2014) Time series modelling and forecasting of emergency department overcrowding. J Med Syst 38(9):1–20. https://doi.org/10.1007/s10916-014-0107-0
52. Khaldi R, El AA, Chiheb R (2019) Forecasting of weekly patient visits to emergency department: real case study. Proc Comput Sci 1(148):532–541
53. Sun Y, Heng BH, Seow YT, Seow E (2009) Forecasting daily attendances at an emergency department to aid resource planning. BMC Emerg Med 9(1):1–9. https://doi.org/10.1186/1471-227X-9-1
54. Whitt W, Zhang X (2019) Forecasting arrivals and occupancy levels in an emergency department. Oper Res Health Care. 21:1–18
55. Xu M, Wong TC, Chin KS (2013) Modeling daily patient arrivals at Emergency Department and quantifying the relative importance of contributing variables using artificial neural network. Decis Support Syst. 54(3):1488–1498
56. Zhang Y, Zhang J, Tao M, Shu J, Zhu D (2022) Forecasting patient arrivals at emergency department using calendar and meteorological information. Appl Intell 52(10):11232–11243. https://doi.org/10.1007/s10489-021-03085-9
57. Becerra M, Jerez A, Aballay B, Garcés HO, Fuentes A (2020) Forecasting emergency admissions due to respiratory diseases in high variability scenarios using time series: a case study in Chile. Sci Total Environ 706:134978
58. Rosychuk RJ, Youngson E, Rowe BH (2015) Presentations to Alberta emergency departments for asthma: a time series analysis. Acad Emerg Med 22(8):942–949. https://doi.org/10.1111/acem.12725
59. Hossein A, Rouhi J, Sardashti S, Taghizadieh A, Soleimanpour H, Barzegar M (2013) Emergency Severity Index (ESI): a triage tool for emergency department. Agency for Healthcare Research and Quality (AHRQ). Int J Emerg Med 1–5. https://www.ahrq.gov/patient-safety/settings/emergency-dept/esi.html
60. Elalouf A, Wachtel G (2021) Queueing problems in emergency departments: a review of practical approaches and research methodologies. Oper Res Forum 3(1):1–46. https://doi.org/10.1007/s43069-021-00114-8
61. Ahsan KB, Alam MR, Morel DG, Karim MA (2019) Emergency department resource optimisation for improved performance: a review. J Indus Eng Int 15(1):253–266. https://doi.org/10.1007/s40092-019-00335-x

62. Connelly LG, Bair AE (2004) Discrete event simulation of emergency department activity: a platform for system-level operations research. Acad Emerg Med 11(11):1177–1185. https://doi.org/10.1197/j.aem.2004.08.021
63. Sir MY, Nestler D, Hellmich T, Das D, Laughlin MJ, Dohlman MC et al (2017) Optimization of multidisciplinary staffing improves patient experiences at the mayo clinic. INFORMS J Appl Anal 47(5):425–441. https://doi.org/10.1287/inte.2017.0912
64. Ghanes K, Jouini O, Diakogiannis A, Wargon M, Jemai Z, Hellmann R et al (2015) Simulation-based optimization of staffing levels in an emergency department. Simulation 91(10):942–953. https://doi.org/10.1177/0037549715606808
65. Maughan BC, Lei L, Cydulka RK (2011) ED handoffs: observed practices and communication errors. Am J Emerg Med 29(5):502–511
66. Venkatesh AK, Curley D, Chang Y, Liu SW (2015) Communication of vital signs at emergency department handoff: opportunities for improvement. Ann Emerg Med 66(2):125–130
67. Alvarez R, Sandoval G, Quijada S, Brown AD (2009) A simulation study to analyze the impact of different emergency physician shift structures in an emergency department. In: Proceedings of the 35th International Conference on Operational Research Applied to Health Services ORAHS, Leuven, Belgium. http://www.econ.kuleuven.be/eng/tew/academic/prodbel/ORAHS2009//1b.pdf
68. Füchtbauer LM, Nørgaard B, Mogensen CB (2013) Emergency department physicians spend only 25% of their working time on direct patient care. Dan Med J. 60(1):A4558
69. Girishan Prabhu V, Taaffe K, Pirrallo R, Shvorin D (2020) Stress and burnout among attending and resident physicians in the ED: a comparative study. IISE Trans Healthc Syst Eng. https://doi.org/10.1080/24725579.2020.1814456
70. Woodworth L, Holmes JF (2020) Just a minute: the effect of emergency department wait time on the cost of care. Econ Inq 58(2):698–716
71. Salary.com (2021) Physician—Emergency Room Salary. https://www.salary.com/research/salary/benchmark/er-doctor-salary
72. Girishan Prabhu V, Taaffe K, Pirrallo RG, Jackson W, Ramsay M (2022) Overlapping shifts to improve patient safety and patient flow in emergency departments. Simulation 11:961–978

Chapter 13
Dependent Demand Forecasting Models in Airline Revenue Management: Parametric Estimation Using Simulation

Kavitha Balaiyan, R. K. Amit, Amit Agarwal, and T. V. Krishna Mohan

Abstract Joint forecasting models (JFMs) in airline revenue management are dependent on demand forecasting models that predict demand volume and consumer choice behaviour simultaneously. JFMs consider various factors such as the booking curve, seasonality, "maximum-willingness-to-pay" (MWTP) of the customer, and attributes of the available products. JFMs are parametric models that use mixed logit models to study customer choice behaviour, where the estimation of model parameters is a challenging problem. We propose a sequential two-stage hybrid modelling approach (a simulation-based heuristic algorithm) for parameter estimation of joint forecasting models using the Airline Planning and Operations Simulator (APOS) on actual airline data. In the first stage, parameters of demand volume are estimated by nonlinear least squares approximation using the Levenberg-Marquardt algorithm. In the second stage, MWTP parameters and choice-logit parameters are estimated simultaneously in two steps involving clustering and a simulation-based heuristic algorithm using Quai Monte-Carlo Simulation. Forecast predictions obtained using this method provide accurate results with lesser computational effort when compared with the results obtained using complete simulation through pseudo-Monte-Carlo methods. We recommend this algorithm for enabling faster computation in the

---

K. Balaiyan
Ford Motor Company, Chennai, India
e-mail: bkavith6@ford.com

R. K. Amit
Decision Engineering and Pricing (DEEP) Lab, Department of Management Studies, Indian Institute of Technology Madras, Chennai, India
e-mail: rkamit@iitm.ac.in

A. Agarwal
MeraPashu 360, Gurugram, India
e-mail: amit.agarwal@merapashu360.com

T. V. Krishna Mohan (✉)
Centre for Simulation, Analytics, and Modelling, University of Exeter Business School, Exeter, UK
e-mail: k.m.tv@exeter.ac.uk

© The Author(s), under exclusive license to Springer Nature Switzerland AG 2024
M. Fakhimi and N. Mustafee (eds.), *Hybrid Modeling and Simulation*, Simulation Foundations, Methods and Applications, https://doi.org/10.1007/978-3-031-59999-6_13

estimation of parameters and optimal seat allocation and pricing in choice-based networks in airline revenue management.

Keywords Revenue management · Demand forecasting · Customer choice model · Mixed logit model · Maximum willingness-to-pay · Monte-Carlo · APOS

13.1 Introduction

"Revenue management (RM) refers to the collection of strategies and tactics firms use to scientifically manage demand for their products and services" [1]. The practice of RM started in the airline industry in the USA as an aftermath of the Deregulation Act in 1978. To compete with low-cost carrier *PeopleExpress*, *American Airlines* performed demand management by segmenting their seat sales in the form of capacity control through fares and purchase restrictions, eventually leading to higher profits for them [2]. The success of RM in the airlines persuaded other industries such as hotels, cruises, car rentals, transport, telecommunication, etc., to follow suit.

Airlines segment their seats into multiple fare classes to attract customers with varying preferences. The customer demand for seats depends on a lot of factors such as the day and time of departure of flights, single or network flight type, price, product type, customer demographics, etc. The availability of seats at a given time affects the pricing and allocation decisions of the airline, thus making the demand dependent on the capacity.

Demands for different fare classes depend on the choices at hand for the customer at the time of making a purchase. Independent demand models assume that the demand is independent for different fare classes; while dependent demand models consider that the choice of the customer depends on the available options at the time of booking. Joint forecasting models (JFMs) couple the prediction of demand volume and customer choice behaviour, by considering the booking curve, seasonality, "maximum-willingness-to-pay" (MWTP) of the customer, and attributes of the available products. These parametric models use mixed logit models to study customer choice behaviour.

Estimation of model parameters for dependent demand models that utilise mixed logit formulation is a challenging problem. Joint forecasting models require simultaneous estimation of the parameters of demand volume and product attributes. Ghalehkhondabi et al. [3] conduct a review of various demand forecasting models in transportation to demonstrate the development of various modelling approaches in the field. They identify that traditional time-series or econometric models lack accuracy and are unable to estimate nonlinear and irregular demand patterns. They identify an increasing trend as the combination of forecasting methods such as the integration of econometric methods with metaheuristics, artificial intelligence, or simulation that help in better estimation.

Hence, we propose a two-stage hybrid method that involves clustering and simulation and clustering. This method combines the advantages of using closed-form methods wherever possible and applying simulation methods for simultaneous estimation of MWTP and customer choice, where closed-form methods are not available. Forecast predictions obtained using this method provide better results when compared with the results obtained using complete simulation using pseudo-Monte-Carlo methods.

This chapter is organised as follows: Sect. 13.2 discusses the related literature briefly describing the theoretical background and applications. Section 13.3 elaborates on the hybrid model and Sect. 13.4 discusses the results. Section 13.5 concludes the chapter.

13.2 Related Literature

We broadly classify the literature related to demand models in airline RM as independent and dependent models. The independent demand modelling approach treats the demand for each fare class separately. Various independent demand models developed include Littlewood's rule [4], expected marginal seat revenue model heuristic [5], discrete-time dynamic programming model [6], etc. Dependent models model the demand for fare classes considering various attributes price, schedule, travel time, etc. Examples of discrete choice models are MWTP models [7], "Fare adjustment (FA) theory" [8, 9], discrete choice models [10], etc. Parametric estimation is performed by linear regression methods such as estimators and nonlinear regression estimators such as Quasi-Newton methods [11], Levenberg-Marquardt algorithm [12, 13], etc.

In this section, we detail the literature into three subsections. Section 13.2.1 discusses the logit demand models. Section 13.2.2 discusses the estimation of mixed logit models detailing maximum simulated likelihood estimation, Monte-Carlo methods, and Halton sequences. Section 13.3 discusses the application of hybrid and simulation models.

13.2.1 Logit Models

Discrete choice models (DCMs) are applied in transportation, airline RM, and other industries to study the "choice of a customer" from a "given set of alternatives", which are mutually exclusive. It is assumed that the customer chooses an alternative that maximises his/her utility. A part of the utility can be observed and represented by a utility function, usually denoted as the representative utility in literature. The unobserved utility is added as a random component, whose distribution specification leads to different models in the family of logit models.

Discrete choice analysis describes the choice-making process of a decision-maker with a given set of mutually exclusive alternatives. The discrete choice analysis comprises two major exercises—model specification and estimation of parameters. Joint forecasting models (JFMs) are formulated using a mixed logit framework specifying the underlying behavioural process for choice analysis.

In a discrete choice framework, the decision-maker selects his option from a "set of alternatives". This choice set must be finite, mutually exclusive, and exhaustive. In cases where the sets may not be mutually exclusive or exhaustive, the choice set may be expanded to satisfy these requirements. The condition of finite alternatives is restrictive. Discrete choice models (DCMs) cannot be applied to infinite number of alternatives. DCMs assume that the decision-maker maximises his utility in his choice-making process [14, 15]. Random utility models can be viewed as methods developed to describe the relationship between "the explanatory variables and the choice-outcome". Parameter estimation of these models helps to identify how best the underlying behavioural choice process can be explained.

The utility of a customer n selecting an alternative j, from a "set of alternatives" J is U_{nj}. Option j resulting in the highest utility is opted by the customer. The choice outcome and the product attributes of the choices are observable, but the utility is not observable by the researcher. A function that can relate the observed factors to the utility of the decision-maker, V_{nj}, explains part of the utility function. The unobserved part of the utility ε_{nj} $\forall j$ is random, with density $f(\varepsilon_n)$.

$$U_{nj} = V_{nj} + \varepsilon_{nj} \qquad (13.1)$$

Assuming that ε_{nj} is an "independent and identically distributed (i.i.d) extreme value", β is a "vector of parameters", and x_{nj} is a "vector of observed variables", the utility function becomes:

$$U_{nj} = \beta'_n x_{nj} + \varepsilon_{nj}. \qquad (13.2)$$

The binary logit probability of picking out an option i from a set of alternatives J is then expressed as

$$P_{ni} = \frac{e^{\beta' x_{ni}}}{\sum_j e^{\beta' x_{nj}}}. \qquad (13.3)$$

McFadden [16] proves that "the log-likelihood function for these choice probabilities is globally concave" and the parameters β can be numerically maximised.

Different assumptions about the specifications of the density f(en) lead to different DCMs. Logit, multinomial logit, and nested logit models have closed-form solutions. These models assume that the "distribution of the random portion of the utility is i.i.d extreme value". The probit model assumes that the function is multivariate normal. Mixed logit assumes that a part of the unobserved portion $f(.)$ is "i.i.d extreme value" and the distribution for the other part is defined by the researcher. Train [17] notes that "Probit and Mixed logit integrals do not have a closed-form solution". The integral forms are evaluated numerically using simulation. See Train [17] and Hensher et al. [18] for details on DCMs.

Koppelman and Sethi [19] provide an overview of logit models that have closed-form solutions. The basic binary logit model is applied in simple scenarios for estimating the probability of whether an alternative is chosen or not. The multinomial logit model (MNL) [17] and the nested logit model (NL) are further extensions to the logit family that are applied in scenarios where the probability of selecting an option from a set of mutually exclusive options, or different subsets of nests of choice sets has to be studied.

Advanced choice models like multinomial probit and generalised extreme value and mixed logit existed in the literature for many years [17, 20]. Mixed logit models address the limitations of logit models such as random taste variation, which is more relevant in choice analysis in airline RM. These advanced models are analytically complex to solve, often represented with no closed-form expressions. With developments in simulation methods, the estimation became tractable, and the capabilities of mixed logit models could be utilised in many practical applications.

13.2.2 Estimation of Mixed Logit Models

"Mixed logit probabilities are the integrals of standard logit probabilities over a density of parameters" [17]. It is a "highly flexible model that can approximate any random utility model" [21]. The choice probability of the mixed logit model is the "weighted average of the logit evaluated at different β, with the weights given by the density function $f(\beta)$", and is expressed as

$$P_{ni} = \int \frac{e^{\beta' x_{ni}}}{\sum_j e^{\beta' x_{nj}}}. \tag{13.4}$$

In airline RM, products are differentiated for customers in different segments. It is assumed that customers differentiated by segments have varying "maximum willingness to pay". When the segment in a population is finite, as applicable in airline RM, the density function $f(\beta)$ can take discrete values. Parameters of the product attributes in the utility function can be estimated for each of these segments. Train [17] expresses the choice probability when the β in the mixing distribution $f(\beta)$ takes "M possible values" with probability S_m as

$$P_{ni} = \sum_{m=1}^{M} S_m \left(\frac{e^{b'_m x_{ni}}}{\sum_j e^{b'_m x_{nj}}} \right). \tag{13.5}$$

JFMs [22], incorporate customer segmentation in the demand models by utilising the customer's *maximum willingness to pay* and random utility maximisation of available products. Mixed logit model for modelling customer choice in joint forecasting model without price attribute (JFM-WPA), joint forecasting model with price attribute (JFM-PA), and joint forecasting model with combined WTP and choice utility (JFM-WTPU) is expressed as shown below, where the customer segments are modelled as the share of the customers that have the MWTP between the price points of the products offered. The logit probabilities for the choices are computed by estimating the parameters of the product attributes.

$$\text{JFM-WPA} \Rightarrow \left[\sum_{C_j \in C_{kM}} \overbrace{(F(p_j+1) - F(p_j))}^{\text{Segment share: MWTP}} \overbrace{\frac{e^{\sum_{a_k \in A} \beta_a a_k}}{\sum_{l \in C_j} e^{\sum_{a_l \in A} \beta_a a_l}}}^{\text{logit probabilities}} \right] \quad (13.6)$$

$$\text{JFM-PA} \Rightarrow \left[\sum_{C_j \in C_{kM}} \overbrace{\left(e^{-\alpha(p_j - p_0)} - e^{-\alpha(p_{j+1} - p_0)}\right)}^{\text{Segment share: MWTP}} \overbrace{\frac{e^{\sum_{a_k \in A'} \beta_a a_k}}{\sum_{l \in C_j} e^{\sum_{a_l \in A'} \beta_a a_l}}}^{\text{logit probabilities}} \right] \quad (13.7)$$

$$\text{JFM-WTPU} \Rightarrow \left[\sum_{C_j \in C_{kM}} \frac{\overbrace{(1 - F'(p_k))}^{\text{Segment share: MWTP}} \overbrace{e^{\sum_{a_k \in A} \beta_a a_k}}^{\text{utility}}}{\sum_{l \int C_j} (1 - F'(p_l)) e^{\sum_{a_l \in A} \beta_a a_l}} \right] \quad (13.8)$$

Estimation of parameters for MWTP and choice attributes in the mixed logit formulation of JFMs that consider the available choices at the time of purchase is computationally complex. Simulation methods are befitting the simultaneous estimation of parameters of the MWTP and the choice model.

13.2.2.1 Maximum Simulated Likelihood Estimation

Mixed logit models are estimated with relative simplicity with the advent of developments in simulation methods. Maximum simulated likelihood estimation is widely used to estimate parameters for mixed logit formulations. Simulated choice probabilities are used in the "log-likelihood function" which is maximised to obtain the parameter values. Simulation-based estimation and the properties of maximum simulated likelihood estimators are discussed in various articles [23–25]. Hensher et al. [18], Train [17], and Stern [26] provide a review of mixed logit models and simulation-based estimation methods.

The general concept of MSL estimation is approximating the probabilities through simulation and maximising the "log-likelihood function" over the parameters to be estimated. The functional form of choice probability in a mixed logit formulation is expressed as shown in Eq. 13.4.

Choice probabilities are simulated by using the following steps.

(a) Values for β are drawn from $f(\beta)$
(b) The logit formula, $\left(\frac{e^{\beta' x_{ni}}}{\sum_j e^{\beta' x_{nj}}}\right)$ is calculated with the drawn value of β
(c) Steps (a) and (b) are repeated several times (R draws)
(d) The average of all the probabilities calculated in step (b) is calculated.

The simulated probability is then given by,

$$P^{\vee}_{ni} = \frac{1}{R} \sum_{r=1}^{R} \left(\frac{e^{\beta' x_{ni}}}{\sum_j e^{\beta' x_{nj}}} \right) \qquad (13.9)$$

$P^{\vee}_{ni}$ is an unbiased estimator of P_{ni} expressed in Eq. 13.4. As R increases, the variance decreases. The "simulated log-likelihood function" is expressed as

$$\text{SLL} = \sum_{n=1}^{N} \sum_{j=1}^{J} d_{nj} \ln P^{\vee}_{nj}, \quad d_{nj} = \begin{cases} 1, & \text{if } n \text{ chooses } j \\ 0, & \text{otherwise} \end{cases}. \qquad (13.10)$$

Stern [26] notes that the MSL estimators are consistent and efficient for a very large class of problems. Train [17] states that this procedure described above helps in maintaining independence over decision-makers when calculating the simulated probabilities that are used in the SLL.

13.2.2.2 Monte-Carlo Methods

The Monte-Carlo simulation deals with computing the integrand values at *random* points and calculating the mean of the integrand values. Halton [27] provides a detailed review of Monte-Carlo methods. The *random* sequences of points for the evaluation of the integrands are not truly random. Pseudo-random sequences are used in the actual implementation of Monte-Carlo methods. Bhat [28] states that the "*Pseudo Monte-Carlo* (PMC) method has a slow asymptotic convergence rate" and "to obtain an added decimal digit of accuracy, the number of draws needs to be increased 100-fold." (p. 679).

The *Quasi-Monte-Carlo method* (QMC) uses a non-random sequence based on logic, which is more evenly distributed in the integration space. QMC sequences aim to provide a wider spread in the selection of the random points for evaluation of the function values. The construction of quasi-random sequences is not random. QMC methods provide a faster convergence rate, when compared with PMC methods. Refer to Bhat [28] and Train [17] for discussion on the convergence rates, integration errors of PMC and QMC methods. For a specified tolerance level, the QMC methods generally converge faster with a smaller number of *draws* and simulation points for choice probabilities. PMC and QMC models are widely used in demand models. Liu and Cirillo [29] uses PMC for vehicle demand estimation. The use of QMC model for various demand estimations can be found in Bhat [30], Bhat and Gossen [31] and Bastin et al. [32].

Drawing values from joint densities for multinomial distributions may be difficult at times when compared with drawing values from the conditional density, when the values of other parameters are given. Algorithms like *Gibbs Sampling* [33] and *Metropolis-Hastings* (MH) [34, 35] have been discussed in many articles in the literature. The random draws in MH and Gibbs sampling methods are dependent on the previous draws. These methods are called "*Markov Chain Monte Carlo*" (MCMC) methods. MCMC methods received significant attention in Bayesian methods of estimation. Stern [26] states that the Gibbs sampling methods are expensive when compared with other methods like MSL. For the estimation of parameters for practical airline applications using JFM models, Bayesian approaches are not preferred.

Monte-Carlo simulation using pseudo-random draws (Pseudo-MC) is used to estimate the parameters for JFMs simultaneously (Sect. 13.3.2.2.1). To address the issues related to computation time, simulation-based heuristic algorithm is proposed (Sect. 13.3.2.2.2), in which Quasi-Monte-Carlo simulation method is applied. As QMC requires a high-quality quasi-random sequence for the successful implementation, we use Halton sequences. Halton sequences are low-discrepancy sequences that are widely used in Monte-Carlo simulations [36]. Halton sequence is comparatively easier than other low-discrepancy measures to implement because of the ease of the implementation of the radical inverse function.

13.2.2.3 Halton Sequences

"*Halton sequences*" [37] is generally defined in terms of a prime number. For instance, the Halton sequence generated using the prime number 3 becomes:

$$\frac{1}{3}, \frac{2}{3}, \frac{1}{9}, \frac{4}{9}, \frac{7}{9}, \frac{2}{9}, \frac{5}{9}, \frac{8}{9}, \frac{1}{27}, \frac{10}{27}, \frac{19}{27}, \frac{4}{27}, \frac{13}{27},$$

and so on. The sequence is intuitive in providing a wide coverage within the space for integration. The pattern of generation of points in a Halton sequence is in such a way that the subsequence covers the spaces not covered by the previous subsequence. This "*filling-in property*" of Halton sequences can help induce a negative correlation over observations. The simulated probabilities computed using the quasi-random points generated by Halton sequences minimise errors in the SLL function, resulting in faster convergence. The coverage improves when the number of draws increases. The accuracy increases with increasing number of draws.

Halton sequences for multiple dimensions can be created with different prime numbers for each dimension. Better coverage and negative correlation over observations make Halton sequences more effective and preferred than the pseudo-Monte-Carlo method of random draws. Spanier and Maize [38] show that fewer Halton draws can provide good accuracy when compared with random draws. Bhat [28] showed that "100 Halton draws resulted in better convergence than 1000 random draws" for a mixed logit model. Train [39] and Hensher [40] prove that the convergence is faster when Halton sequences are used for draws when compared with random draws. Train [17] shows that the computing time can be "reduced by a factor of ten by using Halton draws" compared to random draws, while the accuracy is reported to be increasing, not reducing. Train [17] also discusses the anomalies observed in some experiments, which need further studies regarding the characteristics of Halton sequences. For JFMs in airline RM applications, Halton sequences prove to be highly effective.

13.2.3 Hybrid and Simulation Approaches

Hybrid models are increasingly developed for demand forecasting models for better accuracy [3]. Many of the hybrid models used are a combination of various combination of econometric models [41–43], statistical and machine learning models [44, 45], hybrid simulation models [46], etc. These models are used for various applications such as water demand forecasting [41, 42], pharmaceutical industry [43], energy sector [44], supply chain management [45], healthcare [46], etc. Hybrid simulations have employed combining other approaches with simulation such as Harper et al. [46] use a combination of demand forecasting and discrete event system simulation for healthcare diagnostic service. Simulation models are also used for demand forecasting such as discrete event system simulation model in acute healthcare services [47], agent-based simulation models in travel demand modelling [48], and system dynamics simulation models in water demand estimation [49], air passenger demand [50], and shipbuilding demand [51].

Juan et al. [52] propose the concept of Simheuristics, an optimisation algorithm formed as a combination of simulation and metaheuristics to solve combinatorial optimisation problems. Juan et al. [52] further classify Simheuristics are a special case of hybrid simulation optimisation technique that can be either simulation optimisation problems that prioritise finding optimal parameter values or hybrid simulation analytic problems that use simulation to refine the parameters of a problem-specific analytical model. Various Simheuristics have been applied for different problems such as: routing heuristics combined with Monte-Carlo simulation for inventory routing problem [53]; Monte-Carlo simulation iterated local search metaheuristic for solving the permutation flow shop problem with stochastic processing times [54]; and genetic algorithm combined with discrete event system simulation [55]. The simulation-based heuristic algorithm in the proposed hybrid model is a simulation optimisation class of Simheuristics as explained in Sect. 13.3.2.2.2.

13.3 Research Design

As the estimation of mixed logit models is complex and requires simulation methods, we propose the hybrid model involving clustering to over any bias and a simulation-based heuristic algorithm using QMC to reduce computational time. Figure 13.1 represents the hybrid modelling approach. The approach has two stages to estimate the parameters of "demand volume" and "customer choice". In stage 1, the parameters of demand volume are estimated by nonlinear least squares approximation using the Levenberg-Marquardt algorithm. In stage 2, the parameters for the MWTP and the choice-logit parameters are estimated simultaneously. MWTP of customers varies during the booking horizon. Clustering is used to identify demand patterns concerning days before the departure of a flight. The parameters for the MWTP and choice-logit parameters are estimated for each cluster in two steps. Sampling of error terms of the model is performed in the first step, assuming a normal distribution. Keeping all other parameters constant, the first parameter is varied within a chosen range, simulated log-likelihood function (SLL) is computed and plotted. The range of the parameter is narrowed, and the process is repeated until the convergence of the SLL. The value of the parameter that maximises the SLL function is chosen. This step is repeated sequentially to estimate the rest of the parameters. In the second step, these parameters are used as initial values and set narrower bounds for quasi-Monte-Carlo simulation. Halton sequences are used for sampling from the intervals specified for the parameters. Data generated by APOS is used for estimating the parameters and testing the models. The stages are detailed in further subsections.

13.3.1 Stage 1: Demand Volume Estimation

Joint forecasting models for airline pricing and revenue management are formulated to estimate the demand for a product between any two reading days. The dependent demand model estimates the share of the demand for the host airline between two reading days, derived from the market demand and booking build-up curves. Estimation of parameters for this demand volume can be handled by closed-form methods like nonlinear least squares approximation. The estimation of demand for the product is modelled in JFMs by considering the proportion of customers in the estimated demand volume between reading days, with a certain MWTP, who would consider the product and the choice attributes of the product. Simultaneous estimation of parameters for the attribute vectors used in this mixed logit model part of the formulation requires the application of simulation methods.

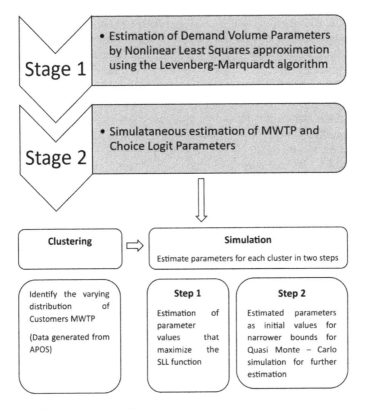

Fig. 13.1 Hybrid modelling approach

13.3.2 Stage 2: Customer Choice Estimation

Customer choice estimation is done through clustering and simulation, detailed as follows.

13.3.2.1 Clustering

Data used for the estimation of JFMs is generated by APOS from real airline booking and availability information. Customer arrival pattern varies at different times in the booking period. Exploration of data for an Origin-Destination (O-D) market can provide information on the sales pattern of different products during the booking horizon. Insights of the proportion of different customer segments at different reading days during the booking horizon can be obtained by investigating the data. Reading days are the days before departure when airlines collect or update their data. For example, reading days (0–20) denote the 20 days before departure of the trip.

A snapshot of booking data for one O-D market for an airline is given in Table 13.1. The distribution of bookings for different fare classes varies with reading days. The sale of products with higher fares is generally observed when the departure date is closer. Clustering data helps in a more precise estimation of parameters for the distribution of the MWTP of the customers. K-means clustering algorithm [56, 57] is used for this experiment. The choice of the clustering algorithm can be at the discretion of the researcher, depending upon the characteristics of the dataset to be analysed and the business needs. Three distinct clusters are identified for the data used for the estimation of JFMs. Reading days (0–20), (21–97), and (98–220) form clusters with well-defined booking patterns. The distribution of fare of the products (booking classes) over reading days is shown in Fig. 13.2. Parameters for MWTP and choice attributes are estimated for each of these clusters using random utility maximisation, to describe the choice process and to produce better forecasts.

A sample of the distribution of different fare classes over the entire booking period for an origin-destination market is given in Table 13.2. The distribution of bookings for different fare classes at different reading days before the departure date is given in Table 13.3. The assumptions on the customers' MWTP distribution can be deduced from the distribution of the booking patterns. All bookings for products with fares greater than $1000 can be observed in the cluster of reading days (0–20), which is near the departure date. (RD 0) indicates the departure date for the flight. No booking for products with fares greater than $750 is observed in the cluster with reading days (98–220), which is far from the departure date. About 56% of total bookings for this O-D route for the airline are observed for the fare classes with a price less than $750. Fare classes with a price greater than $1000 constitute about 25% of the total bookings. Flights in this O-D market are available for booking 330 days prior to departure. Bookings are observed only at 220 days prior to departure and started picking up after 123 days prior to departure.

Exploration of the data used for the estimation of JFMs indicates that the MWTP distributions of the passengers vary across reading days. It also provides information for intelligent assumptions on the MWTP distribution for the total demand for an OD-market for a departure date. Given the data for an O-D market with multiple departure dates, a reasonable assumption for the MWTP distribution of passengers can be inferred. For the dataset used for this research, it is observed that 82% of the total bookings happened when the flight was close to departure (RD 0–20). For practical purposes, the calibration of parameters and optimisation of seat allocation for this market can be more focused during the reading day cluster (0–20). For scenarios where the data is more evenly distributed over the reading days, parameter estimation and optimal allocation of seats must be performed at more frequent intervals in the booking horizon for better accuracy.

13.3.2.2 Simulation

Customer choice formulated using the mixed logit model in JFMs, coupled with the inclusion of "maximum willingness-to-pay" of the customers and available choices at

13 Dependent Demand Forecasting Models in Airline Revenue …   331

Table 13.1 Sample data of an O-D market of an airline for one departure date

FARE →	222.77	736.007	761.024	859.749	878.01	888.807	1092.54	1175.3	1324.61	1377.38	1475.67	1722.1	1800.66	Total
Reading day ↓	Class: E	Class: S	Class: O	Class: L	Class: K	Class: M	Class: N	Class: B	Class: W	Class: H	Class: V	Class: T	Class: Q	
0	7	41	5	7	0	0	19	10	14	15	32	68	13	231
1	37	57	13	5	3	10	45	42	47	3	81	31	13	387
2	63	82	21	21	8	2	91	43	40	12	105	2		490
3	9	15	0	0	1	0	0	1	1	0	0			27
4	10	15	5	0	3	0	1	2	0	0	0			36
5	129	90	11	38	19	1	22	26	39	3	2			380
6	93	57	24	17	18	2	20	19	16	1	0			267
7	175	32	23	17	15	7	20	20	4	15	0			328
8	119	71	32	32	16	3	35	18	80	1	15			422
9	85	19	20	12	12	3	9	5						165
10	6	6	2	0	0	0	1							15
11	38	12	3	1	0	0	1							55
12	136	39	15	9	4	0	1							204
13	60	13	18	4	3	1	1							100
14	64	14	12	0	3	0	0							93
15	72	17	18	6	5	0	0							118
20	126	33	43	8	8	2	1							221
27	153	21	57	21	4									256
34	97	30	74	24										225

(continued)

Table 13.1 (continued)

FARE → Reading day ↓	222.77 Class: E	736.007 Class: S	761.024 Class: O	859.749 Class: L	878.01 Class: K	888.807 Class: M	1092.54 Class: N	1175.3 Class: B	1324.61 Class: W	1377.38 Class: H	1475.67 Class: V	1722.1 Class: T	1800.66 Class: Q	Total
41	61	29	10	8										108
48	28	4	1											33
55	36	34	7											77
62	2	11	1											14
69	8	5	1											14
83	4	4	1											9
97	23	1												24
140	**0**	**4**												**4**
160	**5**													**5**
220	**7**													**7**
Total	1653	756	417	230	122	31	267	186	241	50	235	101	26	4315

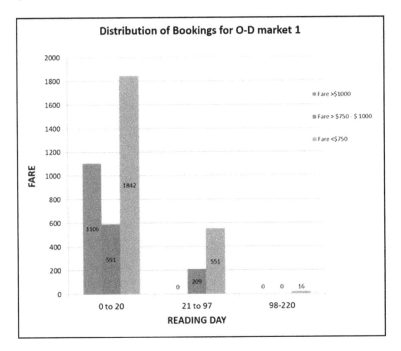

Fig. 13.2 Distribution of bookings for OD-market 1

the time of booking, makes the computations complex. The application of simulation methods assists in addressing the complexity of estimating parameters used in the JFMs. In this section, we explore the estimation of parameters using a complete Monte-Carlo simulation. To address the computation time, a two-stage heuristic algorithm (hybrid model) is proposed. This heuristic uses simulation methods for jointly estimating the parameters for the customer's MWTP distribution and choice attributes. Parameters for demand volume are estimated using nonlinear regression methods.

Monte-Carlo Simulation

Monte-Carlo simulations are used to estimate the parameters for JFMs simultaneously. The parameters for the volume of demand are calibrated using nonlinear least square methods. Random draws are used to compute the simulated log-likelihood estimation assuming uniform distributions for all parameters. Experiments using truncated data pertaining to limited fare classes are conducted to study the forecast prediction accuracy and time taken for computations. The customer's choice set is calculated based on the MWTP assumptions and the available choices at the time of booking. Within these choice sets, the utilities for the choices are calculated using random draws for parameters. The probability of the customer choosing an

Table 13.2 Sample data of fare class distribution over the booking period for O-D market 1

Fare: 222.77		Fare: 736.007		Fare: 761.024		Fare: 859.749		Fare: 878.01		Fare: 888.807		Fare: 1092.54	
RD	Class: E	RD	Class: S	RD	Class: O	RD	Class: L	RD	Class: K	RD	Class: M	RD	Class: N
0	7	0	41	0	5	0	7	0	0	0	0	0	19
1	37	1	57	1	13	1	5	1	3	1	10	1	45
2	63	2	82	2	21	2	21	2	8	2	2	2	91
3	9	3	15	3	0	3	0	3	1	3	0	3	0
4	10	4	15	4	5	4	0	4	3	4	0	4	1
5	129	5	90	5	11	5	38	5	19	5	1	5	22
6	93	6	57	6	24	6	17	6	18	6	2	6	20
7	175	7	32	7	23	7	17	7	15	7	7	7	20
8	119	8	71	8	32	8	32	8	16	8	3	8	35
9	85	9	19	9	20	9	12	9	12	9	3	9	9
10	6	10	6	10	2	10	0	10	0	10	0	10	1
11	38	11	12	11	3	11	1	11	0	11	0	11	1
12	136	12	39	12	15	12	9	12	4	12	0	12	1
13	60	13	13	13	18	13	4	13	3	13	1	13	1
14	64	14	14	14	12	14	0	14	3	14	0	14	0
15	72	15	17	15	18	15	6	15	5	15	0	15	0
20	126	20	33	20	43	20	8	20	8	20	2	20	1
27	153	27	21	27	57	27	21	27	4	27		27	
34	97	34	30	34	74	34	24	34		34		34	
41	61	41	29	41	10	41	8	41		41		41	

(continued)

Table 13.2 (continued)

Fare: 222.77		Fare: 736.007		Fare: 761.024		Fare: 859.749		Fare: 878.01		Fare: 888.807		Fare: 1092.54	
RD	Class: E	RD	Class: S	RD	Class: O	RD	Class: L	RD	Class: K	RD	Class: M	RD	Class: N
48	28	48	4	48	1	48		48		48		48	
55	36	55	34	55	7	55		55		55		55	
62	2	62	11	62	1	62		62		62		62	
69	8	69	5	69	1	69		69		69		69	
83	4	83	4	83	1	83		83		83		83	
97	23	97	1	97		97		97		97		97	
140	0	140	4	140		140		140		140		140	
160	5	160		160		160		160		160		160	
220	7	220		220		220		220		220		220	
Total	1653		756		417		230		122		31		267

Fare: 1175.3		Fare: 1324.61		Fare: 1377.38		Fare: 1475.67		Fare: 1722.1		Fare: 1800.66	
RD	Class: B	RD	Class: W	RD	Class: H	RD	Class: V	RD	Class: T	RD	Class: Q
0	10	0	14	0	15	0	32	0	68	0	13
1	42	1	47	1	3	1	81	1	31	1	13
2	43	2	40	2	12	2	105	2	2	2	
3	1	3	1	3	0	3	0	3		3	
4	2	4	0	4	0	4	0	4		4	
5	26	5	39	5	3	5	2	5		5	
6	19	6	16	6	1	6	0	6		6	
7	20	7	4	7	15	7	0	7		7	
8	18	8	80	8	1	8	15	8		8	
9	5	9		9		9		9		9	
10		10		10		10		10		10	
11		11		11		11		11		11	
12		12		12		12		12		12	

(continued)

Table 13.2 (continued)

Fare: 1175.3	Class: B	Fare: 1324.61	Class: W	Fare: 1377.38	Class: H	Fare: 1475.67	Class: V	Fare: 1722.1	Class: T	Fare: 1800.66	Class: Q
RD		RD		RD		RD		RD		RD	
13		13		13		13		13		13	
14		14		14		14		14		14	
15		15		15		15		15		15	
20		20		20		20		20		20	
27		27		27		27		27		27	
34		34		34		34		34		34	
41		41		41		41		41		41	
48		48		48		48		48		48	
55		55		55		55		55		55	
62		62		62		62		62		62	
69		69		69		69		69		69	
83		83		83		83		83		83	
97		97		97		97		97		97	
140		140		140		140		140		140	
160		160		160		160		160		160	
220		220		220		220		220		220	
Total	186		241		50		235		101		26

Table 13.3 Sample data for OD market 1—distribution of bookings

RD	Fare > $1000	Fare $750–$1000	Fare < $750	Total bookings	Percentage
	Class: (N, B, W, H, V, T, Q)	Class: (O, L, K, M)	Class: (E, S)		
0–20	1106	591	1842	3539	**0.8202**
21–97	0	209	551	760	0.1761
98–220	**0**	**0**	16	16	0.0037
Total Bookings	1106	800	2409	*4315*	
Percentage	**0.2563**	0.1854	**0.5583**		

option is computed using these utilities and used as an input for SLL calculations. This computation is complex and increases the time for convergence of SLL in a complete Monte-Carlo simulation. To address the issues related to computation time and to improve the accuracy of prediction, a simulation-based approach is proposed. This method uses partial application of closed-form solutions using existing algorithms and partial simulation for calibration of parameters. For Monte-Carlo simulation using pseudo-random draws (Pseudo-MC), 20,000 pseudo-random draws were required for convergence of the simulated log-likelihood function.

Simulation-Based Heuristic Algorithm

The estimation method using simulation proposed for calibration of parameters in JFMs has two stages for computation. In stage 1, the parameters of the demand volume are calibrated using nonlinear least squares estimation the by Levenberg-Marquardt algorithm. In stage 2, the parameters for MWTP and choice attributes are estimated simultaneously using simulation. Clustering is used to capture the varying distribution of customers' MWTP with respect to the days before departure of a flight. For each cluster, the parameters for the MWTP and choice-logit parameters are calibrated in two steps. In the first step, sampling of error terms of the model is performed, assuming normal distribution. Keeping all other parameters fixed, the first parameter is varied within a chosen range, and SLL is computed and plotted. The range of the parameter is narrowed, and the process is repeated until SLL converges. The value of the parameter that maximises the SLL function is chosen. This step is repeated sequentially to estimate the rest of the parameters. In the second step, these parameters are utilised as initial values and narrower bounds are set for the quasi-Monte-Carlo simulation. Halton sequences are used for sampling from the intervals specified for the parameters. Simulation-based heuristic approach (Quasi-random MC) resulted in a convergence of the simulated log-likelihood function with 100 Halton draws. As the parameters that maximise the value of the SLL are chosen finally, the proposed simulation-based heuristic algorithm aligns with the

Algorithm Steps for estimation of parameters for JFMs using simulation

Stage 1 :
 Calculate the average market demand from host share
 Estimate parameters for demand volume using LM algorithm
 	
 Stage 2 :
 Identify clusters in data using any clustering technique (eg. K-means algorithm)
 	
 step 1 :
 for *parameter* $= 1 \to n$ **do**
     Sample using random draws, vary *parameter* for a chosen range
     Keep all other parameters fixed
     Compute simulated probability and simulated log-likelihood function
     Choose parameter value that maximizes SLL function
 end for
 step 2 :
 Set the parameter values estimated in *step 1* as initial values
 Set narrower bounds for Quasi Monte - Carlo simulation
 Use Halton sequences for sampling
 Simulate for all parameters simultaneously, compute SLL
 Choose parameter value that maximizes SLL

Fig. 13.3 Simulation-based heuristic algorithm

simulation optimisation class of Simheuristics of Juan et al. [52]. Details of the proposed simulation method and the steps are given as an algorithm (Fig. 13.3).

13.4 Results and Discussion

In this section, the incremental demand forecasts for fare products at different reading days during the booking horizon for an origin-destination market for an airline are computed and discussed. Model JFM-PA [22] is chosen for simultaneous estimation using pseudo-Monte-Carlo simulation method and the proposed hybrid approach (we state a simulation-based heuristic algorithm). This model includes price as a choice attribute in calculating the utility of products that fall in the consideration set of customers with a certain MWTP, based on the choices available at the time of booking decision. This model is chosen as it best represents the complexity and covers a wide range of practical scenarios. Data used for computation, results, and analysis is OD-market 1 for an airline, generated by APOS, Sabre Inc.

The dependent demand forecast is computed at the reading day level using Monte-Carlo simulation, and the simulation heuristic algorithm (hybrid approach) is compared with the actual bookings at all reading days during the booking period, as shown in Fig. 13.4.

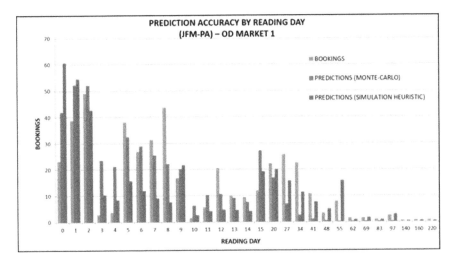

Fig. 13.4 Accuracy of prediction by reading day—estimation by simulation

The accuracy of predictions made by the hybrid approach (simulation-based heuristic algorithm) on a reading day level is higher than the complete Monte-Carlo simulation method when the whole booking horizon is considered. The number of bookings predicted by the Monte-Carlo method compared to actual bookings is better on reading days when more bookings are observed. This method does not produce predictions with good accuracy on reading days with fewer bookings. For reading days RD 220 to RD 41, the Monte-Carlo method predicts almost no bookings. The trend of booking patterns is captured better by estimation using the simulation-based heuristic algorithm, for the entire booking horizon from RD 220 to RD 0. This heuristic produces results of higher accuracy even in cases where the bookings are less. Origin-destination markets with fewer bookings (*lean markets*) can apply the simulation-based heuristic algorithm for parameter estimation, which brings benefits to the market in addressing enhancements of product attributes based on the utilities preferred by the customers in that market. For OD-market 1 taken for this experiment, it is noted in Sect. 13.3.1 that 82% of the bookings are observed between RD 20 and RD 0. The accuracy of predictions for specific reading days between RD 20 and RD 0 can further be improved by clustering and estimating the parameters of the model within this period. Airlines run optimisation models more frequently when the flight is close to departure. Frequent runs of joint forecasting and optimisation models during reading days closer to the departure date increase the accuracy of prediction at the reading day level.

The JFMs formulate the customer choice using a mixed logic model. Analysing the predictions based on the fare classes offered helps to compare the performance of the estimation methods in predicting customer choices. The choices of the customers are predicted by estimating the parameters related to fare, product attributes like elapsed travel time, departure time, and the customer's "maximum willingness-to-pay". A

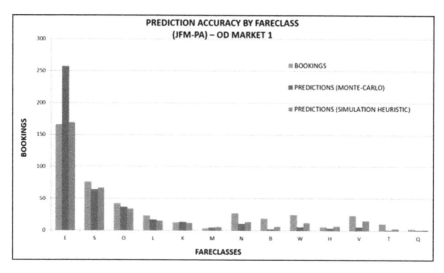

Fig. 13.5 Accuracy of prediction by fare class—estimation by simulation

comparison of simulation-based estimation methods in predicting customer choices based on fare classes is shown in Fig. 13.5.

The fare classes are shown in the order of increasing fares in Fig. 13.5. As discussed in Sect. 13.3.1, around 74% of the total bookings are observed for fare classes with fares less than $750, which constitutes fare classes E to M. It can be noted that the parameter estimation made by Monte-Carlo simulation provides inflated predictions for the lowest fare class, which has the highest number of bookings. Simulation-based heuristic algorithm (hybrid approach) uses clustering to analyse the distribution of customers with maximum willingness to pay. This adds to the benefits of this algorithm to overcome the bias induced by the bookings for fare class E in the data. Parametric estimation by simulation-based heuristic algorithm presents better accuracy of prediction of customer choices based on product characteristics. This method shows good prediction power for all fare classes when compared with Monte-Carlo simulation methods. This method can be applied in markets where only certain booking options are preferred by the customers. Airlines can apply this method to understand the distribution of customers and their utilities to help implement their sales strategies and optimise their seat allocation and pricing.

13.4.1 Comparison of Estimation Methods

Sequential estimation of parameters requires the introduction of a normalisation factor to reduce the build-up of errors in the sequential steps [22]. Simulation methods for parameter estimation in Sect. 13.3.2 help in estimating the parameters of the models simultaneously. In this section, we compare the accuracy of the prediction of the proposed estimation methods with Balaiyan et al. [22].

Parameters calibrated using sequential estimation, Monte-Carlo simulation, and simulation-based heuristic algorithms are listed in Table 13.4. In Monte-Carlo simulation method, the parameters γ and α are estimated by nonlinear least squares estimation, as described in the sequential estimation method. The β parameters pertaining to the choice attributes of the products are estimated using random draws within a chosen range, assuming a uniform distribution. The three methods of estimation are implemented using R programming language. The calculation of the set of options considered by the proportion of customers with a certain MWTP, the utilities for all options considered using the random draws of the parameters, and the simulated log-likelihood functions are coded. Estimations are performed using Intel(R) Core™ i5-8400H CPU @ 2.50 GHz processor with 16 GB RAM. The initial range of parameters chosen is identical between the simulation methods. This supports the comparison of simulation-based estimation methods regarding convergence and prediction accuracy. For Monte-Carlo simulations, 20,000 random draws are used within the given range and the parameters are chosen, which results in the maximum value of the SLL function. It is observed that the convergence of the maximum simulated log-likelihood function happens with 100 Halton draws within the given range for all parameters.

In the simulation-based heuristic algorithm stage 2, random draws are sampled within a chosen range for a parameter, keeping all other parameters fixed. SLL function is computed for this parameter with random draws, until the convergence of the SLL. This step is repeated for every parameter, keeping all other parameters fixed. The parameter value that maximises the SLL function is chosen at every step. Figure 13.6 describes the SLL function for each parameter. It is observed that the SLL function converges with 100 random draws within the chosen range.

The mean absolute deviation (*MAD*) is a good measure of performance indication in the airline industry. To analyse the performance of the parameter estimation methods, considering the distribution of bookings over the reading days and the fare class distribution in an origin-destination market, the weighted mean absolute percentage error (*WMAPE*) is chosen as a performance measurement indicator.

The average deviation of predicted bookings over the booking horizon for all reading days is given by *MAD*. This measure does not capture the variation in prediction considering the distribution of bookings over the reading days. WMAPE explained in Eq. 13.11 captures the weighted average of the deviations in bookings, providing a good measure of the accuracy, which can be compared between different estimation methods. The weight for each reading day is calculated using the percentage of bookings for a reading day.

$$\text{WMAPE} = \frac{\sum_{\text{rd}=0}^{n}(w_{\text{rd}}|A_{\text{rd}} - F_{\text{rd}}|)}{\sum_{\text{rd}=0}^{n}(w_{\text{rd}}|A_{\text{rd}}|)}, \quad (13.11)$$

where

Table 13.4 Parameters calibrated using different estimation methods

Parameters → Estimation method ↓		γ	α	β DFARE	β ETIME	β DEPSLOT1	β DEPSLOT2	β DEPSLOT3	β DEPSLOT4	β DEPSLOT6	β DEPSLOT7
Sequential estimation		0.08428613	0.001251	−0.004	−0.006	0	0	−0.944	−0.106	−0.501	−1.003
Monte-Carlo estimation		0.08428613	0.001251	−1	0	2	3	−1	0	−1	−1
Simulation-based heuristic (hybrid approach)	Cluster 1 (RD 0–20)	0.25131	0.0011	−0.7	0	1	1	−1	0	−1	−1
	Cluster 2 (RD 21–97)	0.0432031	0.00215	−33	0	5	−10	−1	−1	−1	−1
	Cluster 3 (RD 98–220)	0.0173259	0.0040742	−113	0	3	2	−1	3	−1	−1

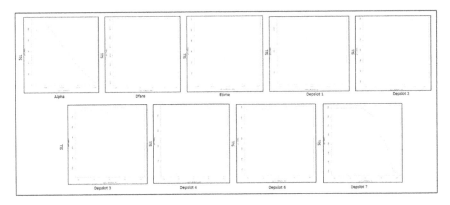

Fig. 13.6 Accuracy of prediction by fare class—estimation by simulation

rd   reading day
w_{rd} weights for reading day
A_{rd} actual bookings at reading day rd
F_{rd} predicted bookings at reading day rd.

The accuracy of prediction at the fare class level is also computed as a weighted mean absolute percentage error. The calculation of WMAPE for the prediction of fare classes in bookings is expressed by Eq. 13.12. The weight for each fare class is calculated using the percentage of bookings for a fare class in the origin-destination market during the booking period. This measure aids in analysing the performance of different estimation methods in calibrating the parameters of the choice attributes of the options available at the time of purchase. Prediction accuracies at the reading day level illustrate the predicting power of the model with reference to the number of bookings observed on a reading day. Computing WMAPE regarding the fare classes gives an outlook of the predicting power of the model with reference to customer choices based on fare and other product attributes.

$$\text{WMAPE} = \frac{\sum_{c=1}^{n}(w_{rd}|A_c - F_c|)}{\sum_{c=1}^{n}(w_c|A_c)}, \quad (13.12)$$

where

c fare class differentiated by price
w_c weights for a fare class
A_c actual bookings for a fare class
F_c predicted bookings for a fare class.

The WMAPE values are computed and shown in Table 13.5.

The WMAPE computed on the *reading day* level is comparable to the simulation methods, approximately 1.9%. It can be observed that the WMAPE calculated on the *reading day* level for the sequential estimation is less than 1%. This can

Table 13.5 WMAPE for predictions based on different estimation methods

Estimation method	WMAPE (reading day)	WMAPE (fare class)
Sequential estimation	0.0076	0.0371
Monte-Carlo simulations	0.0192	0.0358
Hybrid approach (simulation-based heuristic)	0.0197	0.0152

be attributed to the fact that the normalisation step used in the sequential method assists in fine-tuning the predictions on a reading day level. The WMAPE computed on the *fare class* level is comparable between the sequential estimation and the Monte-Carlo estimation methods, indicating the percentage errors to be approximately 3.7% and 3.6%, respectively. The hybrid approach m is observed to be reporting a WMAPE value of 1.5%. This can be attributed to the clustering used in this heuristic assisting to refine the parameter estimation regarding the distribution of the "maximum willingness-to-pay" of the customers.

In summary, the use of a hybrid approach helps in improving accuracy, estimating parameters in less time, and is robust to variation in demand. Compared to the Monte-Carlo simulation, the use of clustering in the hybrid method helped to overcome the bias in bookings. This encourages us to recommend the use of hybrid methods for better demand estimation.

13.5 Conclusion

This chapter proposes the estimation of parameters for joint forecasting models using simulation methods. The Monte-Carlo simulation method is used for parameter estimation, with pseudo-random draws. A hybrid method (simulation-based heuristic algorithm) is proposed, which utilises nonlinear least squares estimation for demand volume parameters and simulation methods for choice parameters. The simulation-based heuristic algorithm uses Halton sequences for sampling for parameters, resulting in faster convergence.

Performances of the sequential estimation, Monte-Carlo estimation, and the hybrid approach are compared. Prediction accuracies of the number of bookings observed on a reading day are comparable for the Monte-Carlo simulation method and the hybrid approach. Monte-Carlo simulation method does not exhibit good predicting power for the prediction of the number of bookings for different fare classes. The hybrid approach performs better in a choice-modelling perspective and displays good prediction accuracy at the fare class level. The performance of sequential estimation and Monte-Carlo method is comparable when the accuracy of prediction for the number of bookings at the fare class level is analysed. The hybrid approach provides better prediction at the fare class level with 57% less WMAPE than the Monte-Carlo simulation and 59% less WMAPE than the sequential estimation [22].

The hybrid approach proves to be a good alternative for complex estimation problems in joint forecasting models. This method is also suitable for some origin-destination markets where the number of bookings observed is very low. In an RM setting, the forecasting and optimisation modules are run frequently when the departure date is near. The hybrid approach assists in faster computation times for the estimation of parameters and optimal seat allocation and pricing in a choice-based network, considering the available options at the time of booking. The forecast parameters are built on the customer's choice based on product attributes, resulting in revenue management in a classless environment.

Our research has some limitations. We have not considered the effects of competition on demand estimation. Time and market-dependent conditions for estimation in the market demand models can improve the accuracy of the research. As our model is limited only to the airline industry, we would like to generalise our model for other settings with perishable assets and substitutable products.

Acknowledgements We thank SABRE Inc. for their support. APOS is an in-house simulation tool developed by SABRE Inc. Amit Agarwal contributed when working for SABRE Inc.

References

1. Van Ryzin GJ, Talluri KT (2005) Emerging theory, methods, and applications. An introduction to revenue management. INFORMS, Catonsville, MD, USA, pp 142–94
2. Smith BC, Leimkuhler JF, Darrow RM (1992) Yield management at American airlines. Interfaces 22(1):8–31
3. Ghalehkhondabi I, Ardjmand E, Young WA, Weckman GR (2019) A review of demand forecasting models and methodological developments within tourism and passenger transportation industry. J Tourism Futures 5(1):75–93
4. Littlewood K (2005) Special issue papers: forecasting and control of passenger bookings. J Revenue Pricing Manag 1(4):111–123
5. Belobaba PP (1989) OR practice—application of a probabilistic decision model to airline seat inventory control. Oper Res 37(2):183–197
6. Lee TC, Hersh M (1993) A model for dynamic airline seat inventory control with multiple seat bookings. Transp Sci 27(3):252–265
7. Garrow LA, Jones SP, Parker RA (2007) How much airline customers are willing to pay: an analysis of price sensitivity in online distribution channels. J Revenue Pricing Manag 1(5):271–290
8. Fiig T, Isler K, Hopperstad C, Belobaba P (2010) Optimization of mixed fare structures: theory and applications. J Revenue Pricing Manag 1(9):152–170
9. Fiig T, Isler K, Hopperstad C, Olsen SS (2012) Forecasting and optimization of fare families. J Revenue Pricing Manag 1(11):322–342
10. Ben-Akiva ME, Lerman SR, Lerman SR (1985) Discrete choice analysis: theory and application to travel demand. MIT Press
11. Nocedal J, Wright SJ (2006) Quadratic programming. In: Numerical optimization, pp 448–492
12. Levenberg K (1944) A method for the solution of certain non-linear problems in least squares. Q Appl Math 2(2):164–168
13. Marquardt DW (1963) An algorithm for least-squares estimation of nonlinear parameters. J Soc Ind Appl Math 11(2):431–441

14. Thurstone LL (1927) A law of comparative judgment. Psychol Rev 34(4):273
15. Marschak J (1974) Binary-choice constraints and random utility indicators (1960). In: Economic information, decision, and prediction: selected essays: volume I Part I Economics of decision. Springer, Dordrecht, The Netherlands, pp 218–239
16. McFadden D (1973) Conditional logit analysis of qualitative choice behaviour. In: Zarembka P (ed) Frontiers in econometrics. Academic Press, New York, NY, USA, pp 105–142
17. Train K (2003) Discrete choice methods with simulation. Cambridge University Press, New York
18. Hensher DA, Rose JM, Greene WH (2005) Applied choice analysis: a primer. Cambridge University Press
19. Koppelman FS, Sethi V (2000) Closed-form discrete-choice models. In: Handbook of transport modelling
20. Hensher DA, Greene WH (2003) The mixed logit model: the state of practice. Transportation 30:133–176
21. McFadden D, Train K (2000) Mixed MNL models for discrete response. J Appl Economet 15(5):447–470
22. Balaiyan K, Amit RK, Malik AK, Luo X, Agarwal A (2019) Joint forecasting for airline pricing and revenue management. J Revenue Pricing Manag 18:465–482
23. Gourieroux C, Monfort A (1993) Simulation-based inference: a survey with special reference to panel data models. J Economet 59(1–2):5–33
24. Lee LF (1995) Asymptotic bias in simulated maximum likelihood estimation of discrete choice models. Economet Theor 11(3):437–483
25. Hajivassiliou VA, Ruud PA (1994) Classical estimation methods for LDV models using simulation. Handb Econ 1(4):2383–2441
26. Stern S (1997) Simulation-based estimation. J Econ Lit 35(4):2006–2039
27. Halton JH (1970) A retrospective and prospective survey of the Monte Carlo method. SIAM Rev 12(1):1–63
28. Bhat CR (2001) Quasi-random maximum simulated likelihood estimation of the mixed multinomial logit model. Transp Res Part B Methodol 35(7):677–693
29. Liu Y, Cirillo C (2016) Small area estimation of vehicle ownership and use. Transp Res Part D Transp Environ 1(47):136–148
30. Bhat CR (2003) Simulation estimation of mixed discrete choice models using randomized and scrambled Halton sequences. Transp Res Part B: Methodol 37(9):837–855
31. Bhat CR, Gossen R (2004) A mixed multinomial logit model analysis of weekend recreational episode type choice. Transp Res Part B: Methodol 38(9):767–787
32. Bastin F, Cirillo C, Toint PL (2006) An adaptive Monte Carlo algorithm for computing mixed logit estimators. CMS 3:55–79
33. Geman S, Geman D (1984) Stochastic relaxation, Gibbs distributions, and the Bayesian restoration of images. IEEE Trans Pattern Anal Mach Intell 6:721–741
34. Metropolis N, Rosenbluth AW, Rosenbluth MN, Teller AH, Teller E (1953) Equation of state calculations by fast computing machines. J Chem Phys 21(6):1087–1092
35. Hastings WK (1970) Monte Carlo sampling methods using Markov chains and their applications. Biometrika 57(1):97–109. ISSN 0006-3444
36. Mascagni M, Chi H (2004) On the scrambled Halton sequence. Monte Carlo Methods Appl 10(3–4):435–442. https://doi.org/10.1515/mcma.2004.10.3-4.435
37. Halton JH (1960) On the efficiency of certain quasi-random sequences of points in evaluating multi-dimensional integrals. Numerische Mathematik 2(1):84–90. https://doi.org/10.1007/BF01386213
38. Spanier J, Maize EH (1994) Quasi-random methods for estimating integrals using relatively small samples. SIAM Rev 36(1):18–44
39. Train K (2000) Halton sequences for mixed logit
40. Hensher DA (2001) Measurement of the valuation of travel time savings. J Transp Econ Policy (JTEP) 35(1):71–98

41. Wu S, Han H, Hou B, Diao K (2020) Hybrid model for short-term water demand forecasting based on error correction using chaotic time series. Water 12(6):1683
42. Pandey P, Bokde ND, Dongre S, Gupta R (2021) Hybrid models for water demand forecasting. J Water Resour Plan Manag 147(2):04020106
43. Siddiqui R, Azmat M, Ahmed S, Kummer S (2022) A hybrid demand forecasting model for greater forecasting accuracy: the case of the pharmaceutical industry. Supply Chain Forum: Int J 23(2):124–134
44. Chreng K, Lee HS, Tuy S (2022) A hybrid model for electricity demand forecast using improved ensemble empirical mode decomposition and recurrent neural networks with ERA5 climate variables. Energies 15(19):7434
45. Mitra A, Jain A, Kishore A, Kumar P (2022) A comparative study of demand forecasting models for a multi-channel retail company: a novel hybrid machine learning approach. Oper Res Forum 3(4):58
46. Harper A, Mustafee N, Feeney M (2017) A hybrid approach using forecasting and discrete-event simulation for endoscopy services. In: Winter simulation conference (WSC), 3 Dec 2017. IEEE, pp 1583–1594
47. Demir E, Gunal MM, Southern D (2017) Demand and capacity modelling for acute services using discrete event simulation. Health Syst 6(1):33–40
48. Zhang L, Levinson D (2004) Agent-based approach to travel demand modeling: exploratory analysis. Transp Res Rec 1898(1):28–36
49. Qi C, Chang NB (2011) System dynamics modeling for municipal water demand estimation in an urban region under uncertain economic impacts. J Environ Manage 92(6):1628–1641
50. Suryani E, Chou SY, Chen CH (2010) Air passenger demand forecasting and passenger terminal capacity expansion: a system dynamics framework. Expert Syst Appl 37(3):2324–2339
51. Wada Y, Hamada K, Hirata N, Seki K, Yamada S (2018) A system dynamics model for shipbuilding demand forecasting. J Mar Sci Technol 23:236–252
52. Juan AA, Faulin J, Grasman SE, Rabe M, Figueira G (2015) A review of simheuristics: extending metaheuristics to deal with stochastic combinatorial optimization problems. Oper Res Perspect 1(2):62–72
53. Juan AA, Grasman SE, Caceres-Cruz J, Bektaş T (2014) A simheuristic algorithm for the single-period stochastic inventory-routing problem with stock-outs. Simul Model Pract Theory 1(46):40–52
54. Juan AA, Barrios BB, Vallada E, Riera D, Jorba J (2014) A simheuristic algorithm for solving the permutation flow shop problem with stochastic processing times. Simul Model Pract Theory 1(46):101–117
55. Rabe M, Deininger M, Juan AA (2020) Speeding up computational times in simheuristics combining genetic algorithms with discrete-event simulation. Simul Model Pract Theory 1(103):102089
56. Lloyd S (1982) Least squares quantization in PCM. IEEE Trans Inf Theory 28(2):129–137
57. MacQueen J (1967) Classification and analysis of multivariate observations. In: 5th Berkeley symposium on mathematical statistics and probability, 21 June 1967. University of California, Los Angeles, LA, USA, pp 281–297

Index

A

Agent-based Modelling (ABM), 10–12, 28, 31–34, 36, 37, 51, 81, 83, 84, 90, 92, 104, 121, 133, 165, 170, 213–239, 250, 307
Agent-based Simulation (ABS), 3–14, 16, 18, 81, 186, 249, 271–278, 281–284, 288, 290, 291, 295–297, 300, 307, 327
Algorithm, 5, 14, 30, 32, 85, 94, 128, 159, 160–164, 166, 167, 169–177, 189, 191, 193–196, 298, 304, 313, 314, 319, 321, 326–328, 330, 333, 337–341, 344
Analytics, 6, 23, 25–27, 29, 30, 40, 87, 89, 176, 327
AnyLogic, 214, 237, 273, 314
Artificial Intelligence (AI), 61, 85, 89, 90, 94, 161, 164, 173, 175, 238, 239, 320
Autonomous, 34, 74, 164, 174, 175, 216, 272

B

Bayesian, 168, 170, 326
Behaviour, 9, 11, 13, 32, 33, 35, 47, 48, 50, 61, 74, 76, 89, 90, 92, 108, 113, 122, 125–129, 131–133, 135, 137, 139, 142–144, 146, 148–150, 152, 155, 170, 171, 173, 189, 214–221, 225, 227, 228, 230, 231, 235, 273, 274, 283, 285, 286, 296, 319, 320
Behavioural modelling, 11, 90, 216, 219, 228, 322

C

CELL-DEVS, 125, 126, 129–133, 139, 141, 155
Cellular automata, 85, 125, 128, 130, 131, 133
Cloud, 6, 9, 16, 82, 248, 249, 251
Clustering, 163, 164, 167, 173, 301, 319, 321, 328, 329, 330, 337, 339, 340, 344
Coding, 24, 38, 50, 51, 53, 68, 84, 177, 214, 237, 238, 251, 297
Cognitive diversity, 75, 93, 94
Cognitive mapping, 6, 13, 15, 18
Computational, 4, 7, 9, 10, 12, 31, 34, 35, 47, 48, 51, 53, 59, 60, 64, 65, 67, 68, 73, 81–84, 86–88, 90, 92, 132, 199, 205, 222, 231–233, 246, 262, 319, 328
Computational Fluid Dynamics (CFD), 7–13, 16, 18, 84
Conceptual modelling, 5, 6, 9, 14, 15, 23, 24, 36, 39, 41, 47–54, 59–61, 63, 65–68, 166
Containerisation, 245, 247–251, 255, 262–264
Continuous, 7–11, 13, 15, 17, 18, 82–85, 103–109, 111, 116, 121, 159, 202, 203, 215, 218, 221, 234, 245, 247, 249, 252, 255, 256, 258, 259, 260, 262, 264, 271, 299, 308
Continuous integration, 245, 247, 249, 252, 255, 256, 258, 259, 260, 262, 264
Covid-19, 92, 126, 130, 226, 228, 249, 253, 275, 283, 284, 286–290, 302
Cross-cultural research, 84, 85
Cross-disciplinary research, 7, 82, 83, 85

Customer choice model, 319
Cyber-physical systems, 74, 82

D
Dashboard, 249, 290, 291
Data-driven, 29, 89, 161, 164, 167, 170, 175, 176, 186, 191, 192
Data farming, 171–173
Data mining, 191, 193
Decision-making, 4, 7, 39, 61, 65, 77–82, 88, 90, 161, 187, 192, 216
Decision tree, 162, 170, 172, 193, 203, 204, 206, 207, 304
Demand forecasting, 319, 320, 327
Deployment, 15, 19, 190, 208, 245–256, 260, 262–264, 274
Digital twins, 16, 89, 175, 176
Discrete, 7–11, 13, 15–18, 82–85, 103–107, 109, 111, 113, 114, 116, 121, 125, 126, 131, 167, 198, 216, 218, 234, 249, 273, 274, 303, 321–323, 327
Discrete-Event Simulation (DES), 3–18, 18, 28, 29, 31–34, 36, 37, 51, 53, 59–61, 64, 65, 67, 104, 121, 165, 185, 186, 190–192, 194, 195, 200–203, 208, 209, 214, 220, 235, 239, 247, 249–253, 255, 262, 264, 271–277, 280–284, 287, 288, 290, 291, 295–297, 300, 307, 308
Distributed simulation, 5, 6, 16, 19, 273, 277
Documentation, 26, 32, 37, 39, 50, 235, 238, 260, 309

E
Emergency medical services, 12, 16, 271, 274, 276, 281, 291
Energy, 11, 13, 62, 65, 66, 116, 121, 170, 171, 327

F
Fluid dynamics, 7, 10, 12, 84
Forecasting, 5, 7, 12, 14, 16, 18, 191, 272, 295, 297–301, 303, 310, 311, 313, 314, 319, 320, 322, 324, 327, 328, 339, 344, 345
Formalism, 5, 6, 85, 86, 103–106, 119, 121, 122, 125, 126, 131–133, 139, 155
Framework, 3–7, 15–17, 19, 26–28, 30, 36, 41, 51, 53, 59, 60, 64, 65, 67, 73, 83, 85, 93, 103–106, 120, 122, 128, 168, 173, 174, 192, 213–215, 218–220, 223, 224, 235, 237, 238, 245–248, 250, 255, 256, 262–264, 303, 322

G
Geographic Information System (GIS), 274, 291

H
Healthcare, 4, 5, 12–16, 51, 62, 64–66, 82, 94, 167, 185–187, 190–192, 214, 216, 217, 232, 239, 245–250, 253, 255, 262–264, 271–275, 283, 287, 290, 295–298, 300, 307, 315, 327
Healthcare Systems Modelling, 296
Heuristic, 5, 30–33, 35, 91, 121, 219, 296, 297, 319, 321, 326–328, 333, 337–342, 344
High Level Architecture (HLA), 16, 104, 121, 277, 281, 282, 290, 291
Hybrid conceptual modelling, 47, 49, 52, 61, 66, 68
Hybrid M&S, 3, 6, 8, 15, 17–19, 74, 75, 77, 81–83, 85–94, 133
Hybrid modelling, 3, 5, 6, 15, 17, 19, 25–29, 32, 35, 37, 40, 48, 52, 67, 73, 75, 82, 83, 88, 165, 213, 236, 238, 239, 245, 246, 262, 295, 297, 303, 319, 328, 329
Hybrid modelling and simulation, 52, 73, 75, 82, 88, 165, 295, 303
Hybrid models, 3, 4, 7–9, 13, 14–16, 18, 24, 26, 32, 83, 103, 104, 121, 126, 171, 187, 213, 214, 218, 220, 221–223, 225, 226, 228, 232, 234, 236–239, 246, 248, 256, 263, 272, 284, 288, 313, 321, 327, 328, 333
Hybrid simulation, 3, 4, 6, 8–11, 15, 17–19, 23–25, 28, 29, 32, 36, 39, 40, 48, 51, 83, 85, 91, 131, 160, 161, 165, 170, 174, 185, 186, 189, 191, 213–215, 221, 222–224, 227, 229–232, 235, 239, 246, 248, 271–273, 276, 277, 281, 283, 292, 295, 297, 307–309, 313, 327
Hybrid simulation designs, 224
Hybrid simulation framework, 26

I
Information systems, 49, 62, 65, 274, 291

Index   351

Inheritance of topology, 103, 105, 112, 115, 122

L
Linear programming, 29, 84, 295, 297–300, 303, 313
Logistic regression, 162, 204–206
Logistics, 13, 67, 162, 167, 171, 173, 174, 204–206
Logit Model, 319–321, 323, 324, 327, 328, 330

M
Machine Learning (ML), 12, 14, 16, 18, 29, 90, 155, 159–177, 185, 186, 188–193, 203, 208, 209, 274, 275, 292, 299, 304, 311, 313, 314, 327
Management, 4, 9, 11, 18, 49, 59, 60, 79–81, 86, 187, 214, 216, 222, 245, 246, 248–250, 256, 262, 271, 273–275, 282–284, 287–291, 295–297, 319, 320, 327, 328, 345
Manufacturing, 4, 11, 13, 59, 62, 66, 67, 167, 170, 173
Mathematical model, 298–300, 311–313
Maximum willingness-to-pay, 319–321, 324, 328, 330, 333, 337–339, 341, 344
Meta-model, 5
Mixed-integer programming, 295, 297, 299, 300, 303
Mixed-method, 62, 66, 68, 234
Modelling frame, 23, 25–28, 31–41, 51, 53, 59, 60, 67, 248
Model product lines, 103–105, 112, 120, 122
Monte Carlo Simulation (MCS), 6, 16, 165, 273, 292, 296, 319, 325–328, 333, 337–341, 344
Multi-agent Systems (MAS), 275

N
Neural network, 162, 163, 168, 170, 171, 193, 203, 204, 206
National Health Service (NHS), 251–253, 255, 262
Nonlinear, 215, 319, 320, 321, 328, 333, 337, 341, 344

O
Open-source, 159, 168, 239, 245–250, 262

Operational Research (OR), 3, 4, 28, 47, 217, 274
Operations, 4, 9, 18, 13, 31, 33, 50, 53, 59, 77, 82–86, 90, 91, 95, 104–106, 114, 115, 122, 161, 168, 186, 187, 190, 191, 250, 260, 274, 275, 287, 295, 296, 298, 300, 302, 309, 311, 319
Optimisation, 5, 12, 14, 16, 30, 34, 35, 75, 84, 91, 160, 164, 166, 171–174, 176, 192, 272, 274, 275, 327, 330, 338, 339, 345
Overview, Design Concepts, Details (ODD) Protocol, 217, 218, 235, 236, 290

P
Pandemic crisis management, 271, 273, 275, 283, 288, 290
Participatory modelling, 73, 75–79, 81, 82, 86–89, 91–95
Process mining, 12, 14, 191
Programming, 12, 29, 38, 84, 120, 159, 192, 193, 203, 217, 235, 237, 238, 250, 274, 290, 295, 297–300, 303, 313, 321, 341
Python, 199, 203, 238, 245, 250, 251, 253, 256, 258, 263, 264, 275, 288, 314

Q
Qualitative, 5, 7, 9, 13, 30, 47–51, 61, 62, 66, 68, 82, 84, 85, 215, 218
Queueing system, 186–192, 203, 208, 209
Queueing theory, 84, 191, 193

R
Random forest, 163, 174, 203, 204, 206, 304, 310, 311
Real-time, 16, 19, 89, 94, 122, 126, 155, 175, 185–193, 200, 202, 203, 206–209, 274, 314
Real-time delay prediction, 185–193, 200, 202, 203, 207–209
Regression, 14, 162, 163, 171, 172, 174, 187, 191, 193, 194, 203–208, 299, 304, 321, 333
Reinforcement learning, 30, 159–161, 164, 174–176, 193, 295, 298, 314, 315
Representation method, 23, 25–30, 32, 36–41
Revenue management, 319, 320, 328, 345

S

Scheduling, 11, 59, 174, 229, 253, 295–297, 299–301, 304
Semi-systematic review, 00
SimPy, 250–253, 258, 264, 288
Simulation analysis, 172, 228
Simulator, 81, 86, 88, 177, 284, 291, 319
Social media, 125, 126, 128, 132, 134, 136, 148, 150–152, 155
Social systems, 48, 62, 66, 127
Socio-technical systems, 47–50, 62, 65–68, 73–77, 83, 86, 87, 92, 93
Soft OR, 5, 6, 9, 13, 15, 18, 19, 85
Stakeholder, 15, 23–25, 29, 30, 32, 35–37, 39–41, 60, 75–77, 79–81, 86–88, 90, 91, 93, 172, 223, 226, 229–231, 233, 235, 236, 245, 272, 296, 299
Stakeholder engagement, 15, 23, 25, 35, 37, 40
Statistical, 66–68, 89, 90, 162, 186, 188, 189, 191, 192, 202, 205, 208, 327
Stock and flow, 7, 37, 61, 66
Supervised learning, 89, 161–163, 170, 172, 173, 193, 304
System Dynamics (SD), 3–13, 18, 28, 34, 36, 37, 51, 67, 83, 84, 92, 103, 104, 121, 165, 167, 186, 192, 213–239, 250, 271, 272, 292, 295, 296, 297, 327

T

Taxonomy, 4, 15, 17–19, 82, 83
Terminology, 91, 220, 297

U

Unified Modelling Language (UML), 5, 6, 61, 62, 65–68, 104, 105, 112, 218, 235
Unsupervised learning, 89, 161, 163

V

Validation, 5, 35–37, 48, 53, 59–61, 142, 155, 173, 200–203, 206, 217, 235–238, 286, 304, 310
Verification, 5, 59, 60, 173, 223, 237
Visualisation, 50, 81, 87, 88, 91, 94, 95, 141, 238, 274, 275, 290

Printed in the USA
CPSIA information can be obtained
at www.ICGtesting.com
CBHW061017090924
14265CB00003B/32